T0269225

THE HUMAN SCIENCES AFTER THE DECADE OF THE BRAIN

THE HUMAN SCIENCES AFTER THE DECADE OF THE BRAIN

Edited by

JON LEEFMANN

Friedrich-Alexander-Universität Erlangen-Nürnberg, Erlangen, Germany

ELISABETH HILDT

Illinois Institute of Technology, Chicago, IL, United States

ACADEMIC PRESS

An imprint of Elsevier
elsevier.com

Academic Press is an imprint of Elsevier
125 London Wall, London EC2Y 5AS, United Kingdom
525 B Street, Suite 1800, San Diego, CA 92101-4495, United States
50 Hampshire Street, 5th Floor, Cambridge, MA 02139, United States
The Boulevard, Langford Lane, Kidlington, Oxford OX5 1GB, United Kingdom

Copyright © 2017 Elsevier Inc. All rights reserved.

No part of this publication may be reproduced or transmitted in any form or by any means, electronic
or mechanical, including photocopying, recording, or any information storage and retrieval system,
without permission in writing from the publisher. Details on how to seek permission, further
information about the Publisher's permissions policies and our arrangements with organizations
such as the Copyright Clearance Center and the Copyright Licensing Agency, can be found at our
website: www.elsevier.com/permissions.

This book and the individual contributions contained in it are protected under copyright by the
Publisher (other than as may be noted herein).

Notices
Knowledge and best practice in this field are constantly changing. As new research and experience
broaden our understanding, changes in research methods, professional practices, or medical treatment
may become necessary.

Practitioners and researchers must always rely on their own experience and knowledge in evaluating
and using any information, methods, compounds, or experiments described herein. In using such
information or methods they should be mindful of their own safety and the safety of others, including
parties for whom they have a professional responsibility.

To the fullest extent of the law, neither the Publisher nor the authors, contributors, or editors, assume
any liability for any injury and/or damage to persons or property as a matter of products liability,
negligence or otherwise, or from any use or operation of any methods, products, instructions, or ideas
contained in the material herein.

British Library Cataloguing-in-Publication Data
A catalogue record for this book is available from the British Library

Library of Congress Cataloging-in-Publication Data
A catalog record for this book is available from the Library of Congress

ISBN: 978-0-12-804205-2

For Information on all Academic Press publications
visit our website at https://www.elsevier.com/books-and-journals

Working together
to grow libraries in
developing countries

www.elsevier.com • www.bookaid.org

Publisher: Mara Connor
Acquisition Editor: April Farr
Editorial Project Manager: Timothy Bennett
Production Project Manager: Chris Wortley
Designer: Mark Rogers

Typeset by MPS Limited, Chennai, India

Contents

I

PROSPECTS AND LIMITATIONS OF NEUROSCIENCE RESEARCH IN THE HUMANITIES AND SOCIAL SCIENCES

II

THE NEUROSCIENCES OF SOCIAL SCIENCES AND ETHICS

III

THE NEUROSCIENCES IN SOCIETY. SOCIAL, CULTURAL, AND ETHICAL IMPLICATIONS OF THE NEURO-TURN

List of Contributors

Alexander Bergmann University of Leipzig, Leipzig, Germany

Anne Biehl University of Leipzig, Leipzig, Germany

Nikola Biller-Andorno University of Zürich, Zürich, Switzerland

Berit Bringedal LEFO - Institute for Studies of the Medical Profession, Oslo, Norway; Columbia University, New York, United States

Markus Christen University of Zürich, Zürich, Switzerland

Philip Clapson Birkbeck, University of London, London, United Kingdom

Guillermo Del Pinal Zentrum für Allgemeine Sprachwissenschaft (ZAS), Berlin, Germany

Mattia Della Rocca University of Pisa, Pisa, Italy

Anna Drozdzewska Université catholique de Louvain, Louvain-la-Neuve, Belgium

Nadia El Eter Paul Valéry University of Montpellier, Montpellier, France

Yvonne Förster Leuphana University of Lüneburg, Lüneburg, Germany

Gerd Grübler Technical University of Dresden, Dresden, Germany

Oonagh Hayes University of Tübingen, Tübingen, Germany

Michael Jungert Friedrich-Alexander-Universität Erlangen-Nürnberg, Erlangen, Germany

Melissa M. Littlefield University of Illinois at Urbana-Champaign, Urbana, IL, United States

Hironori Matsuzaki University of Oldenburg, Oldenburg, Germany

Marco J. Nathan University of Denver, Denver, CO, United States

Alberto Rábano Carlos III Health Institute, Madrid, Spain

João Bettencourt Relvas University of Porto, Porto, Portugal

Işık Sarıhan Central European University, Budapest, Hungary

Alexandros Tillas University of Düsseldorf, Düsseldorf, Germany

Ties van de Werff Maastricht University, Maastricht, the Netherlands

Rui Vieira da Cunha University of Porto, Porto, Portugal

Jörg Zabel University of Leipzig, Leipzig, Germany

Introduction

Knowledge about the brain has affected thinking in the humanities and social sciences in more than one way and not just as recently as the Decade of the Brain. However, 25 years after the proclamation of this successful scientific research program, the influence of neuroscience in the social sciences and the humanities can no longer be denied. Originally intended to deepen knowledge of basic brain mechanisms and to investigate possibilities for medical intervention, the program has produced insights with implications reaching far beyond the life sciences and medicine. The growth of new academic neuro-disciplines, from neuro-philosophy and neuro-ethics to neuro-history, neuro-economics, and neuro-education to name but a few, bears witness to this development. Yet, beyond the purely academic realm, reference to neuroscientific knowledge has gained an explanatory function for many every-day-life phenomena whose interpretation used to be under the authority of the social sciences and humanities. Folk understandings of human behavior, human decision-making, and even human identity evermore often draw on—sometimes ill-conceived—knowledge of the brain and the nervous system.

However, less obvious than these consequences of neuroscientific engagement, neuroscience has also gained valuable input from some of the humanities and social sciences. Philosophy of science, for example, has put much analytical rigor into clarifying what makes a good neuro-scientific explanation, and investigators in social neuroscience and in the neuroscience of moral behavior have modeled their experiments on conceptions and theories from the cognitive psychology of decision-making. Ethical considerations stimulated by the development of new neurotechnologies have also stimulated first thoughts on legal regulations of neuroscientific research and technology development. Because the mutual influences of the neurosciences, the humanities, and society at large are so obvious and far reaching that neuroscience has been granted the role of the new leading discipline of the millennium, this has led some scholars to proclaim a neuroscientific turn in the humanities, the social sciences, and even beyond in the practices of education in schools and legal judgment in courts. Whether this turn mainly affects the practices of the humanities and social sciences via translations of neuroscientific knowledge and methodology to questions so far unrelated to neuroscience or—as for instance Fernando Vidal has argued in his 2009 article "Brainhood, Anthropological Figure of Modernity"—should be interpreted as an overarching (and problematic) anthropological outlook, is open for debate. The papers collected in this volume provide differentiated and diverse answers to this discussion.

One point most scholars seem to agree on is that interactions between neuroscience and the humanities are not always unidirectional. They mostly agree that neuroscience is not to be conceived of as a new universal science able to explain everything from the ion channel to human subjectivity to the symbolic orders that structure human societies. And neither is it neuroscience alone that helps shed new light on problems in the social sciences and humanities, nor is it the only conceptual and theoretical work undertaken in philosophy and the humanities that offers the theoretical frameworks in which neuroscientific research questions can be operationalized and approached experimentally. Interactions are multidirectional and can be fruitful on both sides of the disciplinary divide between the natural and the human sciences.

*

The chapters in the first part of this volume, "Prospects and Limitations of Neuroscience Research in the Humanities and Social Sciences," investigate the forms of possible interactions between neuroscience and the social sciences and humanities. The contributions probe the prospects of neuroscientific investigations in the social sciences and humanities and vice versa, drawing on a variety of concrete examples. In the first chapter "Neurophilosophy or Philosophy of Neuroscience?" *Michael Jungert* compares the two main forms of interaction of philosophy and neuroscience in the current discourse about the neuroscientific turn. While the *philosophy of neuroscience* tries to apply methods and classical approaches from the philosophy of science to neuroscience and is concerned with methodological, conceptual, and explanatory problems of the neurosciences, so-called *neurophilosophy* takes a different approach by applying neuroscientific findings to classical philosophical issues. Jungert criticizes the philosophy of neuroscience for its indifference of the empirical core business of the neurosciences, which only allows for a critique from the outside but not for a mutual and interdisciplinary interaction. As a result, he turns to neurophilosophy as a more eligible candidate for interdisciplinary engagement. Jungert identifies three principle ways neurophilosophy relates philosophical concepts to empirical findings from the neurosciences and chooses an approach that aims at integrating neuroscientific findings into a philosophical theory in order to gain an empirically informed and enriched theory. By analyzing the example of memory in theories of personal identity, he evaluates neurophilosophy's claim to integrate neuroscientific findings into philosophical theory and concludes that this neurophilosophical approach has much to offer for philosophers as well as for neuroscientists.

Against this case for a reasonable interaction of neuroscience and philosophy *Işık Sarıhan* puts forward a more skeptical position. In his chapter "Philosophical Puzzles Evade Empirical Evidence. Some Thoughts and Clarifications Regarding the Relation between Brain Sciences and Philosophy of Mind" he chooses the example of philosophy of mind to take a two-step approach to the investigation of possible interactions with neuroscience. In the first step, he describes philosophy as functioning as an aide and inspector for science, which helps to clarify concepts and to develop theories. The more skeptical stance toward an interaction between neuroscience and philosophy of mind, however, relies on the second type of interaction: the usefulness of empirical neuroscience to philosophy. Based on two examples, Sarıhan argues that attempts to resort to neuroscience in order to settle philosophical questions cannot be successful. Philosophical questions are conceptual questions and differ categorically from empirical questions and hence cannot be solved by empirical neuroscientific evidence.

The following two chapters explore in a more concrete fashion some of the implications of interdisciplinary collaboration between neuroscience and philosophy. In the third chapter, "'Who's afraid of the Big Bad Neuroscience?' Neuroscience's Impact on Our Notions of Self and Free Will," *Rui Vieira da Cunha* and *João Bettencourt Relvas* reflect on the consequences of neuroscientific findings on shared assumptions about rational and self-ruling persons, free will, and responsibility. After delineating some of the promises and perils of neuroscience, they contrast neuroscience-based interpretations of self and free will with the respective folk conceptions. Following an analysis of the impact of neuroscientific findings on notions of self and free will, they conclude that neuroscience alone is not able to change those widely held notions.

In the fourth chapter "Free Will—Between Philosophy and Neuroscience" *Anna Drozdzewska* deepens this discussion, exploring a variety of neuroscience-based approaches to free will. Focusing on the experiments conducted by Benjamin Libet and on the position held by Peter Ulric Tse, she starts from the hypothesis that the problem of mental causation is crucial for an effective account of free will and discusses the role mental causation has both for basic intuitions concerning free will and for an experimental neuroscientific approach. She then goes on to reflect on how neuroscience can influence the philosophical discussion of free will and how philosophical reflection can influence neuroscientific studies on free will.

While theoretical philosophy and especially the philosophy of mind provides widely acknowledged examples for some kind of interaction

between empirical neuroscience and the a priori reasoning of philosophy, the possible interactions of neuroscience with other disciplines
form the humanities is much less investigated. In the last chapter of
Part I, *Mattia Della Rocca* therefore turns to historiography as another
field of research that has been affected by neuroscience in the past
decade. In his contribution he shows how the field neurohistory has
formed itself alongside the more established history of neuroscience
and how both approaches differ from one another. The former presents
itself as a methodology for doing historiography, based on the findings
of current cognitive and brain sciences and aims to explain how historical and cultural changes developed out of the confrontation of the nervous system with an ever-changing physical and cultural environment.
The chapter, in contrast, occupies itself with the framing of neuroscientific triumphs into the chronological collection of historical recording
and with the discovery of "precursors" of the discipline to justify and
celebrate current research in neuroscience. Both approaches, Della
Rocca argues, are, however, one-sided and do not suffice for a fruitful
interaction between history and neuroscience. Neurohistory, by explaining historical shifts based on the dominant knowledge available in neuroscience at a time, has a tendency toward presentism, ahistoricism,
and the neglect of the sociocultural embeddedness of neuroscientific
explanation. The history of neuroscience, instead, is unable to account
for the influence of cognitive foundations guiding human behavior in
history. Referring to the example of neuroplasticity, Della Rocca shows
how a third possibility—a critical interaction of neuroscience and
history—could avoid both kinds of restrictions.

*

Against this first part, which explored the possibilities of mutually
relating neuroscience and the humanities toward one another, the second part of the present volume turns to critical discussions of the idea
that neuroscience has a foundational role in explaining social and moral
phenomena. This part, entitled "The Neurosciences of Social Science
and Ethics," starts out with the delineation of a major theoretical
scheme for the role of neuroscience in the explanation of social phenomena. Arguing for an eliminativist approach in the philosophy of
mind, in his contribution "The Theory of Brain-Sign," *Philip Clapson*
reconstructs consciousness as "the brain phenomenon" and shows how
brain-signs, which he conceives of as means of interneural communication, are generated by the brain from its causal orientation toward the
world. This allows him to explain cooperative actions of organisms in
physical terms and to argue that human beings cannot be conceived of
as mental subjects with immediate access to themselves and the world.
If the brain is explained as a causal organ, instead, human beings can

be fully explained in terms of the natural sciences. As a result, Clapson claims that the natural sciences and the social sciences gain a uniform scientific and practical foundation. It is the language of the natural sciences that can account for everything there is to say about social phenomena.

In the following chapter, "On the Redundancies of 'Social Agency,'" *Alexandros Tillas* presents an example of how a neuroscientifically informed approach to the nature of thinking can shed light on a long-standing sociological debate. Starting from the debate over the priority of structure versus agency for the explanation of social behavior, a sociological debate that cuts across many of the most fundamental aspects of human cultural life, he shows how results from cognitive science propose fruitful reinterpretations of the concept of agency. In the "structure versus agency" debate, agency usually refers to the capacity of individuals to exercise free will and to act upon their goals, whereas structure comprises the social arrangements and scaffoldings presumed to influence choices of individuals, as well as biological factors. Drawing on findings from cognitive science and neuroscience, Tillas argues that agency is itself "structured" and that, because of this, agents cannot act independently once they have internalized existing norms and principles. This leads him to claim that the traditional "structure versus agency" debate as well as the notion of "social agency" are redundant.

Bridging levels of explanation is, however, theoretically problematic not only between the coarsely construed levels of cognitive processes and social theory. The more crucial step is the explanation of hypotheses over cognitive processes with reference to neuroscientific data. One method in cognitive neuroscience to derive explanations of cognitive processes from neuroscientific data is reverse inference. In reverse inference the presence of a cognitive process is inferred from measured brain states. This is often used to discriminate between competing psychological hypotheses to explain cognitive processes. In their contribution "Two Kinds of Reverse Inference" *Guillermo Del Pinal* and *Marco Jacob Nathan* analyze this method of explanation from a philosophy of neuroscience perspective. They distinguish between location-based reverse inference (LRI) and pattern-based reverse inference (PRI), and discuss several methodological problems of reverse inference. They argue that a key assumption for reverse inference is the establishment of a linking condition between neuroscientific data and cognitive processes. Because of the multifunctionality of brain regions and the threat of circular explanation, they doubt that such a linking condition can be fulfilled by LRI, but show that instead PRI does not face similar methodological problems.

The second part of the volume closes with a contribution by *Nadia El Eter*. El Eter is concerned with the different interpretations given to the problem of the "naturalistic fallacy." She argues that the distinction between the descriptive—or the realm of what "is"—and the normative—or the realm of what "ought to be"—in the branch of neuroethics that Adina Roskies has dubbed the "neuroscience of ethics" is construed differently from the classical arguments in Hume's *Treatise* and Moore's *Principia Ethica*, as well as from its contextually abstracted orthodox reading. In neuroethics, neuroimaging studies investigating human morality draw on presuppositions about the relation between the explanatory levels of cognitive neuroscience and cognitive psychology that cannot be isolated from epistemological considerations. Drawing among others on the problem underlying reverse inference (RI) and the problematic concept of cognition, El Eter establishes an argument that aims to show how accusing neuroscientific investigations into the basis of morality of committing a "naturalistic fallacy" does not do justice to the multifaceted contextual presuppositions underlying the current research field of neuroscience of ethics. In this branch of neuroethics the distinction between the descriptive and the normative realm cannot be drawn consistently.

*

The chapters in the third part of the volume transgress the border of theorizing the possible relations of neuroscience and the humanities and of investigating the specific theoretical problems underlying mutual interactions between social sciences and natural science. They put neuroscience in the broader perspective of cultural uses and interpretations. Under the title "The Neurosciences in Society. Social, Cultural, and Ethical Implications of the Neuro-Turn," these chapters investigate the broad spectrum of symbolic and metaphorical meanings connected to neuroscientific concepts and assess the implications of neuroscience for educational purposes. Finally, the last two contributions address ethical concerns with regard to the organization of large-scale neuroscience research projects and the representation of deep brain stimulation in the media.

At the opening of part three the contribution by *Yvonne Förster* called "The Neural Network as a Paradigm for Human Self-Understanding," explicates from an interdisciplinary perspective of philosophy of mind and media studies how the neuroscientific turn helps to stir the imagination of a new metaphysical dimension for (post-)human consciousness. Using the successful Hollywood movies *Her* and *Transcendence* as examples, she shows how the turn from phenomenological to neuroscientific and cognitive science approaches in the philosophy of mind has popularized the view that subjective first person experience is derivative of cognitive processes of the neural networks that make up the human brain. Many current science-fiction movies display the idea of neural networks without picturing the human brain or body, but by imagining information that can be embodied by multiple different substrates and that transcends strict physical boundaries. This evokes the image of posthuman conscious lifeforms, which in turn can serve as a foil for human self-understanding.

The following chapter, "Brain, Art, Salvation. On the Traditional Character of the Neuro-Hype," by *Gerd Grübler* takes up the approach of investigating the cultural meanings of neuroscience, but focuses on a much more general pattern of interpretations neuroscience shares with interpretations of other, earlier scientific endeavors and achievements. Grübler defends the thesis that neurologization of society is a rather conservative and traditional endeavor. He draws from several examples from the history of the European sciences (such as using a clock or a mechanical machine as a metaphor) in order to delineate that the enormous role currently ascribed to neuroscience can be seen in the context of a search for salvation. In this way, understanding the brain and building on this knowledge is seen as a way to allow mankind to escape finite existence. He then goes on to explore the role his analysis may have for the field of applied ethics.

Besides the danger of furnishing neuroscientific knowledge with metaphysical expectations about infinite human existence there are, however, other more mundane threats to a fruitful development of neuroscience under the current circumstances of western society. Even though the neuroscientific turn is typically conceptualized as a phenomenon that primarily impacts the humanities and social sciences, the contribution of *Melissa M. Littlefield* "A Mind Plague on Both Your Houses" explores the consequences of the neuroscientific turn for the neurosciences themselves. Analyzing the discourses surrounding the most recent backlash on the neurosciences and the restructuring of Europe's currently most prestigious neuroscientific research project—the Human Brain Project (HBP)—she finds that premature and ill-conceived translations of neuroscientific findings to the social sciences and humanities and pressure for marketability of scientific results might threaten the respectability and credibility of the neurosciences in the future. These imagined repercussions of the neuroscientific turn on the neurosciences have great potential to determine the future of neuroscientific research.

Some of these imagined repercussions might also come from experiences with neuroeducation. As *Ties van de Werff* reveals in his chapter "Being a Good External Frontal Lobe: Parenting Teenage Brains," knowledge about adolescent brain development used by pedagogues, family coaches, and other parenting professionals to educate parents on how to deal with their teenage children often comes in ambiguous versions. Van de Werff explicates how the topic of adolescent brain development goes along with the idea that good parents should act as external frontal cortex of their adolescents' brains in order to complement the developing teenage brain. Based on an analysis of a variety of media in Dutch popular culture, he reveals two different normative discourses surrounding the idea of parenting-as-external-frontal-lobe, which demand different courses of parental action: in the first discourse, parents are seen as guardians of external stimuli, whereas in the second, parents are considered to be stimulating coaches.

Before using neuroscientific knowledge for developing new educational paradigms, knowledge about the brain needs to be encountered. Hence, in their chapter "Toward Neuroscience Literacy?—Theoretical and Practical Considerations," *Alexander Bergmann, Anne Biehl*, and *Jörg Zabel* investigate the status quo of neuroscience education. Starting from the concept of scientific literacy they discuss neuroscience literacy, a skill to understand basic neuroscientific assumptions, explanations, and methods. They discuss what it means to be neuroscientifically educated and reflect on how neuroscience education can be improved. Starting from an analysis of current science classroom conditions, which focuses on science teaching in Germany, they reflect on ways to improve neuroscientific learning. In particular, they suggest a critical-reflective teaching method that involves students' "everyday myths" concerning neuroscientific research and neuroethical discourses.

The last two chapters of this book, finally, are dedicated to ethical implications of neuroscience research and the use of neuro-technologies in the clinical context. In this sense, both contributions can be seen as being typical neuroethical analyses.

The currently most prestigious neuroscientific research project in Europe, the Human Brain Project (HBP), serves as an example in many of the contributions in this volume. However, the chapter "'Strangers' in Neuroscientific Research" by *Berit Bringedal* and colleagues describes the structure and unclear role of ethics consultant in the HBP and offers a first rapprochement to the ethical, social, and institutional challenges surrounding huge research projects in the neurosciences. Having been involved in the ethical advisory board of HBP, the authors identify the most pressing ethical challenges of the project from a research ethics perspective. Taking into account the recent criticism of the project from inside the neuroscience community, they identify a precautionary attitude, the weighing of harm and benefits, and transparency as the most important ethical principles to tackle the structural problems of big science projects such as the HBP. The role of ethical advisors in big science projects, they conclude, should be to work toward an implementation of institutional structures that support these principles.

Finally, *Oonagh Hayes* takes up the issue of public imaginations about the power of neuro-technology drawing on the example of deep brain stimulation. Analyzing media reports and film sequences, her contribution reveals the rhetorical and discursive mechanisms producing the image of an easy and simple quick-fix technology that allows changing motional and psychological states "at the push of a button." This imagination is ethically problematic as far as it distorts the actual risks and timeframe of the intervention and obscures the difficulties that often arise during treatment.

This book summarizes experiences from presentations and discussions at the interdisciplinary conference "The Human Sciences after the Decade of the Brain," which was held in March 2015 at the Department of Philosophy at Johannes Gutenberg-University of Mainz. The conference and this publication would not have been possible without a generous research grant from the German Research Foundation (DFG) for the project "The 'Neuro-Turn' in European Social Sciences and Humanities" (NESSHI).

Erlangen and Chicago, July 2016

Jon Leefmann
Friedrich-Alexander-Universität Erlangen-Nürnberg, Erlangen, Germany

Elisabeth Hildt
Illinois Institute of Technology, Chicago, IL, United States

PROSPECTS AND LIMITATIONS OF NEUROSCIENCE RESEARCH IN THE HUMANITIES AND SOCIAL SCIENCES

1

Neurophilosophy or Philosophy of Neuroscience? What Neuroscience and Philosophy Can and Cannot Do for Each Other

M. Jungert

Friedrich-Alexander-Universität Erlangen-Nürnberg, Erlangen, Germany

INTRODUCTION

Ever since the rise of modern neuroscience in the 1980s, there has been controversial discussion about its potential influence on topics that have been traditionally seen as part of the domain of social sciences and humanities (see, e.g., Gold & Stoljar, 1999; Satel & Lilienfeld, 2013; Tallis, 2014).[a] The heated public and scientific debate on proposed neuroscientific solutions to the problem of the freedom of the will might be considered as the most prominent example (see, e.g., Mele, 2010, 2015; Schlosser, 2014; Walter, 2001). Moreover, the formation of a large number of neuro-hyphenated disciplines in the field of social sciences and the humanities such as neuro-theology,

---

[a] One result of this controversy is the formation of so-called "critical neuroscience" (see, e.g., Choudhury & Slaby, 2012; Slaby, 2010; Wolfe, 2014).

© 2017 Elsevier Inc. All rights reserved.

neuro-psychoanalysis, neuro-education, or neuro-economics, to name just a few, shows the appeal and attraction of applying neuroscientific methods to traditional scientific fields.

In *philosophy*, two distinct ways of dealing with the problems and prospects of neuroscience have been developed in recent decades: On the one hand, the *philosophy of neuroscience* tries to apply methods and classical approaches from the philosophy of science to neuroscience, e.g., to shed light on its specific explanatory strategies. While this view is sometimes considered to be a more skeptical, critical, or even destructive one, so-called *neurophilosophy* takes a different approach. Here, neuroscientific findings are applied to classical philosophical issues such as the nature of emotions, the concept of morality, or the nature of consciousness, in order to develop empirically informed philosophical concepts and theories.

In this chapter, I am going to evaluate the premises and prospects of both approaches by discussing the following issues: I will start by reviewing the methods, theoretical assumptions, and explanatory aims of both the philosophy of neuroscience and neurophilosophy. In the next step, I will look into neurophilosophy's claim to integrate neuroscientific findings into philosophical theory by analyzing the relation between memory and personal identity. Based on this analysis, I aim to shed light on the more general question of what philosophy and neuroscience can and cannot do for each other.

WHAT IS THE PHILOSOPHY OF NEUROSCIENCE?

The so-called "philosophy of neuroscience" can be considered a branch of philosophy of science representing the ongoing trend to move from very general questions about science to more detailed discussions of particular issues of specialized disciplines. It applies classical concepts and questions from the general philosophy of science to the field of neuroscience. Research questions of the philosophy of neuroscience include: Is there a specific scientific method in neuroscience (Machamer, McLaughlin, & Grush, 2001)? Are there special kinds of explanations in neuroscience that differ from the types of explanations in other fields of science (Bechtel, 1994)? What is the impact of neuroscience on theories of human agency (Runyan, 2014)? Which concepts of causality or reduction are involved in neuroscientific explanation (Bickle, 2003)?[b]

[b] For a detailed general overview of the philosophy of neuroscience (see Bickle, Mandik, & Landreth, 2012).

One way of pursuing those questions is purely descriptive. If done that way, the agenda of the philosophy of neuroscience equals the approach of other specialized branches of the philosophy of science, e.g., like the philosophy of biology, physics, or psychology. In all those cases the main aim of philosophical investigation is to illuminate the field-specific ways of research and argumentation of an empirical discipline. Regarding neuroscience, one famous debate is the discussion about reductionism (see, e.g., Bickle, 2008; Craver, 2005; Schouten & Looren de Jong, 2007).

Generally speaking, the task of the philosophy of neuroscience is threefold:

First, it is the philosopher's job to discover and explicate the theoretical assumptions that are often more or less implicitly "woven into the fabric of empirical research" (Hyman, 1989, p. XIV). For example, Max Bennett and Peter Hacker state in their seminal work *Philosophical Foundations of Neuroscience*, "Many brain-neuroscientists have an implicit belief in reductionism. Few try to articulate what exactly they mean by this term of art" (Bennett & Hacker, 2003, p. 355).

Some philosophers of neuroscience see it as their task to make such implicit beliefs explicit in order to make them an object of investigation in the philosophy of science. This kind of explication work on background concepts can be considered a manifestation of the philosopher's general aim to dissolve conceptual puzzles and to confront others with their unquestioned or unconscious beliefs and assumptions. As in other fields of empirical research, neuroscientists mostly do not regard such basic questions as a matter of concern for themselves as they do not feel that these issues belong to their empirical core business. Therefore one main task of the philosophy of neuroscience in this context is to show that the proposed distinction between empirical core business and nonempirical sideline work is illusive as it ignores the fact that concepts, theoretical framework, and empirical investigation are intimately connected.

The second task is the distinction of different meanings of concepts that are either explicitly stated or implicitly used by neuroscientists. One example is the difference between ontological and explanatory reductionism in neuroscientific theories (Bennett & Hacker, 2003, pp. 355–366). Another one concerns the distinction between different meanings of "decision," which is one of the key terms in the debate about the freedom of human will (Walter, 2001, pp. 28–37). In those cases, the philosopher's job is to clarify the meaning of terms and the different ways of using concepts in order to make sure that discussions are really based on common concepts and not just circling around mock debates due to conceptual confusion. The heated free will debate between some neuroscientists and philosophers shows that many misunderstandings and fruitless debates are due to conceptual confusion and could be avoided by clarification of concepts and by creating a common conceptual ground for fruitful interdisciplinary discussion.[c]

Finally, the third task of the philosophy of neuroscience is to discuss the plausibility of conclusions drawn from empirical data. For philosophers of science, one of the most irritating assumptions defended by some neuroscientists is the idea that far-reaching conclusions about human thinking and behavior can be more or less directly drawn from measurement results or brain imaging studies. Therefore it is the philosopher's task to analyze the structure of neuroscientific arguments and theories and to identify conclusions that are logically unsound or not supported by the presented data. Recent neuroscientific claims, among others made by Francis Crick, Gerald Edelman, or Antonio Damasio, offer plenty of examples (see Bennett & Hacker, 2003, pp. 68–74). One of them is the mereological fallacy that consists of ascribing mental states or complex abilities like deciding, believing, interpreting, perceiving, or thinking to the human brain as a part of a person instead of the person as a whole. In *The Astonishing Hypothesis*, Francis Crick gives a good example of this kind of fallacy: "What you see is not *really* there; it is what your brain *believes* is there [...]. Your brain makes the best interpretation it can according to its previous experience [...]. The brain combines the information provided by the many distinct features of the visual scene [...] and settles on the most plausible interpretation of all these various clues taken together" (Crick, 1995, p. 30). Philosophical analysis shows that this kind of ascription of psychological attributes to the brain simply does not make any sense. As Bennett and Hacker state, "The brain is not a logically appropriate subject for psychological predicates" (Bennett & Hacker, 2003, p. 72).

By categorizing such neuroscientific claims as confusing or even senseless, the philosopher is not just making a descriptive statement about neuroscience. In contrast to, for example, the reconstruction and description of theory formation in neuroscience, he takes a *normative* position toward his object of investigation. In a similarly normative way, he could try to show that certain correlations gained by neuroimaging studies do not reveal anything interesting about causal relations between brain states and behavior. The focus of the philosophy of neuroscience thereby switches from describing the structure of neuroscience to judging certain claims and eventually proposing alternative interpretations or models of explanation.

A survey of possible points of criticism concludes the characterization of the philosophy of neuroscience: Firstly, neuroscientists might complain that the philosophy of neuroscience represents exactly the

[c]See Kane (2011) for a broad overview.

kind of pointless "armchair philosophy" that tries to criticize empirical research from the outside without really knowing anything about its contents or methods.

Secondly, one could object that, while tackling foundational issues like explanatory strategies or concepts of representation, the actual research topics and empirical results are not at the center of attention. Instead of discussing current findings and helping to analyze, interpret, and consolidate the outcome of neuroscientific research, the philosophy of neuroscience only takes an interest in abstract conceptual and logical analysis. Moreover, it tends to lecture empirical scientists about issues that are remote from their core business or even completely irrelevant to their factual doing.

Thirdly, and finally, one could point at the one-sidedness of the philosophy of neuroscience. While it aims at analyzing and sometimes criticizing neuroscience, there is no attempt to consider neuroscience as a potential enrichment for philosophy, especially for the philosophy of mind. This ignorance, so the objection goes, inhibits productive interdisciplinary cooperation that is necessary for extensive research on the human mind and brain.

NEUROPHILOSOPHY: HOW TO COMBINE NEUROSCIENTIFIC FINDINGS WITH PHILOSOPHICAL THEORY

Against the background of this criticism, we can now turn to neurophilosophy as quite a different way of dealing with the prospects and challenges of neuroscience. The publication of Patricia Churchland's much debated book *Neurophilosophy* in 1986 can be seen as a major step in the development of this discipline. Churchland's approach is based on the assumption that close cooperation between neuroscience and philosophy and mutual integration of each other's findings and concepts are crucial for successfully studying brain and mind.

John Bickle, Peter Mandik, and Anthony Landreth describe Churchland's program as follows:

> She was introducing philosophy of science to neuroscientists and neuroscience to philosophers. Nothing could be more obvious, she insisted, than the relevance of empirical facts about how the brain works to concerns in the philosophy of mind. Her term for this interdisciplinary method was "co-evolution" [...]. This method seeks resources and ideas from anywhere on the theory hierarchy above or below the question at issue. Standing on the shoulders of philosophers like Quine and Sellars, Churchland insisted that specifying some point where neuroscience ends and philosophy of science begins is hopeless because the boundaries are poorly

defined. Neurophilosophers would pick and choose resources from both disciplines as they saw fit. (Bickle et al., 2012)

In Churchland's understanding, neurophilosophy strongly differs from the philosophy of neuroscience. Neurophilosophers do not see themselves as critical observers of neuroscience, neither do they draw a sharp line between nonempirical matters of philosophy and empirical matters of science. In fact, they consider the philosophy of mind and neuroscience as closely related, intertwining, or even merging disciplines. However, the label "neurophilosophy" implies a unified concept while in fact it comprises several different approaches regarding methods and leading questions. The case of personal identity might help to illustrate this.

Taking a neurophilosophical approach to personal identity could, among others, be interpreted as follows:

- Firstly, one could attempt to transform philosophical criteria for personal identity into what Georg Northoff calls a "self-rating scale for empirical assessment of personal identity" (Northoff, 2004, p. 92). By doing this, abstract philosophical concepts are thought to be operationalized and converted into empirical concepts that can be applied to practical problems such as personality changes after brain surgery.
- Secondly, one could attempt to identify neural correlates of philosophical concepts, in this case personal identity. The potential results are sometimes thought to be a more precise replacement for allegedly cloudy philosophical concepts.
- Thirdly, one could attempt to integrate neuroscientific findings into a philosophical theory of personal identity in order to gain an empirically informed and enriched theory of personal identity. This approach is driven by the idea that—at least in some philosophical theories—there are elements of the theory that are open to the integration of empirical science.

I will now illustrate this third model by discussing the question of how the integration of memory research can be used to advance the philosophical debate on personal identity.

NEUROPHILOSOPHY IN ACTION: PERSONAL IDENTITY AND MEMORY RESEARCH

Memory is one of the most important features of human beings. Its importance becomes especially apparent in cases of severe amnesia or dementia, where loss of memory often seems equal to loss of identity

(Clark, 2010; Hoerl, 1999; Klein & Nichols, 2012). Although memory is a key element in many theories of personal identity from John Locke to Derek Parfit, the exact role that memory plays for the constitution and preservation of identity remains largely unclear (Schechtman, 1994). I claim that this negligence is, among others, due to diffuse concepts of memory and the lack of application of neuropsychological knowledge to philosophical theory (Jungert, 2013, 2015). I argue that both problems can be solved by integrating neuropsychological findings into philosophical theory, thereby creating an empirically informed new approach (Jungert, 2015).

As a start, it is helpful to consider the most prominent classification of memory systems, introduced by neuroscientist Larry Squire in the 1980s. Squire separates declarative memory from nondeclarative memory. While declarative memory is defined as "the capacity for conscious re-collection about facts and events" (Squire, 2004, p. 173), nondeclarative memory includes, among other things, skills, habits, and different forms of learning and is characterized "through performance rather than recollection" (Squire, 2004, p. 173). Although both systems are relevant to personal identity, I will limit the following discussion to declarative memory and especially to autobiographical memory as one of its subsystems. In contrast to semantic memory, the other subsystem of declarative memory, autobiographical memory is not memory of pure, neutral facts and knowledge. Instead, it contains significant emotionally charged memories about important events in the unique history of a person. Autobiographical memories can be remembered over a long period, often include a high level of detail, and enable the specifically human ability to mentally reexperience episodes from one's own past. By doing so, it becomes possible for human beings to evaluate and anticipate current and future actions based on experiences, personal preferences, and decisions from the past.

By combining the features and mechanisms of autobiographical memory discovered by neuroscience with the philosophical concept of biographical identity, it becomes possible to develop a much richer theory of the relationship between memory and identity compared to classical analytical approaches. I will give only one short example that concerns the emotional dimension of autobiographical memory. In many philosophical theories, memories are seen as countable units whose main purpose is to carry information over time (see Schechtman, 1994). They completely ignore the fact that memories are also a necessary precondition for experiencing one's life as a coherent, meaningful, and ongoing process, as they connect and organize remembered events in a way that allows for threading these different parts and that results in the ability of seeing one's life as a whole (Jungert, 2015, pp. 133–136).

Recent neuroscientific research elucidates this property of memory (Berntsen & Rubin, 2006). The emotional index attached to a certain memory at the time of encoding can become a part of the person's biographical identity. However, this emotional index might also be subject to change over time. The change can be caused by a new evaluation of the emotion and memory in question (Debus, 2007). If, for example, someone has changed his attitude toward smoking over the years, he might end up attaching his current aversion against smoking to his former memories, even though at the time of encoding he had been a passionate smoker and therefore attached positive emotions to these memories originally. Such cases demonstrate the reverse direction of influence from (current) self to (former) emotion and memories: "Not only is our sense of self based on memories of past experiences, [...] but our retrieval, recollection, and reconstruction of the past is, reciprocally, influenced by the self" (Schacter, Chiao, & Mitchell, 2003, p. 227). This modification of memories through the current self-image of a person usually happens without conscious awareness, resulting in implicit harmonization of remembered past and experienced presence.

Empirical findings such as these are extremely helpful for developing an advanced philosophical theory of the importance of memory for personal identity that is able to explain how persons are capable of developing narrative structures. It shows that these structures are necessary to understand one's own life as a more or less coherent story and demonstrates that memories can not only affect persons as carriers of information, but first and foremost as "transmitters of influence" (Wollheim, 1984, p. 101).

WHAT NEUROSCIENCE AND PHILOSOPHY CAN AND CANNOT DO FOR EACH OTHER

This example, although discussed very briefly, shows the enormous potential of this kind of neurophilosophy. The integration of neuroscientific findings into philosophical theory, done in the right way, can be of high value for both disciplines. For philosophy, this value consists of:

• Firstly, the chance to compare philosophical concepts to related concepts from neuroscience. In this context, "related" means that the concepts in question have some common content and are on a similar level of description. In the case of memory—as discussed—both disciplinary perspectives aim at a mental faculty and try to describe its importance for human beings. Therefore the comparison might show the philosopher that his own concepts are too narrow and fail to include important aspects of the object of investigation.

- Secondly, the potential revision of some elements of philosophical theories in reference to neuroscience. For instance, this holds for elements that claim to describe the actual functioning of certain cognitive powers. In the case of memory, such an element is the idea that memory can be considered as a warehouse that stores items of the past. Some philosophical theories of memory and personal identity rely on this idea implicitly or explicitly. Neuroscientific research shows that this idea is wrong in many respects. If the neuroscientific findings about the dynamics and inconstancy of memory are taken seriously by philosophers, this will also lead to new philosophical insights about memory and its role in the formation and structure of personal identity (Jungert, 2015).

For neuroscience, the value of neurophilosophical approaches consists of:

- Firstly, the discovery of implications of neuroscientific findings for topics that were originally outside the disciplinary focus. Examples include personal identity (Mathews, Bok, & Rabins, 2009), the nature of desire (Schroeder, 2004), or phenomenal consciousness (Clark, 1993).
- Secondly, the chance to make use of philosophical tools and methods of investigation. If neuroscientists are willing to engage in serious interdisciplinary dialog with philosophers, this provides the opportunity to benefit from conceptual and logical analysis. In contrast to some philosophers of neuroscience, neurophilosophers will consider the application of philosophical methods to neuroscientific findings not as a way of correction and falsification from a neutral outside perspective. In fact, they will see it as part of mutual learning and exchange that aims at a better understanding of complex mental phenomena by means of close collaboration on equal terms.

While it was my aim to mainly discuss the chances and positive effects of a certain way of understanding neurophilosophy, there are of course also certain problems and limitations. The outlined understanding could be characterized as "weak neurophilosophy," as it preserves the autonomy of the disciplines involved and abstains from strong claims regarding reduction or elimination. For some neuroscientists and neurophilosophers this conception will be way too careful and conservative.

In addition, the proposed way of integrating neuroscientific findings into philosophical theory might work well regarding topics like memory or perception. However, the intense debate on the possibility and meaning of "neuroethics" (see, e.g., Churchland, 2011; Farah, 2011;

Levy, 2009) suggests that it will not or only to a lesser extent work for other fields. To decide which fields can or cannot be an object of neurophilosophy and to determine boundaries and objections, one in turn needs help from the philosophy of neuroscience, and so the circle is complete.

References

Bechtel, W. P. (1994). Levels of description and explanation in cognitive science. *Minds and Machines, 4*(1), 1−25.

Bennett, M. R., & Hacker, P. M. S. (2003). *Philosophical foundations of neuroscience.* Malden, MA: Blackwell.

Berntsen, D., & Rubin, D. C. (2006). Emotion and vantage point in autobiographical memory. *Cognition and Emotion, 20*(8), 1193−1215.

Bickle, J. (2003). *Philosophy and neuroscience: A ruthlessly reductive account.* Norwell, MA: Kluwer Academic Press.

Bickle, J. (2008). Real reduction in real neuroscience: Metascience, not philosophy of science (and certainly not metaphysics!). In J. Hohwy, & J. Kallestrup (Eds.), *Being reduced: New essays on reduction, explanation, and causation* (pp. 34−52). Oxford: Oxford University Press.

Bickle, J., Mandik, P., & Landreth, A. (2012). The philosophy of neuroscience. In E. N. Zalta (Ed.), *Stanford encyclopedia of philosophy* (Summer 2012 ed.). <http://plato.stanford.edu/archives/sum2012/entries/neuroscience/>.

Choudhury, S., & Slaby, J. (2012). *Critical neuroscience: A handbook of the social and cultural contexts of neuroscience.* Malden, MA: Wiley-Blackwell.

Churchland, P. S. (1986). *Neurophilosophy.* Cambridge, MA: MIT Press.

Churchland, P. S. (2011). *Braintrust: What neuroscience tells us about morality.* Princeton, NJ: Princeton University Press.

Clark, A. (1993). *Sensory qualities.* Cambridge: Cambridge University Press.

Clark, A. (2010). Memento's revenge: The extended mind extended. In R. Menary (Ed.), *The extended mind* (pp. 43−66). Cambridge, MA: MIT Press.

Craver, C. F. (2005). Beyond reduction: Mechanisms, multifield integration and the unity of neuroscience. *Studies in History and Philosophy of Science, 36*(2), 373−395.

Crick, F. (1995). *The astonishing hypothesis: The scientific search for the soul.* New York: Touchstone.

Debus, D. (2007). Being emotional about the past: On the nature and role of past-directed emotions. *Noûs, 41*(4), 758−779.

Farah, M. J. (2011). Neuroscience and neuroethics in the 21st century. In J. Illes, & B. J. Sahakian (Eds.), *Oxford handbook of neuroethics* (pp. 761−781). Oxford: Oxford University Press.

Gold, I., & Stoljar, D. (1999). A neuron doctrine in the philosophy of neuroscience. *Behavioral and Brain Sciences, 22*(5), 809−830.

Hoerl, C. (1999). Memory, amnesia, and the past. *Mind and Language, 14*(2), 227−251.

Hyman, J. (1989). *The imitation of nature.* Malden, MA: Blackwell.

Jungert, M. (2013). *Personen und ihre Vergangenheit. Gedächtnis, Erinnerung und personale Identität.* Boston, MA/Berlin: De Gruyter.

Jungert, M. (2015). Memory, personal identity, and memory modification. In R. Ranisch, M. Rockoff, & S. Schuol (Eds.), *Selbstgestaltung des Menschen durch Biotechniken* (pp. 129−140). Tübingen: Francke.

Kane, R. (Ed.). (2011). *The Oxford handbook of free will* (2nd ed.). Oxford: Oxford University Press.

Klein, S., & Nichols, S. (2012). Memory and the sense of personal identity. *Mind*, *121*(483), 677−702.

Levy, N. (2009). Neuroethics: Ethics and the sciences of the mind. *Philosophy Compass*, *4*(1), 69−81.

Machamer, P. K., McLaughlin, P., & Grush, R. (Eds.). (2001). *Theory and method in the neurosciences*. Pittsburgh, PA: University of Pittsburgh Press.

Mathews, D. J. H., Bok, H., & Rabins, P. V. (Eds.). (2009). *Personal identity and fractured selves: Perspectives from philosophy, ethics, and neuroscience*. Baltimore, MD: Johns Hopkins University Press.

Mele, A. R. (2010). Testing free will. *Neuroethics*, *3*(2), 161−172.

Mele, A. R. (Ed.). (2015). *Surrounding free will: Philosophy, psychology, neuroscience*. Oxford: Oxford University Press.

Northoff, G. (2004). What is neurophilosophy? A methodological account. *Journal for General Philosophy of Science*, *35*(1), 91−127.

Runyan, D. (2014). *Human agency and neural causes: Philosophy of action and the neuroscience of voluntary agency*. Basingstoke: Palgrave Macmillan.

Satel, S., & Lilienfeld, S. O. (2013). *Brainwashed: The seductive appeal of mindless neuroscience*. New York: Basic Books.

Schacter, D. L., Chiao, J. Y., & Mitchell, J. P. (2003). The seven sins of memory: Implications for self. In J. E. LeDoux, J. Debiec, & H. Moss (Eds.), *The self: From soul to brain* (pp. 226−239). New York: Wiley.

Schechtman, M. (1994). The truth about memory. *Philosophical Psychology*, *7*, 3−18.

Schlosser, M. E. (2014). The neuroscientific study of free will: A diagnosis of the controversy. *Synthese*, *191*(2), 245−262.

Schouten, M. K., & Looren de Jong, H. (Eds.). (2007). *The matter of the mind: Philosophical essays on psychology, neuroscience, and reduction*. Malden, MA: Blackwell.

Schroeder, T. (2004). *Three faces of desire*. Oxford: Oxford University Press.

Slaby, J. (2010). Steps towards a critical neuroscience. *Phenomenology and the Cognitive Sciences*, *9*(3), 397−416.

Squire, L. R. (2004). Memory systems of the brain: A brief history and current perspective. *Neurobiology of Learning and Memory*, *82*(6), 171−177.

Tallis, R. (2014). *Aping mankind: Neuromania, darwinitis and the misrepresentation of humanity*. London: Routledge.

Walter, H. (2001). *Neurophilosophy of free will*. Cambridge, MA: MIT Press.

Wolfe, C. T. (Ed.). (2014). *Brain theory: Essays in critical neurophilosophy*. Basingstoke: Palgrave Macmillan.

Wollheim, R. (1984). *The thread of life*. Cambridge: Cambridge University Press.

Philosophical Puzzles Evade Empirical Evidence: Some Thoughts and Clarifications Regarding the Relation Between Brain Sciences and Philosophy of Mind

I. Sarıhan

Central European University, Budapest, Hungary

In this chapter, I chart the general territory of interdisciplinary interaction between analytic philosophy and brain sciences and the relevance the two fields have for each other. I differentiate two main styles of interaction, from philosophy-to-neuroscience and from neuroscience-to-philosophy. The first type of interaction is motivated by philosophers' interest in the issues that arise within neuroscience, the findings of the field, and the claims made by neuroscientists in the context of the philosophers' function as a conceptual and logical corrector or aid toward a more accurate science. The other aspect concerns the relevance of the findings of neuroscience for the resolution of philosophical debates. I will make some brief remarks about the first, and then investigate more closely the second type of interaction. I will not discuss the controversial field called "experimental philosophy."

The Human Sciences after the Decade of the Brain.
DOI: http://dx.doi.org/10.1016/B978-0-12-804205-2.00002-1
© 2017 Elsevier Inc. All rights reserved.

Especially since the 1980s and the emergence of "neurophilosophy," more and more philosophers have been closely following the findings of neuroscience and other related sciences and bringing home morals for philosophical questions. I will discuss whether these attempts succeed or not in working toward a resolution of philosophical debates. I will mention two cases—one regarding a puzzle in philosophy of perception and one regarding the mind-body problem—and I will argue that such attempts are not successful in trying to answer questions like whether we see the external world directly in perception or whether psychology can be reduced to neuroscience.

This failure is not due to any problems with the neuroscientific data itself, but results from an ability of the philosophical questions to evade the data, i.e., the data mentioned fails to settle the philosophical debates conclusively. There is a reason for this: If those philosophical questions could be settled by empirical evidence, they wouldn't be philosophical questions in the first place, they could be reframed as scientific, empirical questions. What makes these questions persisting philosophical questions is precisely that there is no way to settle them through empirical evidence. Rather, they are conceptual questions, some of them also with phenomenological aspects, and their solution lies in conceptual analysis and/or phenomenological methods.

PHILOSOPHY-TO-SCIENCE: ETHICS, LOGIC, TERMINOLOGY, AND CLARITY

There are many ways philosophy functions as an aid and an inspector for science. For continental philosophy, this is generally by highlighting and discussing various normative, political, and societal aspects of scientific theories, terminologies, and processes, but this chapter is about analytic philosophy, which I will refer simply as "philosophy" from now on. I will very briefly identify some functions of philosophy in relation to science.

One contribution of analytic philosophy to sciences is through introducing new conceptual and terminological tools to a field (or the criticism of the introduction of such tools). For instance, the use of the term "representation" in cognitive neuroscience comes through a long philosophical tradition, a famous example promoting the use of such a conceptual scheme being Fodor's seminal work (1975). This is, of course, something that is also often done by people who are not necessarily philosophers by profession but who are working on the theoretical levels in a scientific field. Another function common to both philosophers and scientists is asking questions that can inspire scientific work.

Indeed, the philosopher and the theoretical scientist is hard to distinguish at this level. The philosophizing aspect of the philosopher comes into the picture where the contribution to be carried out by the philosopher involves the kind of conceptual analysis and logical argumentation typically found in analytic circles.

One thing a philosopher can do, qua philosopher, is to criticize a certain idea that appears in a scientific context on the grounds that it doesn't logically follow from the evidence, or to criticize a terminology on the grounds that concepts deployed are problematic. To give an example, Bennett and Hacker (2003) criticize the conceptual background of neuroscience from a Wittgensteinian metaphilosophical perspective, arguing that many terms used in neuroscience are confused, in the sense that they rest on misunderstandings of terms in everyday vocabulary (Bennett & Hacker, 2003). Another prominent example of this type of contribution is the debate sparked by the empirical findings of Libet on the basis of which some have claimed that we don't have free will because before we are aware of a mental act of decision-making, a certain unconscious process that factors into which decision is to be made is already going on in our brains (Wegner, 2002). This conclusion was challenged by philosophers in many ways, e.g., by claiming that it relies on a misunderstanding of the term "free will": When one analyzes the concept, it is claimed, one sees that in order to freely perform actions one doesn't need to be aware of a mental act of "deciding" or "willing"; in order for an action to be free other criteria are sufficient, such as the action's being in line with one's beliefs and desires and the decision not being taken under duress (see O'Connor, 2010, for a review of various analyses of free will; also see Dennett, 2003, Chapter 8, for a criticism of Wegner).

We should recognize that many such debates around science are actually spillovers from philosophers' debates. For instance, "qualia" is a term that originally appeared in philosophical literature, and started to appear in certain scientific work especially after the 1990s (see, e.g., Edelman & Tononi, 2000; Ramachandran & Hirstein, 1997). But many philosophers claim that the concept of qualia is confused and attempts to refer to something that doesn't exist, if it is attempting to refer to an internal quality, and according to this view scientists who are trying to solve the problem of qualia are dealing with a pseudo-problem. (Representationalists like Byrne (2006) or Wittgensteinians like Bennett and Hacker (2003) are examples.) This is nothing other than the philosophers' debate carrying itself over to the battleground of science. Similarly, going back to the Libet example above, Dennett (2003) claimed that the confusion regarding free will based on Libet's experiments has a long philosophical background that goes back at least to Descartes.

One last function of philosophy that should be mentioned is the role played by ethics in the context of science. Ethics of neuroscience is a relatively well-established field, and is involved in various important worldly matters such as neuro-marketing, neuro-enhancement, and animal experimentation.

SCIENCE-TO-PHILOSOPHY: CAN SCIENTIFIC DATA SETTLE PHILOSOPHICAL DEBATES?

Now we will look at the other side of the interaction, taking empirical data and trying to come to conclusions regarding the traditional questions of philosophy. By going through two cases, one regarding the question of whether we see the outside world directly or not, and the other regarding the question of whether the mind can be reduced to the brain, I will try to show this direction of interaction is not very fruitful, for systematic reasons due to the nature of philosophical questions. Of course, there are other science-to-philosophy directions of intellectual contribution that I will not talk about, say, some new empirical data providing a good case example to discuss something in a philosophical context or triggering novel philosophical questions. An empirical finding can also have some relevance for ethics. For instance, if it turns out that a creature feels pain and pleasure we are obliged to behave differently toward that creature. However, it should be remembered that we have these obligations already in the context of a background ethical code, itself a matter of philosophical debate untouched by empirical evidence, like not causing unnecessary harm or not terminating otherwise pleasurable lives.

Case One: Philosophy of Perception

One philosophical question that has persisted for centuries is whether or not we see the external world directly; whether the things we seem to be directly acquainted with in perception are objects and properties in the external world or whether they are internal or intermediary items in some sense. These intermediary items have had various names, most famously "sense-data" or sometimes "sensations." More recently, some have adopted the term "qualia" for this terminological purpose (Wright, 2008), even though more often the term has been deployed to mean something slightly different.

Many philosophers and scientists have insisted that the findings of many empirical fields, particularly psychophysics and neuropsychology of conscious experience, have given us proof that "indirect realism," the philosophical view that we don't perceive the external world directly, is the correct view (Revonsuo, 2006; also see science-based articles in Wright, 2008). According to this approach, this centuries-old philosophical question has been waiting for the emergence of scientific evidence to be settled. However, unfortunately the evidence doesn't settle the issue, and even if it did, much simpler evidence could be just enough, without extensive modern-day research.

The science-based approach to settle this question relies mainly on the fact that every type of experience one can have depends on a certain internal, neural state, which seems to be enough to have that type of experience without an external corresponding object, as is the case in illusions, dreams, and hallucinations. One can see a sunset in a dream, one can see an object as blue even if it is some other color, or still have the lingering phantom feeling of an amputated arm. Since the hallucinatory cases and the externally caused veridical counterparts are subjectively the same experiences, and since we are both aware of something in both cases, then, just like we are not directly aware of an external entity in the illusory cases, we are not directly aware of the external world in the veridical case either. Rather, what we are directly aware of are models or internal images constructed by the brain. This idea is also supplemented by the finding that there is a lot of constructive work going on in one's brain during ordinary perception, say, the filling in of the blind spot or various other "inferences" the perceptual system makes to construct an image of the world.

This old philosophical argument that predates contemporary science is known as the "argument from error" (also "argument from hallucination" or "argument from illusion") in the philosophical literature. Note that if the argument could settle the debate, it could have settled it a long time ago without help from the rigorous empirical science of today, since hallucinations, illusions, and dreams have long been well-known phenomena. One doesn't need contemporary neuroscience to know that one can have the experience of seeming to see a particular object or particular quality without there being a corresponding thing in the environment. So, the first point is that, if empirical data could settle the issue, hard science wouldn't be necessary.

The second point is, empirical evidence cannot settle the issue because the conclusion of the argument, that we do not see the world directly, rests on a particular understanding of certain terms like "seeing." A common objection to the argument from error has been that when we hallucinate something, we don't really see anything, rather, we "seem to see" something, i.e., we have a hallucinatory experience of seeing something with our eyes. So one cannot generalize from hallucination to veridical perception and say that what we see in both cases is the same and therefore in veridical perception we don't see the external world directly. In hallucination we are aware of something for sure, the "content" of our experience, but this doesn't make it impossible for us to be in direct empirical touch with the external world. In hallucination, the content doesn't match the outside world, in veridical perception, it matches, and via "having" this content (as opposed to "seeing" or "perceiving" this content) we are aware of the external world directly without being aware of anything else. It is indeed the case that the brain has some constructive role in everyday perception, and the sensory content regarding what is in front of us provided by this constructive process partly distorts the reality out there. However, such distortions need not force us to say that we indirectly see the world via seeing an internal image. The distortion is compatible with the fact that where our perceptual system does not distort reality, we see it directly, and where it is distorted, we siimply fail to see, rather than seeing an internal image. Think of a visual state, which is partly illusory—you experience a scene that represents the world as mainly as it is but that includes the illusory appearance of a bent stick in water. Some would like to analyze this perceptual state as the brain constructing an image of the world that is partly incorrect, and the immediate object of our awareness is this internal image. But the brain's construction of an image, a model, or an appearance of the outside world need not be analyzed in this epistemic way. One can as well say that as a result of some external and internal processes, the world appears to us in a certain way, and if this appearance is truthful, we see the world directly, and the misleading representation of the bent-stick-in-water shouldn't be understood as "seeing" an image of a bent stick, but failing to see a particular spatial property of the stick because of having a misleading visual representation of the world.

This is one among various responses given to the argument from hallucination. I am not interested in logically proving it here, but to demonstrate that one's reliance on empirical data to argue for indirect realism rests on a certain understanding of certain concepts like seeing, awareness, perception, etc., and this conceptual debate cannot be settled by empirical evidence. And if it could, we wouldn't need hard science, since the mere existence of illusions and hallucinations could prove this point.

Case Two: The Mind-Body Problem

The mind-body problem is the problem of understanding what the relation between the mind and body is, or more precisely, whether mental phenomena are a subset of physical phenomena or not. There are many philosophical positions associated with this problem—

substance dualism ("mind and body are two different substances"), property dualism ("there is only one, physical substance, but mental properties of subjects cannot be reduced to their physical properties"), and physicalist reductionism ("mental properties can be identified with, or can be spelled out in terms of, physical properties"), among other positions.

Some philosophers in recent decades have argued that modern neuroscience has already given us an answer to this question: Mental states are nothing other than neural states, and we can talk of mental phenomena through physical vocabulary without any loss of meaning or reference. The founders of what is called "neurophilosophy," Patricia and Paul Churchland, have been among the most famous advocates of this position, even though their views oscillated between reductionism and eliminativism, the latter view being that mentality (or certain aspects of it) is a prescientific construct that will have no place in the scientific understanding of the world once we have a fully developed neuroscience (see, e.g., Churchland, 1988).

This brand of reductionism relies heavily on the findings of neuroscience, often on quite detailed empirical data about the relation between certain psychological states and brain states or how neural processes generate behavior. Well, almost all brands of reductionism have some reference or other to the brain science, but they are often uninterested in detailed data. Rather, they simply point to the general scientific consensus that there is a very direct relation between mental and neural states, but this relation by itself does not play an important role in their arguments, for these arguments generally rely on a conceptual analysis of mental states to see if mental states could be reduced to *any* physical states to begin with, and if such reduction is possible, modern day science tells us which physical states are the reduction base, which turns out to be neural or bodily states. On the other hand, the kind of reductionism we are interested in here takes it that *science has proven* mind-brain identity.

However, the scientific evidence does not settle the philosophical question. No matter what detailed and direct mapping we establish between mental and neural states, there are so many options that remain on the table before we can proclaim that we have reduced mental processes to brain processes. The first obstacle is that correlation does not mean identity, and the reductionist should answer arguments to the effect that the relation is better explained as a correlation. The most important and pervasive argument I will mention here is the argument that, to put it roughly, the reduction of mental phenomena to physical phenomena does not make sense, on conceptual or logical grounds. A phenomenon described by a physical description like "such and such connectivity and firing in this and that brain area" simply cannot be identical to "feeling pain" or "thinking about Budapest" or "having a visual experience of a yellow lemon" as these mental phenomena have certain characteristics that the physical description does not capture, even though antireductionist philosophers disagree on what these characteristics are. Some say it is some subjective aspect, the "feel" or "what-it-is-likeness" associated with the mental state, others say it is "intentionality," the property of "having a content" or "being directed to an object," such as the object experienced or thought about. When we are given a description like "such and such neural firing," we cannot infer whether this state is an experience of the color green or color red or something else. This disparity is termed "the explanatory gap": A successful reduction, it is argued, should tell us not only an identity based on an observation of a correlation between the occurrence of mental and neural states, but also make us understand how is it possible that such and such mental firing could constitute experiencing the color red (Levine, 2001). Another very different antireductionist argument, a behaviorist one, is the argument that the mind cannot be the brain because mental terms do not strictly refer to states or processes, but dispositions to behave, and it doesn't make sense to identify a disposition with a state or process (Ryle, 1949).

Of course, philosophers have answers to these challenges. Some claim that identities don't require an explanation (Block & Stalnaker, 1999), others try to analyze mental phenomena into physical or "topic-neutral" vocabulary (vocabulary that is common both to physical and mental terminology, such as the vocabulary of basic ontological phenomena like causation; the approach was popularized by Smart (1959) and many causal-informational analyses of mental phenomena can be considered as a continuation of this strategy). To give a more specific example, some philosophers like Millikan analyze mental states as natural indicators, e.g., "thinking of X" is being in a state that has the function of indicating the presence of X. So in the case of humans, it is a brain state that has acquired the biological function of signaling the presence of X through the organism's interaction with its environment (Millikan, 1984). Again, it is not important for us here whether this analysis is correct or not. But if this analysis is correct, then there is no obstacle for mental states to be reduced to neural states—as instances at least, if not as scientific or functional kinds, since different types of neural states can realize the same mental states—and one doesn't need to look at detailed empirical data to claim that reduction is possible, given that every mental state will correlate with some neural state. From then on, if Millikan's analysis is accepted, whether mental state M is identical to neural state N is philosophically uninteresting in the context of the mind-body problem, i.e., further empirical evidence does not add anything to the solution of the philosophical problem.

CONCLUSION

The two examples above have shown us that philosophical puzzles cannot be settled by empirical evidence, and it is not due to any specific features of these two cases, it is something that results from the nature of philosophical questions. Philosophical questions are often classified as questions of a logical or conceptual nature. But let's avoid here the task of defining what a philosophical question is; it is enough to mention a negative aspect: Philosophical questions are not empirical questions. If they could be conclusively settled down by empirical evidence, then they would be turned into scientific questions and taken from the hands of philosophers. This is arguably what happened with some philosophical questions in the past, questions that stopped being philosophical questions and turned into questions of physics or biology, such as some questions regarding spatio-temporal behavior of objects, questions regarding basic elements, some questions related to vision, and issues about the origins and diversity of lifeforms. (Think of ancient Greek theories of objects moving toward their "natural place" or the philosophical theory of vision that postulated beams emitted by the eyes.) Answering questions like whether we see the world directly or whether mentality can be reduced to physics requires first of all an agreement on what "seeing the world directly" means or what we mean by "mentality" or "reduction." And when these questions are settled, we still won't be able to say that the "philosophical question" can be answered empirically, because the philosophical part of the question would already be solved.

Another point about this nonempirical nature of philosophical questions is that their solution is of little or no practical concern. If we cannot settle a question empirically, i.e., by intervening with the world and then observing the outcomes, then the settlement of the question won't give us any advantage of intervening with the world to bring about outcomes either. Of course, if a philosophical question has an ethical or normative aspect, then one's converging on one answer or the other has consequences for one's personal behavior or how we conduct science. But otherwise, the resolution of the mind-body problem or the philosophical problems of perception don't seem to add anything to our ability in scientific contexts, in terms of, say, prediction or engineering. No matter whether the mental properties are identical to neural properties or are just irreducible and systematically correlated with neural properties, we still get the same results in neuroimaging or behavioral neuroscience. No matter if we see the world directly or not, we get the same results in psychophysics. Still, one can expect a scientist to be careful about what background philosophical view she takes, if she takes any at all, while using a set of terms and concepts in relation to the scientific work. This means either putting the philosophical questions aside, or, when making big claims, being aware of rich and complicated debates going on in the contemporary philosophical background surrounding such issues. While the very act of looking into the brain and acquiring the data itself rests on certain background philosophical assumptions, many are unimportant in the particular context of the interaction of neuroscience and philosophy of mind. Instead, they are related to general philosophical issues surrounding science and observation, and a scientist cannot be expected to worry about all such philosophical problems before taking up some work. But after the empirical part is done, when one starts making claims like there is no free will, the mind is the brain, or we don't see the external world directly, one has stepped into the domain of philosophy of mind.

References

Bennett, M. R., & Hacker, P. M. S. (2003). *Philosophical foundations of neuroscience*. Hoboken, NJ: Wiley-Blackwell.

Block, N., & Stalnaker, R. (1999). Conceptual analysis, dualism, and the explanatory gap. *Philosophical Review, 108*(1), 1−46.

Byrne, A. (2006). Color and the mind-body problem. *Dialectica, 60*(3), 223−244.

Churchland, P. M. (1988). *Matter and consciousness*. Cambridge, MA: MIT Press.

Dennett, D. C. (2003). *Freedom evolves*. New York: Viking Press.

Edelman, G. M., & Tononi, G. (2000). *A universe of consciousness: How matter becomes imagination*. New York: Basic Books.

Fodor, J. A. (1975). *The language of thought*. Cambridge, MA: Harvard University Press.

Levine, J. (2001). *Purple haze: The puzzle of consciousness*. Oxford: Oxford University Press.

Millikan, R. (1984). *Language, thought and other biological categories*. Cambridge, MA: MIT Press.

O'Connor, T. (2010). Free will. In E. N. Zalta (Ed.), *Stanford encyclopedia of philosophy* (Fall 2014 ed.). <http://plato.stanford.edu/archives/fall2014/entries/freewill/>.

Ramachandran, V. S., & Hirstein, W. (1997). Three laws of qualia: What neurology tells us about the biological functions of consciousness, qualia and the self. *Journal of Consciousness Studies, 4*(5-6), 429−458.

Revonsuo, A. (2006). *Inner presence: Consciousness as a biological phenomenon*. Cambridge, MA: MIT Press.

Ryle, G. (1949). *The concept of mind*. London: Hutchinson.

Smart, J. J. C. (1959). Sensations and brain processes. *Philosophical Review, 68*, 141−156.

Wegner, D. (2002). *The illusion of conscious will*. Cambridge, MA: MIT Press.

Wright, E. (Ed.). (2008). *The case for qualia*. Cambridge, MA: MIT Press.

3

"Who's Afraid of the Big Bad Neuroscience?" Neuroscience's Impact on Our Notions of Self and Free Will

R. Vieira da Cunha and J.B. Relvas

University of Porto, Porto, Portugal

INTRODUCTION

Neuroscience is coming of age but not all of its advancements mean good news for everyone. For some, neuroscience might obliterate the assumption that a society is based on rational and self-ruling persons wholly responsible for their actions while offering in exchange merely a mechanist and reductionist vision of a collective made of irrational and neuron-ruled bodies deprived of free will and moral and legal responsibility.

We begin by establishing the promises and the perils of neuroscience, i.e., the fantastic benefits but also the concerns brought about by neuroscience's advances. After determining the different kinds of concerns that neuroscience poses to our societies, we will focus on one of those concerns and begin by looking into "the selves of neuroscience," i.e., the latest progresses made by neuroscience about the self and the free will it commands, and compare them to "our selves," i.e., our folk conceptions of the self and their free will. An integral part of our

© 2017 Elsevier Inc. All rights reserved.

argument is the premise that there are too many different philosophical notions of self (and free will and responsibility) to be sure of what is actually being challenged by neuroscience and how grounded that supposed picture of rational and self-ruling persons wholly responsible for their actions is. Another important part of the argument are the findings of experimental philosophy and social psychology about the folk conceptions of self and free will, which do not seem to be undermined by new neuroscientific evidence. We conclude that the evidence is subject to many different interpretations and that "the jury is still out" on these matters. Thus perhaps neuroscience alone is unable to change the humanistic worldview and its assumption of the rational and self-ruling person with free will worthy of moral and legal responsibility.

THE PROMISES AND PERILS OF NEUROSCIENCE

One would have to be completely unaware of the scientific discoveries of recent years—and of their growing impact, both in media attention and in other disciplines' research trends (see, e.g., the ever-increasing field of neuro-prefixed disciplines such as neuroeconomics, neurolaw, neuroethics[a])—to deny that neuroscience and neuroscientific explanations[b] have made important progresses in the last few decades.

US President George H.W. Bush designated the 1990s as "the Decade of the Brain"[c] and great advances were made, from imaging technologies to pharmacological development. The biggest one was perhaps the

[a]The trend is omnipresent these days but 10 years ago Colin Blakemore had remarked "'Neuro-', like 'Psycho-', 'Cyber-' and 'Retro-', offers itself promiscuously for the creation of neologisms" (Blakemore, 2005, p. v). A more recent effort by Muzur and Rinčić (2013) criticizes this trend while trying to map the origins of many such neuro-prefixed neologisms.

[b]We use "neuroscience," "neurosciences," and related terms like "neuroscientific findings" in a very broad sense to denote all the scientific endeavors to study mental phenomena in humans from a biological point of view, with a focus on the brain and all of the nervous system.

[c]The Decade of the Brain was a designation for 1990–99 by US president George H.W. Bush as part of a larger effort involving the Library of Congress and the National Institute of Mental Health of the National Institutes of Health "to enhance public awareness of the benefits to be derived from brain research" (Presidential Proclamation 6158, http://www.loc.gov/loc/brain/proclaim.html). As seen by the number of brain research projects in the 2000s (Allen Brain Atlas, Human Connectome Project, BRAIN Initiative, and Human Brain Project, to mention just a few), the 2000s have not lagged behind the 1990s, as far as brain research goes.

realization that our brain continues to grow new cells even in adult-hood, albeit at a slower pace than in earlier years. As it turns out, by the beginning of the 2000s, we were coming to terms with the idea that our brain, the repository of our identities, of our selves, never ceases to change (Blakeslee, 2000).

Neuroscientific findings have continued into the 2000s and the 2010s and we now have ways to relate thoughts to images and the extension of our "mind reading" abilities, if you will, have gone way beyond crude CAT (computed axial tomography) scans to MRI (magnetic reso-nance imaging) and particularly fMRI (functional magnetic resonance imaging). While the first relies on X-ray images to generate computer-processed "slices" of specific areas of the brain, MRI and fMRI depend on magnetic fields to detect brain activity indicated by blood flow—BOLD (blood-oxygen-level dependent) imaging is the most common form of fMRI. Recent studies have allowed us to discern people's prefer-ences, to relate specific thoughts to images and access mental content, to predict visual responses and unconscious vision, and even to exchange information with patients unable to communicate (see Evers & Sigman, 2013; Farah & Wolpe, 2004).

Furthermore, it is not only our abilities to read and interpret activi-ties in the brain that have improved in the last decades. We can now tinker with the brain in ways that seemed mere science fiction some years ago, and it is increasingly harder to draw the line between real and hypothetical technologies when it comes to neuroenhancement.[d]

Individuals with no medical condition may attempt to enhance their cognition from substances used for current treatments of Attention Deficit Hyperactivity Disorder (ADHD) and narcolepsy such as methyl-phenidate and modafinil (Repantis, 2013). Although there is no conclu-sive evidence of their benefits, with efficacy depending on a number of factors and no knowledge of their long-term effects (Smith & Farah, 2011), their use as study aids in higher education seems to be on the rise (Franke & Lieb, 2013; Wilens et al., 2008).

Along with promising enhancement benefits there comes a myriad of technologies that range from the deeply invasive, such as deep brain stimulation (DBS) to the noninvasive, like transcranial direct current stimulation (tDCS) and transcranial magnetic stimulation (TMS). While the first involves the surgical implantation of electrodes in specific regions of the brain to stimulate them and treat, for instance, depression (Mayberg et al., 2005), the latter use either electrical current or weak

---

[d]Throughout the chapter we prefer the expression "neuroenhancement" but see Hildt (2013) on the distinction between related terms such as "cognitive enhancement," "mood enhancement," and other kinds of enhancement.

electromagnetic fields to stimulate specific brain areas without the need for surgery and have been used to treat, for instance, addiction (Conti et al., 2011; Jansen et al., 2013). Lastly, neurofeedback can also be considered as an enhancement technology. It is a form of brainwave training that works by measuring the patterns of electrical current in the brain and presenting sensory stimulus to recondition and retrain the brain (Hammond, 2007).

All of these technologies have therapeutic as well as enhancement potential: DBS has been reported to improve memory (Synofzik & Schlaepfer, 2008); electrical stimulation may improve sensorimotor skills, memory, attention, and problem solving (Krause, Márquez-Ruiz, & Cohen Kadosh, 2013); and neurofeedback has been shown to have positive effects on reaction times, spatial abilities and creativity (Doppelmayr & Weber, 2011), and on mood and cognition (Vernon et al., 2009).

Although those advances infuse us with hope of better clinical solutions and solutions to many medical conditions, they are not met without concern. There are ethical problems, namely privacy issues, arising from the use of the new imaging technologies on which neuroscience so heavily relies (Farah & Wolpe, 2004). Their use as lie detectors or general crime-prevention tools can have many more social than scientific problems (Farah, Hutchinson, Phelps, & Wagner, 2014). Additionally, enhancement technologies bring their own ethical and social concerns. To begin with, what counts as enhancement and what is merely therapeutic? In the case of TMS, for instance, using it for a clinically depressed person is certainly therapy but, not to even mention the problem with threshold diagnosis, what do we make of healthy subjects using it for mood improvement?[e] Will society as a whole benefit from the enhancement of some (but not all) of their members? Would this reduce or increase social inequalities? Would this be fair to those deprived of access to enhancement technologies? Which characteristics should we enhance? And what if those enhancements change our personality and our sense of self? Should we enhance our children? Is it acceptable to demand certain professionals use enhancement technologies? These and many other similar questions have been discussed for some time now and it is far from certain that discussion alone will lead us to consensus.[f]

[e]It is unclear whether TMS, even if applied in a repetitive manner, can have any effect on the mood of healthy subjects—Moulier et al. (2016) find no significant effects but do report improvement in adaptation to daily life.

[f]For an overview of the moral considerations, see Brukamp (2013) or Hauskeller (2013). Savulescu, ter Meulen, and Kahane (2011) reviews some of the problems, and a recent consideration of ethical problems can be found in Nagel (2014).

A second kind of concern raised by neuroscience has to do with its implications for the humanities and the social sciences. For some, neuroscience is ruining the humanities by putting too much emphasis on how ideas are produced rather than on their meaning. The current *zeitgeist* seems to place a very powerful and attractive aura on the reasoning of neuroscience, so much so that it is feared that the growing charm of neuroscience will undermine our attraction for ideas and its meaning, which will be substituted by some sort of irrational fixation on the causal processes and the physical substrates of those ideas (Krystal, 2014). McCabe and Castel (2008) have delved into that allure and argued that at its base lies our affinity for reductionist explanations of cognitive phenomena. That is why they have found that articles with brain images are more convincing than those with other illustrations. Weisberg, Keil, Goodstein, Rawson, and Gray (2008) claimed to have shown that the introduction of neuroscientific material when explaining psychological experiences triggers more public (nonexpert) interest, even if the material is irrelevant and the explanation is a bad one. This effect was only present in nonexpert groups, but it should be noted that this included students of cognitive neuroscience with a semester's worth of instruction. A tentative explanation offered by the authors hinges also on a reductionist bias, with its origin in the general public's familiarity with interlevel explanations and the description of higher-level phenomena through lower-level theories and concepts. "Neuroenchantment"[8] or "neurohype" seems thus to be corelated with reductionism (even if a semantic or epistemic version of it) and so may pose a peril for the humanities and the social sciences, contributing to their neglect in our societies.

However, such a reductionist propensity can be even more concerning if it endangers not only the place of the humanities and the social sciences in our societies but also the latter's humanistic view of the world. It has been suggested that the brain has become the place of the modern self (or person) in industrialized and highly medicalized societies, making it "the anthropological figure inherent to modernity" (Vidal, 2009). This coheres with the "neuro-essentialism" (Racine, Waldman, Rosenberg, & Illes, 2010) that some diagnose in

---

[8]This persuasive power has been confirmed by Keehner, Mayberry, and Fischer (2011), as well as by McCabe and Castel (2008), although Gruber and Dickerson (2012), Michael, Newman, Vuorre, Cumming, and Garry (2013), and Hook and Farah (2013) have thrown some doubt on this claimed power of neuroscientific explanations and brain images over the layperson's mind. Schweitzer, Baker, and Risko (2013) have claimed that there is, however, such a "neuroimage bias," but only within certain limited contexts, while Ali, Lifshitz, and Raz (2014) still argue for the power of what they call "neuroenchantment."

our contemporary societies, at least in the media. "Neuro-essentia-lism"[h] is characterized as the understandings of the brain as the quin-tessence of a person, the secular equivalent to the soul. According to Reiner (2011, p. 161), "for all intents and purposes, we are our brains." Unlike in the second concern, here it is not about which discipline has explanatory legitimacy but really about our ontologies: could it be that many of the assumptions on which we base our view of the world and our place in it are mere illusions and fairytale stories that neuroscience is busting? One might express this by saying that Crick's famous claim that "you are your brain" (Crick, 1994) threatens to dominate our worldviews or by using Flanagan's (2002) shorthand of "the problem of the soul" to refer to the same fear: that neuroscientific findings are incompatible with widespread and foundational beliefs in our socie-ties, such as that we are rational, autonomous, and self-ruling persons, truly and fully structured permanent and immutable selves, governing our lives by ideas and values and, as such wholly responsible for our actions. The concern is that all that neuroscience has to offer in exchange is merely a mechanist and reductionist vision of a collective formed by irrational, nonautonomous, and neuron-ruled bodies, con-ventional selves conducting their lives on illusions of self-hood, free will, and moral and legal responsibility.

We thus have three kinds of concerns—three perils, if you will—brought by neuroscience: that its expanding field of knowledge and technological application raises even more ethical problems, both in therapy and in enhancement (1), that its status among other disciplines threatens the humanities and the social sciences (2), and that its find-ings challenge some of our most important everyday assumptions and our humanistic view of the world (3).[i] In this chapter, having already

---

[h]The identifications of subjectivity or personhood or self-hood with the brain has had many labels: Vidal (2009) uses the term "brainhood" for a very similar understanding of the "neuro-essentialist" position dubbed by Racine. For a famous statement of such identifications, see Michael Gazzaniga's (2005, p. 31) comment (about the possibility of a brain transplant): "This simple fact makes it clear that you are your brain."

[i]We are aware that by setting the question in these terms ("our humanistic view of the world" vs "neuroscientific findings") we may be criticized by embracing or perpetuating some sort of two-culture distinction or by granting the humanities a more prominent role in people's hearts and minds than neuroscience. We believe it is clear from this work that the only reason this oppositional formulation is being used is because it simplifies a matter that would otherwise be too convoluted for such a short chapter. In a brief provisional attempt of defense, we can argue that the new element in the discussion, particularly in the last few years, is precisely the neuroscientific findings and their interpretations and it is by that reason alone that they are pitted against the previous cultural notions, be they popular or academic, theoretical or practical.

described the foundation of these worries and briefly delved into (1) and (2), we intend to focus specifically on (3).

THE SELVES OF NEUROSCIENCE AND OUR SELVES

The working assumption of most neuroscientists is usually said to be that all psychological properties are emergent properties of the brain, and that no change in thought or sense of selfhood could occur without corresponding changes in neurophysiology. Although practically everyone can agree on this, the exact form of tracing mental properties to brain properties—or mental states to brain states—is one of the most difficult problems to solve, for scientists as well as for philosophers.

Even though Dennett famously claimed that "It is a category mistake to stark looking around for [selves] in the brain" (1992, p. 109), some scientists try to pinpoint the specific brain areas or brain functions involved in self-related thoughts or actions.[j] Turk, Heatherton, Macrae, Kelley, and Gazzaniga (2003) group together certain beliefs and knowledge about our selves and relate it to the activity of the left hemisphere, while Platek et al. (2003, p. 147) claim that there "is growing evidence that processing of self-related information (e.g., autobiographical memory, self face identification, theory of mind) is related to activity in the right frontal cortex" (see also Devinsky, 2000; Miller et al., 2001).

According to Vogeley and Gallagher (2011), on which this brief synthesis is based, the entire cortex seems relevant for self-referential processing, echoing Gillihan and Farah's (2005) conclusion that the brain lacks a specific region for self-related representations.

Nevertheless, the same Vogeley and Gallagher conclude that the self, which at first seemed to be spread across all the brain, "now seems to be nowhere" (2011, p. 118), since on some authors' interpretation (namely Legrand & Ruby, 2009), there is no region of the brain whose activation is exclusively for the self. Indeed, these authors (Legrand & Ruby, 2009) show, by reviewing neuroimaging studies, that self-related activities involve a widespread cerebral network that is also engaged during many other states (mind reading, memory recall, and reasoning), which is explained by the involvement of cognitive processes common to all those tasks.[k] In fact, self-referential mental activity,

---

[j]Whether or not scientists studying the brain for self-related processes is the same as looking for the self in the brain is a moot question. The fact is that such a scientific endeavor is very usually presented in that reductionist manner—see, for instance, the title in Feinberg and Keenan (2005).

[k]A previous review (Gillihan & Farah, 2005) of the neuroscientific findings about the self also concluded that there are no clear results for specialized brain regions.

particularly in the resting state, has been repeatedly related to the so-called default mode network (DMN), which consists of a set of areas in the cortex that show an increase in activity when people's mind is left to wander independent of external stimulus or any task-oriented processes (Davey, Pujol, & Harrison, 2016; Qin & Northoff, 2011; Raichle, 2015; Shulman et al., 1997; Spreng and Grady, 2010). However, it is still too soon to call the DMN the place of the self, since it does not seem to be self-specific (there are many other processes that also involve the DMN) nor self-exclusive (there are other areas involved in self-related processes).

But what do we make of all these findings? Do they confirm a neuroessentialist and reductionist view of the person or not? Some authors (Farah & Heberlein, 2007; Illes & Racine, 2005) seem to believe very clearly that they do lay the ground for a reductionist to do his work, while others (Buford & Allhoff, 2007, for instance) maintain that reductionists have no reasons to look for allies in neuroscientists.[1]

This lack of certainty about the self and our sense of self-hood has not prevented the media from blasting such headlines as "Neuroscience backs up the Buddhist belief that 'the self' isn't constant, but ever-changing" (Goldhill, 2015) presenting some form or another of the "neuro-essentialism" alluded above (Racine et al., 2010).[m] Racine et al. (2010) identified not only a lack of details but also a clear optimism in the media (UK and US) when depicting neuroscience. Furthermore, certain popularizations of scientific and philosophical positions, from authors such as V.S. Ramachandran, Thomas Metzinger, or Bruce Hood, may also play their part in creating a certain public view of neuroscience as endorsing certain views about the nonexistence of the self. A similar observation can be made about many fiction works, both books and films, which depict a similar reductionist and neuro-essentialist view: think, for instance, of the many Hollywood movies premised on brain transfers or similar brain-altering situations and its corresponding change of self.

It can certainly be objected that such reductionism and neuro-essentialism is to be found in the interpretations (by the media or by Hollywood, for instance) and not on the neuroscientific findings themselves. This, however, would miss the point that cultural change does not happen in a neutral and ahistorical ground but rather in a

---

[1]This sort of reductionist account can take many different forms, varying in degree in their intensity: for a representative position, see Churchland (1986) or Bickle (2003).

[m]It can still be called a neuro-essentialist position, even if is being claimed that there is no self, since the grounds for such claim is that the self cannot be found in the brain—i.e., if it is not in the brain, it cannot exist. Note that, in the case of Goldhill's newspiece, the original scientific article is actually about meditation, not specifically about the self.

web of influences that is not solely influenced by science. In any case, as many in the social studies of science have shown, scientific knowledge itself is not solely influenced by science. Furthermore, that neuro-essentialism is widespread among academics and laymen alike seems indubitable (McCabe & Castel, 2008; Satel & Lilienfeld, 2013; Weisberg et al., 2008).

Nonetheless, as concluded from the review of neuroscientific findings above, the jury is clearly still out on this issue of the self. Regarding the notions of free will and responsibility, the outlook does not differ greatly: even though there is no scientific consensus, a number of different findings are usually reported in the media in a sensationalist manner and the overall feeling for the layman, probably, is that free will, according to the scientists, is an illusion.

Perhaps Benjamin Libet's experiments in the early 1980s are to blame for such impression. In these experiments, participants were asked to make simple decisions related to motor tasks (flex a wrist) whenever they wanted and then report the moment they became conscious of their intention—which they kept track of by observing a modified clock, where a revolving hand moved faster than normal clock hands (Mele, 2014b). The participants' brain activity was measured during the experiment using electroencephalogram technology, and it was observed that it spiked on areas toward the front of the brain (that prepare actions) about 550 ms before the motion, even though participants' report of their intention was only 200 ms prior to the motion. Libet and many others concluded that free will could not have played a role here, since the "readiness potential" that caused the motion preceded the moment the participants reported having made the decision to move.

Libet's experiments have had great impact on other academic disciplines, with many following in his footsteps. Soon et al. (2008) is one such case, with fMRI being used to measure the participants' brain activity during the making of several simple decisions related to motor tasks (to press "freely," that is, whenever they felt the urge to do so, between one of two buttons) and reporting that the outcome of a decision can be encoded in certain areas (prefrontal and parietal cortex) of the brain activity up to 10 s before one is aware of it. This particular study deserves attention on two levels. On the one hand, because it is not clear, as in many other similar neuroscientific findings (Fried, Mukamel, & Kreiman, 2011 replicated Libet's experiments using single-neuron recording) that it really provides evidence that we don't have free will. As Mele (2014b) has pointed out, the encoding accuracy is only about 60% and, in any case, this test is about arbitrarily picking a button, which is very different from other kinds of decisions where we are usually interested in asserting free will and responsibility: to choose a course, to choose a partner, to have a child, etc.

On the other hand, this study, despite the openness to interpretation of its results, has been presented in the media as a "case closed for free will" (Youngsteadt, 2008), clearly showing how there is certain hype and exaggeration in the depiction of these studies. Similarly, other headlines such as "Free will may be an illusion, study says," or even "Free will could be the result of 'background noise' in the brain, study suggests" (respectively, Matyszczyk (2014) and Molloy (2014)).

Like in the case of the existence or not of a true self, it is arguable that some scientists, such as Sam Harris or Michael Gazzaniga, in their guise of popularizers of science, have contributed to creating this picture of a relentless and consensual scientific advance of the no-free-will hypothesis, although much of the debate actually still continues and we are far from anything like unanimity of the scientific community in these matters. Also, as in the case of the existence of a true self, it is important to keep in mind but not overstress the distinction between the neuroscientific findings and their interpretations: by the media, by the layman, and even by the neuroscientific community itself.

If these are "the selves of neuroscience" and their free will and responsibility, at least as the media depict them, what then are "our selves"? Philosophically, it is highly unlikely that, even constraining our inquiries to the Western societies, that there is something such as "the" concept of self to be found but rather an historical procession of concepts of person and self that stretch from the autobiographical self to the working self (Strawson, 2009, p. 18).[n]

Furthermore, it is arguable that there isn't even a single problem of the self and personal identity but rather a wide range of loosely connected questions,[o] dominated, at least in Anglo-Saxon analytical philosophy, by what Quante (2006) has labeled as the persistence-of-person problem and that we can better express by resorting to Noonan (2003, p. 2): "(...) the problem of giving an account of the logically necessary

---

[n]Galen Strawson is himself the proponent of the minimal self theory (SESMET—subject-as-single mental-thing, the subject as the thing "that remains when one has stripped away everything other than the being of experience") and coincidently (or maybe not) one of the most well-known incompatibilist determinists. The extent to which his deflated and minimal concept of the self is reliant on his determinist position about free will (or vice versa) is an interesting, although outside the scope of this essay.

[o]Quante (2006, pp. 148–150) identified four such questions: (1) The conditions-of-personhood-problem; (2) the unity-of-person-problem; (3) the persistence-of-person-problem; and (4) the structure-of-personality-problem. Olson (2016) identified as many as 10, with the most relevant ones being the personhood question (what are the defining characteristics of a person as opposed to, say, an object?) and the persistence question (what are the conditions for a person to continue existing from one time to another?).

and sufficient conditions for a person identified at one time being the same person as a person identified at another."

Another problem is how the concept of self relates to other concepts: the Platonic soul or the Aristotelian *psyche*, the Lockean *person*, the "thinking intelligent" being (Locke, 1979 [1690], II. 27. 9) or the Leibnizian monad or Hume's bundle of perceptions, or even Kant's transcendental unity of apperception, to name but a few.[P] If this is so, one may ask, how is neuroscience (and its technological applications in the form of neuroenhancement, for instance) supposed to challenge "it," if there is no single conception to be found?

Regarding the notions of free will and responsibility, the stock isn't much easier. The basic terms of the debate are supposedly well-known, according to any basic encyclopedia of philosophy: libertarians oppose hard determinists on the question of whether the universe is indeterministic but both parties agree that free will and a deterministic universe are incompatible, while compatibilists do not (compatibilists that also believe in determinism are also labeled as soft determinists).

This, however, is only part of the story. What exactly is our western conception of free will and responsibility? More often than not, biologists and neuroscientists seem to commit the strawman fallacy, by describing a supposed picture of free will that seems to be almost magical and inherently bounded with magical or ghostly elements and presupposing the existence of something like a soul, while in fact many philosophers do not seem to require any of that in their understanding of free will. Mele (2014a, p. 205) states the problem clearly (regarding the discussion of Soon et al., 2008): "Before we make a judgment about whether particular data threaten the existence of free will, we should ask ourselves how free will would need to be understood in order for the threat to be a genuine one."

All of this, of course, applies to responsibility: some authors seem to think that neuroscience can truly transform people's intuitions and notions of free will and responsibility. Greene and Cohen (2004, p. 1775) believe neuroscientific findings "will challenge and ultimately reshape our intuitive sense(s) of justice," even if they may not affect our everyday practices. Perhaps the main problem with such a view has been best expressed by Vincent (2010), who paralleled Mele's point about free will and argued that not only is neuroscience not unified but also that we can understand responsibility in at least six different ways, which makes it extremely difficult to see what is changing and what is doing the change.

[P]For historical developments of the idea of self, namely its connections with the notion of soul, see Martin and Barresi (2001, 2008) and Crabbe and James (1999).

However, if we want to understand "our selves" and the free will and responsibility involved, we are not limited to philosophy: recent studies seem to show that in our culture, despite some association of the brain with the self and all the neurohype, our ideas about personal identity are chiefly directed by moral traits: we do not seem to ascribe identity to persons by the degree of mental connectedness or any cognitive process but simply by the presence of moral traits (Strohminger and Nichols (2014) call it "the essential moral self hypothesis"). The point is that moral traits are more essential to determining personal identity than any other mental features.

It has also been argued that people have a general tendency to believe that there is something akin to a "true self" and that such true self is fundamentally good. In their studies, Newman, Bloom, and Knobe (2014) found that behaviors deemed morally good are more likely to be "conscripted" as reflecting the person's true self, by comparison with behaviors deemed bad. The interesting point here, in regards to neuroscientific findings, is that the population sample, even if probably aware of many of the neuroscientific findings about the self and the dispute about the existence or not of something like "a true self," still believes in such a notion and has very morally biased ways of determining it, completely unrelated to any neuroscientific findings.

However, being influenced by our own moral values in what we consider to be someone's "true self" is not the only way in which our culture refuses to acknowledge or embrace recent neuroscientific findings about the self. Knobe and Nichols (2011) report that one's viewpoints about the boundaries of the self depend on the context, with broader views being employed when interactions with other agents and the world are the focus, only adopting methods of cognitive science and a truly different conception of the self when they "zoom in," so to speak. What these different perspectives seem to show is that most people are perfectly fine with entertaining two different (and probably, at least for academics, incompatible) notions of self, easily adopting the one they find more adequate to their context.[q]

On the notions of free will and responsibility, we can't help but start by noting that here also we have many survey studies to help us scrutinize the folk notions. Some of these analyses aim to confirm that people are libertarians: Nichols (2004) has argued that young children deploy a notion of agent causation, for instance. However, it has also been shown

[q]Additionally, Nichols and Bruno's (2010) studies on framing about the persistence of selves led them to conclude that people are more prone to adopt a psychological theory of personal identity when they are faced with open-ended, abstract questions.

that people tend to have incompatibilist intuitions when thinking about the free will problem in a more abstract, cognitive way, but compatibilist intuitions when reasoning in a more concrete, emotional way (Nichols & Knobe, 2007).

On the other hand, whether a person believes in free will clearly affects the way responsibility is faced and how punishment is dealt with: Krueger, Hoffman, Walter, and Grafman (2014) have shown that people with a strong belief in free will punish more harshly than people with weak belief in free will, but only in low affective cases. More concluding is the evidence from Vohs and Schooler (2008) pointing out that immoral behavior, such as cheating, can be primed by reading sentences where scientists denied the existence of free will, or from Baumeister, Mele, and Vohs (2010), where subjects exhibited more aggressive behavior after reading similar sentences.

However, and despite whatever beliefs and evidences about free will one has, the notion of responsibility does not seem to fade. Knobe and Nichols (2011) have claimed that people affirm responsibility even when told that the agent's actions are the inevitable result of his genes and environment, that the agents lived in a completely deterministic universe, and when they are told the agent has a neurological disorder and that if anyone else had this illness he or she would behave in the same way.[r]

A good description of the problem can be found in Vargas (2014). The problem, it is said, is that responsibility has social and normative dimensions that should be understood as a practice supporting and extending our concept of morality, with the question of whether such practice is justified being a clear normative question, unapproachable by neuroscience.

CONCLUSION

Neuroscience is coming to age and is providing us with increasingly refined and more accurate explanations and predictions of how the brain regulates behavior and cognition. We have looked into recent neuroscientific findings about self and free will to determine whether they change our philosophical and folk notions. We concluded that philosophical notions of self and free will are numerous and so it is a difficult question to answer. The folk notions, however, seem mostly unaffected: people still believe in a "true self," in an "essential moral

---

[r]Dweck and Molden (2008) have also suggested that personal well-being increases with the belief in free will.

self," and still believe in free will and responsibility, even when confronted with deterministic views of the world and the brain.

First of all, we must remember that what we are really dealing with in those studies are only experimental setups, which may have many procedural weaknesses and whose extrapolation to everyday human life is uncertain. In a sense, we could say that it is highly hopeful to believe that neuroscientific findings can change those rooted notions: after all, decades of knowledge about evolution and climate change hasn't changed everyone's beliefs. The public uptake of science is not a straightforward matter. Moreover, these findings are not as uncontroversial as evolution theory and climate change: as we've seen, there is room for interpretation from the scientists themselves to the media.

Likewise, philosophers have struggled for centuries with different theories taking aim at the self and its free will, without managing to change our notions of it in any way. Perhaps neuroscience can at most illuminate some of the competitor's claims and discard theories but probably not much more than this. Roskies (2006), for instance, argues convincingly that, even though neuroscientific advances may worry us that people will abandon their belief in free will, neuroscience can only remain silent about the (in)deterministic nature of our universe, because only our best physical theory can answer that question.

We need to recognize the self's embodiment, the body's influence, and the ever-present sociocultural constraints as a way of qualifying and moderating many exaggerated claims about neuroscience's powers, while all the while admitting that our shared assumptions about self, free will, and responsibility might have been too optimistic in the first place.[5] Maybe neuroscience can contribute to (but not determine) a more doubting view of persons as rational and autonomous selves, wholly responsible for their actions, without necessarily dispelling our humanistic vision of the world.

References

Ali, S. S., Lifshitz, M., & Raz, A. (2014). Empirical neuroenchantment: From reading minds to thinking critically. *Frontiers in Human Neuroscience, 8*, 357.
Baumeister, R., Mele, A., & Vohs, K. (2010). *Free will and consciousness: How might they work?* Oxford: Oxford University Press.
Bickle, J. (2003). *Philosophy and neuroscience: A ruthless reductive account*. New York: Springer.

[5]Rose and Abi-Rached (2013) take that approach, all the while urging us to accept the opportunity to use neuroscientific findings to move beyond the illusion of a rational and autonomous subject.

Blakemore, C. (2005). Foreword. In J. Illes (Ed.), *Neuroethics: Defining the issues in theory, practice and policy* (pp. v–vi). Oxford: Oxford University Press.

Blakeslee, S. (2000, January 4). A decade of discovery yields a shock about the brain. *New York Times.* Retrieved from <http://www.nytimes.com/2000/01/04/science/a-decade-of-discovery-yields-a-shock-about-the-brain.html?pagewanted = all> Accessed 22.11.15.

Brukamp, K. (2013). Better brains or bitter brains? The ethics of neuroenhancement. In Elisabeth Hildt, & Andreas G. Franke (Eds.), *Cognitive enhancement: An interdisciplinary perspective* (pp. 99–112). Dordrecht: Springer.

Buford, C., & Allhoff, F. (2007). Neuroscience and metaphysics (redux). *The American Journal of Bioethics: AJOB, 7,* 34–36.

Churchland, P. S. (1986). *Neurophilosophy.* Cambridge, MA: MIT Press.

Conti, C. L., Nakamura-Palacios, E. M., Goldstein, R. Z., Volkow, N. D., Nitsche, M. A., Cohen, L. G., et al. (2011). Bilateral transcranial direct current stimulation over dorsolateral prefrontal cortex changes the drug-cued reactivity in the anterior cingulate cortex of crack-cocaine addicts. *Brain Stimulation, 7*(1), 130–132.

Crabbe, M., & James, C. (1999). *From soul to self.* London: Routledge.

Crick, F. (1994). *The astonishing hypothesis: The scientific search for the soul.* New York: Touchstone Press.

Davey, C. G., Pujol, J., & Harrison, B. J. (2016). Mapping the self in the brain's default mode network. *NeuroImage, 132,* 390–397.

Dennett, D. C. (1992). The self as a center of narrative gravity. In F. S. Kessel, P. M. Cole, & D. L. Johnson (Eds.), *Self and consciousness: Multiple perspectives* (pp. 103–115). Hillsdale, NJ: Lawrence Erlbaum Associates.

Devinsky, O. (2000). Right cerebral hemisphere dominance for a sense of corporeal and emotional self. *Epilepsy & Behavior, 1*(1), 60–73.

Doppelmayr, M., & Weber, E. (2011). Effects of SMR and theta/beta neurofeedback on reaction times, spatial abilities, and creativity. *Journal of Neurotherapy, 15*(2), 115–129.

Dweck, C. S., & Molden, D. C. (2008). Self-theories: The construction of free will. In John Baer, James C. Kaufman, & Roy F. Baumeister (Eds.), *Are we free? Psychology and free will* (pp. 44–64). New York: Oxford University Press.

Evers, K., & Sigman, M. (2013). Possibilities and limits of mind-reading: A neurophilosophical perspective. *Consciousness and Cognition, 22,* 887–897.

Farah, M. J., & Heberlein, A. S. (2007). Personhood and neuroscience: Naturalizing or nihilating? *The American Journal of Bioethics AJOB, 7*(1), 37–48.

Farah, M. J., Hutchinson, J. B., Phelps, E. A., & Wagner, A. D. (2014). Functional MRI-based lie detection: Scientific and societal challenges. *Nature Reviews Neuroscience, 15* (2), 123–131.

Farah, M. J., & Wolpe, P. R. (2004). Monitoring and manipulating brain function: New neuroscience technologies and their ethical implications. *The Hastings Center Report, 34*(3), 35–45.

Feinberg, T. E., & Keenan, J. P. (2005). Where in the brain is the self? *Consciousness and Cognition, 14*(4), 661–678.

Flanagan, O. (2002). *The problem of the soul: Two visions of mind and how to reconcile them.* New York: Basic Books.

Franke, A. G., & Lieb, K. (2013). Pharmacological neuroenhancement: Substances and epidemiology. In Elisabeth Hildt, & Andreas G. Franke (Eds.), *Cognitive enhancement: An interdisciplinary perspective* (pp. 17–27). Dordrecht: Springer.

Fried, I., Mukamel, R., & Kreiman, G. (2011). Internally generated preactivation of single neurons in human medial frontal cortex predicts volition. *Neuron, 69*(3), 548–562.

Gazzaniga, M. S. (2005). *The ethical brain.* New York: Dana Press.

Gillihan, S. J., & Farah, M. J. (2005). Is self special? A critical review of evidence from experimental psychology and cognitive neuroscience. *Psychological Bulletin, 131*(1), 76–97.

Goldhill, O. (2015, September 20). Neuroscience backs up the Buddhist belief that "the self" isn't constant, but ever-changing. *Quartz*. Retrieved from <http://qz.com/506229/neuroscience-backs-up-the-buddhist-belief-that-the-self-isnt-constant-but-ever-changing> Accessed 22.11.15.

Greene, J., & Cohen, J. (2004). For the law, neuroscience changes nothing and everything. *Philosophical Transactions of the Royal Society of London. Series B, Biological Sciences, 359* (1451), 1775–1785.

Gruber, D., & Dickerson, J. A. (2012). Persuasive images in popular science: Testing judgments of scientific reasoning and credibility. *Public Understanding of Science, 21*(8), 938–948.

Hammond, D. C. (2007). What is neurofeedback? *Journal of Neurotherapy: Investigations in Neuromodulation, Neurofeedback and Applied Neuroscience, 10*(4), 25–36.

Hauskeller, M. (2013). Cognitive enhancement – To what end? In Elisabeth Hildt, & Andreas G. Franke (Eds.), *Cognitive enhancement: An interdisciplinary perspective* (pp. 113–123). Dordrecht: Springer.

Hildt, E. (2013). Cognitive enhancement – A critical look at the recent debate. In Elisabeth Hildt, & Andreas G. Franke (Eds.), *Cognitive enhancement: An interdisciplinary perspective* (pp. 1–14). Dordrecht: Springer.

Hook, C. J., & Farah, M. J. (2013). Look again: Effects of brain images and mind–brain dualism on lay evaluations of research. *Journal of Cognitive Neuroscience, 25*(9), 1397–1405.

Illes, J., & Racine, E. (2005). Imaging or imagining? A neuroethics challenge informed by genetics. *The American Journal of Bioethics, 5*(2), 5–18.

Jansen, J. M., Daams, J. G., Koeter, M. W. J., Veltman, D. J., van den Brink, W., & Goudriaan, A. E. (2013). Effects of non-invasive neurostimulation on craving: A meta-analysis. *Neuroscience & Biobehavioral Reviews, 37*(10), 2472–2480.

Keehner, M., Mayberry, L., & Fischer, M. H. (2011). Different clues from different views: The role of image format in public perceptions of neuroimaging results. *Psychonomic Bulletin & Review, 18*(2), 422–428.

Knobe, J., & Nichols, S. (2011). Free will and the bounds of the self. In Robert Kane (Ed.), *The Oxford handbook of free will* (pp. 530–554). Oxford: Oxford University Press.

Krause, B., Márquez-Ruiz, J., & Cohen Kadosh, R. (2013). The effect of transcranial direct current stimulation: A role for cortical excitation/inhibition balance? *Frontiers in Human Neuroscience, 7*, 602.

Krueger, F., Hoffman, M., Walter, H., & Grafman, J. (2014). An fMRI investigation of the effects of belief in free will on third-party punishment. *Social Cognitive and Affective Neuroscience, 9*(8), 1143–1149.

Krystal, A. (2014, November 21). The shrinking world of ideas. The Chronicle of Higher Education. Retrieved from <http://chronicle.com/article/The-Shrinking-World-of-Ideas/150141> Accessed 22.11.15.

Legrand, D., & Ruby, P. (2009). What is self-specific? Theoretical investigation and critical review of neuroimaging results. *Psychological Review, 116*(1), 252–282.

Locke, J., & Nidditch, P. H. (1979). *An essay concerning human understanding*. Oxford: Oxford University Press.

Martin, R., & Barresi, J. (2001). *Naturalization of the soul: Self and personal identity in the eighteenth century*. New York and London: Routledge.

Martin, R., & Barresi, J. (2008). *The rise and fall of soul and self: An intellectual history of personal identity*. New York: Columbia University Press.

Matyszczyk, C. (2014, June 22). Free will may be an illusion, study says. CNET. Retrieved from <http://www.cnet.com/news/free-will-may-be-an-illusion-study-says> Accessed 22.11.15.

Mayberg, H. S., Lozano, A. M., Voon, V., McNeely, H. E., Seminowicz, D., Hamani, C., et al. (2005). Deep brain stimulation for treatment-resistant depression. *Neuron, 45*(5), 651–660.

McCabe, D. P., & Castel, A. D. (2008). Seeing is believing: The effect of brain images on judgments of scientific reasoning. *Cognition*, *107*(1), 343–352.

Mele, A. (2014a). Free will and substance dualism: The real scientific threat to free will? In Walter Sinnott-Armstrong (Ed.), *Moral psychology, volume 4: Free will and responsibility* (pp. 195–207). Cambridge: MIT Press.

Mele, A. (2014b). *Free: Why science hasn't disproved free will*. Oxford: Oxford University Press.

Michael, R. B., Newman, E. J., Vuorre, M., Cumming, G., & Garry, M. (2013). On the (non) persuasive power of a brain image. *Psychonomic Bulletin & Review*, *20*(4), 720–725.

Miller, B. L., Seeley, W. W., Mychack, P., Rosen, H. J., Mena, I., & Boone, K. (2001). Neuroanatomy of the self: Evidence from patients with frontotemporal dementia. *Neurology*, *57*(5), 817–821.

Molloy, A. (2014, June 21). Free will could be the result of "background noise" in the brain, study suggests. The Independent. Retrieved from <http://www.independent.co.uk/news/science/free-will-could-be-the-result-of-background-noise-in-the-brain-study-suggests-9553678.html> Accessed 20.11.15.

Moulier, V., Gaudeau-Bosma, C., Isaac, C., Allard, A. C., Bouaziz, N., Sidhoumi, D., et al. (2016). Effect of repetitive transcranial magnetic stimulation on mood in healthy subjects. *Socioaffective Neuroscience & Psychology*, *6*, 29672.

Muzur, A., & Rinčić, I. (2013). Neurocriticism: A contribution to the study of the etiology, phenomenology, and ethics of the use and abuse of the prefix neuro-. *JAHR—European Journal of Bioethics*, *4*(7), 545–555.

Nagel, S. K. (2014). Enhancement for well-being is still ethically challenging. *Frontiers in Systems Neuroscience*, *8*, 72.

Newman, G. E., Bloom, P., & Knobe, J. (2014). Value judgments and the true self. *Personality and Social Psychology Bulletin*, *40*(402), 203–216.

Nichols, S. (2004). The folk psychology of free will: Fits and starts. *Mind and Language*, *19*, 473–502.

Nichols, S., & Bruno, M. (2010). Intuitions about personal identity: An empirical study. *Philosophical Psychology*, *23*(3), 293–312.

Nichols, S., & Knobe, J. (2007). Moral responsibility and determinism: The cognitive science of folk intuitions. *Nous*, *41*(4), 663–685.

Noonan, H. W. (2003). *Personal identity*. New York: Routledge.

Olson, E. T. (2016). Personal identity. In E. N. Zalta (Ed.), *The Stanford encyclopedia of philosophy* (Spring 2016 ed.). Retrieved from <http://plato.stanford.edu/archives/spr2016/entries/identity-personal/>.

Platek, S., Myers, T. E., Critton, S. R., Gallup, G. G., Anderson, N., Baron-Cohen, S., et al. (2003). A left-hand advantage for self-description: The impact of schizotypal personality traits. *Schizophrenia Research*, *65*(2-3), 147–151.

Qin, P., & Northoff, G. (2011). How is our self related to midline regions and the default-mode network? *NeuroImage*, *57*(3), 1221–1233.

Quante, M. (2006). Personal identity between survival and integrity. *Poiesis & Praxis*, *4*(2), 145–161.

Racine, E., Waldman, S., Rosenberg, J., & Illes, J. (2010). Contemporary neuroscience in the media. *Social Science & Medicine*, *71*(4), 725–733.

Raichle, M. E. (2015). The brain's default mode network. *Annual Reviews*, *38*(1), 433–447.

Reiner, P. (2011). The rise of neuroessentialism. In J. Illes, & B. J. Sahakian (Eds.), *The oxford handbook of neuroethics* (pp. 161–175). Oxford: Oxford University Press.

Repantis, D. (2013). Psychopharmacological neuroenhancement: Evidence on safety and efficacy. In Elisabeth Hildt, & Andreas G. Franke (Eds.), *Cognitive enhancement: An interdisciplinary perspective* (pp. 29–38). Dordrecht: Springer.

Rose, N. S., & Abi-Rached, J. M. (2013). *Neuro the new brain sciences and the management of the mind*. Princeton, NJ: Princeton University Press.
Roskies, A. (2006). Neuroscientific challenges to free will and responsibility. *Trends in Cognitive Sciences, 10*(9), 419–423.
Satel, S., & Lilienfeld, S. O. (2013). *Brainwashed: The seductive appeal of mindless neuroscience.* New York: Basic Books.
Savulescu, J., ter Meulen, R., & Kahane, G. (2011). *Enhancing human capacities.* Oxford: Wiley-Blackwell.
Schweitzer, N. J., Baker, D. A., & Risko, E. F. (2013). Fooled by the brain: Re-examining the influence of neuroimages. *Cognition, 129*(3), 501–511.
Shulman, G. L., Fiez, J. A., Corbetta, M., Buckner, R. L., Miezin, F. M., Raichle, M. E., & Petersen, S. E. (1997). Common blood flow changes across visual tasks: II. Decreases in cerebral cortex. *Journal of Cognitive Neuroscience, 9*(5), 648–663.
Smith, M. E., & Farah, M. J. (2011). Are prescription stimulants "smart pills"? The epidemiology and cognitive neuroscience of prescription stimulant use by normal healthy individuals. *Psychological Bulletin, 137*(5), 717–741.
Soon, C. S., Brass, M., Heinze, H. J., & Haynes, J. D. (2008). Unconscious determinants of free decisions in the human brain. *Nature Neuroscience, 11*(5), 543–545.
Strawson, G. (2009). *Selves: An essay in revisionary metaphysics.* New York: Oxford University Press.
Strohminger, N., & Nichols, S. (2014). The essential moral self. *Cognition, 131*, 159–171.
Synofzik, M., & Schlaepfer, T. E. (2008). Stimulating personality: Ethical criteria for deep brain stimulation in psychiatric patients and for enhancement purposes. *Biotechnology Journal, 3*(12), 1511–1520.
Turk, D. J., Heatherton, T. F., Macrae, C. N., Kelley, W. M., & Gazzaniga, M. S. (2003). Out of contact, out of mind. *Annals of the New York Academy of Sciences, 1001*(1), 65–78.
Vargas, M. R. (2014, June 1). Neuroscience and the exaggerated death of responsibility. *The Philosophers' Magazine* (University of San Francisco Law Research Paper No. 2014-20, Vol. 64). San Francisco, CA: University of San Francisco.
Vernon, D., Dempster, T., Bazanova, O., Rutterford, N., Pasqualini, M., & Andersen, S. (2009). Alpha neurofeedback training for performance enhancement: Reviewing the methodology. *Journal of Neurotherapy, 13*(4), 214–227.
Vidal, F. (2009). Brainhood, anthropological figure of modernity. *History of the Human Sciences, 22*(1), 5–36.
Vincent, N. (2010). On the relevance of neuroscience to criminal responsibility. *Criminal Law and Philosophy, 4*(1), 77–98.
Vogeley, K., & Gallagher, S. (2011). The self in the brain. In Shaun Gallagher (Ed.), *The Oxford handbook of the self* (pp. 111–136). Oxford: Oxford University Press.
Vohs, K. D., & Schooler, J. W. (2008). The value of believing in free will: Encouraging a belief in determinism increases cheating. *Psychological Science, 19*(1), 49–54.
Weisberg, D. S., Keil, F. C., Goodstein, J., Rawson, E., & Gray, J. R. (2008). The seductive allure of neuroscience explanations. *Journal of Cognitive Neuroscience, 20*(3), 470–477.
Wilens, T. E., Adler, L. A., Adams, J., Sgambati, S., Rotrosen, J., Sawtelle, R., et al. (2008). Misuse and diversion of stimulants prescribed for ADHD: A systematic review of the literature. *Journal of the American Academy of Child and Adolescent Psychiatry, 47*(1), 21–31.
Youngsteadt, E. (2008, April 14). Case closed for free will? *Science Magazine*. Retrieved from <http://www.sciencemag.org/news/2008/04/case-closed-free-will> Accessed 21.11.15.

Free Will—Between Philosophy and Neuroscience

A. *Drozdzewska*

Université catholique de Louvain, Louvain-la-Neuve, Belgium

In recent decades, some of the questions usually associated with the domain of human science have slowly started making their way through to the experimental realm, especially with advances in neuroscience. Free will is an example of one of those issues, initially discussed almost exclusively by philosophers and other human scientists, but with the developments in neuroscience and an increasing understanding of how the brain works, it is slowly becoming a topic of interest of neuroscientific experiments. This shift can bring a new perspective to this old problem and help us develop new answers as well as new questions. However, as with other interdisciplinary research, some aspects of the problem become more complicated the more domains are involved. In this regard the problem of free will is no different.

The goal of this chapter is to show that, in fact, both in philosophy and in neuroscience, free will has the same issue at its core, namely mental causation. In the first section of this chapter I first try to establish what is usually hiding behind the term free will. The notion, although frequently used in discussions in various fields is not homogenous, and has several possible interpretations. I will proceed to explain the problem of causal efficacy of mental events, focusing on the Causal Exclusion argument.

In the sections "Free Will and Neuroscience—Libet's Approach" and "Free Will and Neuroscience—Tse's Approach" I will analyze the different experimental approaches that can be found in neuroscience,

The Human Sciences after the Decade of the Brain.
DOI: http://dx.doi.org/10.1016/B978-0-12-804205-2.00004-5

42

© 2017 Elsevier Inc. All rights reserved.

focusing mainly on two main lines of inquiry—Libet's (Libet, 1985; Libet, Gleason, Wright, & Pearl, 1983) and Tse's (Tse, 2013). I will explain why mental causation is also essential here and how it fits into the experimental paradigm. I will conclude this chapter by discussing how joint forces of philosophy and neuroscience can, in fact, greatly improve the way we do research on topics that were, until recently, almost exclusively discussed outside the hard sciences.

FREE WILL AND PHILOSOPHY

Defining Free Will

Finding a precise definition of free will, given the complexity and multidimensional nature of the term, is not a straightforward task. But in the multitude of approaches both in philosophy and in neuroscience, we can identify three main categories of definitions (Mele, 2014).

The first group connects free will to having a soul or another supernatural aspect of life. Here all the religious and theological explanations of free will and what agents are would fit. Only this group assumes as necessary a soul that is possibly made out of a different substance is necessary; the other two groups deem the supernatural nature irrelevant. Since in this chapter I mainly focus on (nonreductive) physicalist approaches, this understanding of what is required for free will is automatically excluded, as it assumes a form of dualism.

In the second group of definitions for an individual to have free will is to be able to make rational decisions, which are in no way forced. Those decisions are then a basis for at least some of our actions. This definition is probably the broadest one, and can be accepted by most researchers working on the topic. The notions of rational decision and reasons for action are the core of the argument on free will constructed by J. Habermas (2008), for whom both are essential marks of voluntary action. According to this position, the same movements executed in different contexts and for different reasons are distinct and possibly incomparable. The movement of a hand by a cyclist signaling his intent to turn has a different meaning than the same gesture in the experimental environment, like in Libet's experiments.

For the third group of definitions, the ability to make rational choices is not sufficient. What is needed is an actual choice between different options: for free will to exist in any moment, given the state of the universe at the moment, understood both physically as well as psychologically, there has to be more than one possible action. For John Searle (2010), a free action is one not determined by antecedently sufficient conditions and thus we are not in a grip of some overwhelming force of deterministic sufficient conditions, which would mean actions (though they admit of causal explanations given in terms of the agent's reasons), motives, beliefs, desires, and so forth, are not determined. The totality of the agent's prior states, including all the neurobiological states, was not sufficient to determine that a particular action had to occur. The agent might have performed a different action than the one actually executed. As Mele (2014) suggested this assumption seems to require indeterminism, which explicitly states that we cannot predict the outcome of every action given the precise state of the universe at a specific point of time combined with the laws of nature.

Most of the present approaches fit into one of the three categories, although major differences can be seen in each group. On the other hand, the discussions on what is needed for free will usually revolve around five focal points: (1) determinism vs indeterminism; in this case, the discussion usually focuses not on the actual state of the universe but rather on the possibility of free will, given its potential nature. In other words, whether under the assumption of either determinism or indeterminism, free will is possible; (2) the principle of alternate possibilities—being able to choose between at least two different options. The arguments are connected to the indeterminism/determinism debate, since being able to choose between different paths might mean that any given action is not completely determined by the totality of the previous states. The discussions also point out that different options would also be compatible with determinism or that an illusion of choice would be sufficient; (3) mental causation—probably the least discussed issue of the five points, assumes that for free will to be possible we need mental events to be causally efficacious. The topic has gained more attention recently and will be discussed in detail in later sections of this chapter; (4) free will and moral responsibility—many of the authors bind free will to moral responsibility; and (5) rational choice, the feeling of agency and the context of action puts free will in the space of reason—connecting the problem of free will again to the causes of our actions.

In this chapter, I will focus solely on the issue of mental causation and its connection to the free-will debate. I will show that the problem of causal efficacy of the mental lies at the core of both what we usually understand as being free as well as in the most basic assumptions of the neuroscientific approach to free will. I will hypothesize that, without solving the puzzle of how the mental can influence the physical, the experiments do not tell us as much about free will as we would like them to.

Free Will and Mental Causation

I have argued above that the problem of free will has several different angles, and three of them seem to be interconnected, namely the widely discussed determinism/indeterminism debate and the problem of alternative possibilities, with the focus of this chapter—mental causation.

By mental causation I mean that a mental event or a mental property can be a cause in the physical world, or, can be a cause of a physical event. However, I do not want to state that mental events are different from physical events in terms of substance (thus no substance dualism), but rather, I assume the world is comprised of all physical events with some of them also being mental, with the latter not being reduced to the former (nonreductive physicalism). The main challenge is to show how an event can be a cause through its mental properties rather than the physical ones. It is largely accepted that the two are connected with one another through the relationship of *supervenience*.

Supervenience has become a term used to describe a broad variety of correlations, and it relates two sets of properties (x and z) in such a manner that there can be no difference in one set (x) without there being a difference in the other (z). When applied to the mental causation debate, supervenience means that "there cannot be two events alike in all physical respects but differing in some mental respect, or that an object cannot alter in some mental respect without altering in some physical respect" (Davidson, 2001, p. 214). Since Davidson's formulation the concept has been redefined several times, yet the core intuitions have remained the same—two events cannot be physically identical and at the same time differ in some mental aspect.

Another crucial notion that plays a prominent role in the debate on free will and mental causation is *multiple realizability*. It is also the core notion for functionalism, a version of nonreductive physicalism. It is broadly agreed that the same mental event might supervene on a variety of different physical events. The most commonly evoked example of this is pain; the same sensation can be felt by different physical beings, where different physical states correspond to the feeling of pain. This notion is at the heart of why some (e.g., interventionists, discussed below) think that, in cases like voluntary actions, it is the mental that is the cause of our actions. Since the two— the mental and the physical—are related through supervenience, we are sure that there is a physical basis for every mental event. Because the mental is multiple realizable we know that the same mental event can supervene on different physical events. Hence, it is not certain that if we manipulate the physical event we will always get a different mental event, since the physical event after the manipulation can still be the supervenience basis for the same mental event.

The problem of mental causation as the problem of causes of our actions is sensitive to the conclusions of the famous *exclusion argument* aimed at all versions of nonreductive physicalism, and broadly discussed by Jaegwon Kim (1989, 1998, 2005). The argument is based on three general principles and through them shows how all the causal work is in fact done by the physical and the mental can at best be epiphenomenal (it exists but has no causal influence in the world). It starts with the supervenience claim that the mental property (M) always has a physical basis (P), and the mental is not substantially different from the physical. Two additional principles are the *causal closure of the physical world* and the *no overdetermination principle*. Given the causal closure of the physical, every physical event has a sufficient physical cause. No overdetermination means that each event has only one cause, except from the rare causes of genuine overdetermination, where, for instance, two balls hit a window at the same exact time, and it is impossible to determine which one of those caused the glass to break. However, it is commonly agreed on that mental events are not examples of genuine overdetermination. The argument can be clarified through a simple example of wanting a glass of water. We usually think that the wanting was causally responsible for us going to the kitchen and pouring ourselves a glass of water, but the causal exclusion argument shows us that this cannot be the case. The mental event here would be the wanting, and the physical event is the underlying brain activation, connected through supervenience. Therefore, there is no change in the "wanting to drink water" without the change in the pattern of brain activity. Since the world is causally closed and all physical events have sufficient physical causes, the mental events are excluded as causes, and the physical brain activation is seen as the sole cause of going to the kitchen to pour ourselves a glass.

The problem of mental causation is also connected to a broader issue of causation itself. As often in philosophy, there is no one definition of the concept. The two most popular approaches are those of causation as production and causation as difference making, and the distinction between them is also central in the free-will problem. In the production accounts, or transmission accounts, the relation between the cause and effect is often based on something being transmitted between the two, e.g., energy (Dowe, 1995; Salmon, 1984). On the other hand, causation as difference making (also called interventionism, Menzies & Price, 1993; Woodward, 2003) focuses on the intuitions that are the basis of most neuroscientific experiments, namely that if we manipulate the cause, the effect will also change. In this chapter we will focus on causation as making a difference, as it provides a solution for the free-will problem in the framework of the argument on mental causation and is the closest to the intuitions on the nature of causes in neuroscientific experiments.

As mentioned previously, the connection between mental causation and free will is not often discussed. A comprehensive account of the problem was provided by Peter Menzies and Christian List (List & Menzies, in press), based on three main theses:

1. **"The causal-source thesis**: Someone's action is free *only if* it is caused by the agent, particularly by the agent's mental states, as

distinct from the physical states of the agent's brain and body" (List & Menzies, in press, p. 2)

2. **"The purported implication of physicalism**: Physicalism rules out any agential or mental causation, as distinct from causation by physical states of the agent's brain and body" (List & Menzies, in press, p. 2)

3. **"The source-incompatibilist conclusion**: There can be no free actions in a physicalist world" (List & Menzies, in press, p. 2)

Menzies and List offer a solution to the problem based on the specific understanding of causation as difference making, where the cause for an effect is defined as the event that makes the difference; if the cause was manipulated the effect would change or not occur at all. In the case of mental causation, according to List and Menzies only the mental can be seen as the cause, a claim based on the multiple realizability of the mental. The same mental event can supervene on different physical events, but manipulating the physical basis of the mental would have no effect on it, since the changed physical event might still be the supervenient basis for the same mental event. On the other hand, manipulating the mental will most definitely change the physical.

Further discussion on the approach of Menzies and List requires a chapter of its own, but here I would like to focus on a neuroscientific account that follows the same intuitions, both in terms of seeing the importance of mental causation as well as the manipulation approach to causation.

FREE WILL AND NEUROSCIENCE—LIBET'S APPROACH

In recent decades, problems usually seen as highly theoretical have entered into the experimental domain. Neuroscientists have started to analyze problems such as consciousness or free will using new tools and assumptions (discussion of the latter entered the broader debate with the experiments conducted by Benjamin Libet (Libet, 1985; Libet et al., 1983)). Since then Libet's approach has become a standard in research on the topic, and versions of the experiment are still being conducted.

The Definition of Free Will in Libet's Experiments

According to Benjamin Libet, an action can be defined as voluntary and an exercise of the subject's free will when three conditions are met, namely: "(a) it arises endogenously, not in direct response to an external stimulus or cue; (b) there are no externally imposed restrictions or compulsions that directly or immediately control the subject's initiation and performance of the act; and (c) most important, subjects feel introspectively that they are performing the act on their own initiative and that they are free to start or not to start the act as they wish" (Libet, 1985, pp. 529–530). For Libet the starting point was the assumption that the overt movement under investigation can be caused by either a, presumably mental, conscious decision or a physical brain activation. Either of the two could potentially be the cause but only if it was the former the movement could be considered free. The movement itself was necessary for technical reasons- the data was analyzed by backtracking from the movement to see what was happening in the brain just prior to it. Thus the experimental design, where in analyzing the data, Libet starts from the effect—the movement—and investigates the competing causes, the conscious decision and the brain activation.

Libet's Experiments

The pioneering work done by Libet provided a first experimental look at the causes of human action, as it focused on action initiation and aimed to investigate, if an action is in fact caused by a brain process or rather by our conscious decision to act. Their conclusion was that, contrary to our common convictions, the action might not be started by us (understood as the agent making a decision) but rather by a neuronal process (Libet, 1985; Libet et al., 1983). For Libet himself, his experiment still left space for free will, although with a more limited scope, as a veto power. It can be exercised between the conscious decision and the action and is used to effectively stop the action under execution. Yet the experiment is often used as an example of a clear answer to the free-will problem—free will does not exist.

In order to determine how a voluntary action arises, Libet asked participants to move their fingers or flex their wrist while looking at a spot of light rotating clockwise. The participants were asked to remember the position of the spot of light when they first felt the urge to flex their fingers. Throughout the experiment the participants were connected to an electroencephalogram (EEG) recording the brain activity. The reported time of the conscious intention to move (W judgment—the moment in which the participant reported feeling the urge or decision to act) was then compared with the EEG readings to establish if there was any brain activity preceding the reported urge. The reported time of W judgment, on average, preceded the actual movement by 200 ms, yet in all the subjects a visible activation of the brain (the readiness potential, or RP) was observed much earlier than that. In the fully spontaneous acts, without any reported preplanning, the RP (called type II RP) was observed already 500 ms before the movement. The difference in trials with reported preplanning (type I RP) was even greater, averaging about 800 ms prior to the movement. The W judgments' timing was roughly the same for all of the trials, including those with reported preplanning. This led Libet to conclude that "the brain evidently 'decides' to initiate or at least prepare to initiate the act at a time before there is any reportable subjective awareness that such a decision has taken place. It is concluded that cerebral initiation even of a spontaneous voluntary act of the kind studied here can, and usually does, begin *unconsciously*" (Libet et al., 1983, p. 640). In other words, the beginning of the movement precedes the moment of conscious volition, as reported by the participant. Libet, however, does not immediately conclude that free will is completely an illusion—he saves it while limiting its scope. He states that free will is the veto power that can be exercised between the moment when the conscious decision to move was taken and the execution of the movement itself. Nonetheless, the scientific heirs of Libet's research who repeated, usually with some changes, the experiments (e.g., Haggard & Eimer, 1999; Soon, Braas, Heinze, & Haynes, 2008; discussed here) think that the obtained results offer something else—a final proof that free will as we understand it is just an illusion. In other words, the conscious decision we think was made was in fact preceded by the neuronal activation, which in turn means that the decision was not made consciously but "by the brain" (often reiterated as the phrase "my brain made me do it").

Problems With Libet's Approach

From the moment they were published, Libet's experiments gave rise to a very lively discussion between specialists from both fields: neuroscience and philosophy. The results as well as the basic assumptions of the experiment were put into question and the discussion on their validity is still going on. There are various ways to criticize both the setup as well as the basic assumptions of Libet's experiment and in this section I will focus on what I consider the core of the problem.

As suggested in the previous parts of this chapter, I hypothesize that the core issue for the problem of free will is mental causation; in Libet's experiments this was represented by the conscious decision to act. It is shown as opposite to the brain activation, and no physical process is investigated in connection to it. This, however, might suggest that the decision comes "from nowhere" in two different senses of the word, physically and psychologically. Physically because the physical process that would underlie the decision, if we want to remain in the realm of physicalism, is not accounted for. Psychologically because it appears without accounting for intentions that usually are active in the formation of the decision. For now I would like to focus on the second understanding.

According to Alfred Mele "to decide to do A is to perform a momentary action of forming an intention to A" (Mele, 2010, p. 44), where intentions are executive attitudes toward plans, which can be either very simple or very complex. Those intentions can be divided into proximal (which are the closest ones to the action) and distal (which are formed about actions we are planning to perform later). A decision is then a result of forming an intention to act, probably connected somehow to the process of deliberation. According to Mele, the RP that Libet is analyzing might be just a preparation for the intention forming and not for the action itself.

Contrasting the physical brain activation with the, presumably, conscious mental decision also suggests that, possibly, the conscious decision has no physical basis, as it is not accounted for in the experiments. If we have to remain in the realm of nonreductive physicalism, for every mental event there will be a physical basis connected through supervenience (for dualism this is not an issue; physicalism, on the other hand, would have to provide rules under which the mental is reduced to the physical). The experiment implicitly assumes that our actions are caused by either the conscious decision, or the (physical) brain activation. If it is the former, we as agents are the cause of our actions and we have free will; if the latter, our actions are caused by mostly unconscious brain states and we do not have free will. The lack of any physical basis for the conscious decision can potentially lead to several different outcomes. For example, we could (1) accept substance dualism, (2) agree that there could be uncaused causes, (3) assume that the conscious decision is based on brain activation itself, or (4) assume that the mental is epiphenomenal and does no causal work. In case of Libet's experiment, if we accept the third option, we return to what Mele suggested—the RP that is being investigated might be connected to the conscious decision and may as well be the brain's preparation for forming of either the intention or the decision. The first suggestion, accepting substance dualism, would stand against the scientific approach of Libet. The second one, accepting the existence of uncaused causes, might also be difficult to accept under normal assumptions in an experimental domain where everything has a cause. We could also conclude that the conscious decision is an epiphenomenon, which would make it irrelevant to the process because the brain activation is doing all the work. A number of possible replies to epiphenomenalism have been offered in philosophy, which we will not analyze here, but if we limit ourselves to Libet's own account, mental events have to be causal if we want to preserve free will as a veto power. Libet assumes, in fact, that although the conscious mental decision does not play a role in initiating the action, it can still be causal in later stages of the action execution.

Accepting the third assumption and reassessing the experiments seems most promising, and it might lead to the conclusion that the RP is in fact the physical basis of either an intention or conscious decision itself. This would, however, mean that the conclusions are not valid anymore, as what was earlier treated as the physical initiation of the action is, in fact, the physical basis of the conscious decision. Putting the conscious decision and the physical activation at two opposite sides of the spectrum in order to investigate the causes for our actions might be mistaken, as the two are intertwined and do not exist completely independently from one another (except if we accept substance dualism). Additionally, if we want to conclude that the decision is somehow causal, so that the mental part of the event is the cause, even as the veto power over an action already underway, we have to provide a solution for the *exclusion argument*, and ultimately show how the causality of the mental could be compatible with the causal closure of the physical.

Libet's Legacy in Neuroscience

Libet's experiments inspired multiple researchers to conduct their own, often modified versions; the most famous have been conducted by Patrick Haggard and Chun Siong Soon. Although both of them applied several modifications to the original, some of the basic assumptions, including the problem of causal efficacy of the mental, remained the same.

Patrick Haggard defines freely willed actions as a combination of two components—the ability to do otherwise and the existence of conscious agents who initiate the action (Haggard, 2011). His understanding of the ability to do otherwise differs from the one often discussed in philosophy. Haggard believes that neuroscience is inherently deterministic in its methods, and the free will that is investigated by neuroscience, in order to exist, has to be compatible with determinism. The ability to do otherwise is limited to action selection, where two different action alternatives are developed in parallel, up to late stages of processing, with one of them being chosen in the end. The mechanisms of how they are chosen can be deterministic. This understanding of "could have done otherwise" is not necessarily in line with the philosophical interpretation, as it has little to do with being determined by the antecedently sufficient conditions, but the actual existence of two possible paths for the brain activation might already require some form of indeterminism. Haggard focuses on the representation of the possibilities in the brain, which are different processes leading to different actions being developed simultaneously, regardless of the actual possibility to choose from actions (which would mean that the action was not determined by the previous states, both physical as well as mental). The actual existence of two or more different options to choose from is irrelevant for the experimental setup.

Haggard's definition of free action is closer to philosophical debates on the topic in comparison to Libet's, because it acknowledges the necessity of different possibilities to choose from. He recognizes the dualism in our day-to-day language (Haggard, 2008), but believes it implies the distinctiveness of the mental from both the body and the brain. This led him to think that the approach in which voluntary action is opposed to a stimulus-driven one can provide us with most data about free will. Additionally, he assumes common origin for the conscious intentions as well as the action—in the neural firings that lead to the action. This, however, is not a complete picture. Firstly, dualism and physicalism are not the only two possible positions; the fact that mental events are distinct and not reducible to physical events does not entail them being made from a different substance. Secondly, the way we define causality is not straightforward, and it is not clear which understanding Haggard has in mind when discussing the causality of both the neural findings as well as, or maybe even more importantly, the mental events.

In an updated version of Libet's experiment, Haggard, along with Eimer (Haggard & Eimer, 1999; also discussed in Haggard & Libet, 2001) investigated the possibility roughly suggested by Mele, that the RP is not in fact the cause of the action, but rather is the cause of our experience of intention for moving. Two versions of the experiment were conducted—one using Libet's method and another where the subjects had to choose between making a movement with their right or left hand. Both the RP as well as the lateralized RP (connected to the choice of one of the hands for execution of the movement) were analyzed. If in fact RP is connected to the intention rather than the movement itself, it would appear earlier for actions stemming from distal intentions and later for actions initiated by proximal intentions. Of course, the scope of the investigation into the distal intentions has to be limited given the time constraints of a normal neuroscientific experiment (e.g., time limitiations). No such relation has been found either for the version of the experiment similar to the one conducted by Libet, nor in Haggard and Eimer's variation, in which the main difference was lateralization of the RP. They concluded that Libet's RP cannot be identified with the basis for the conscious intention, but the lateralized RP potentially could, since when the intentions occurred early, so did the lateralization of RP.

Another influential variation of Libet's experiment was conducted by Soon et al. (2008) and the main differences also lie in the structure of the task individuals were asked to perform. The participants, instead of simply flexing their wrist, were asked to decide between pressing one of two buttons instead. The change was applied in order to solve one of the potential problems with the original experiments. It was questioned if, in the case of Libet's experiment, participants are actually making a choice, since it is not the case that different options are available to them. Choosing between different moments to conduct one action might not be considered an example of a choice between different alternatives, in the sense of different movements. Moreover, in this version the recordings were not limited to the motor areas, as the authors questioned whether this is in fact where the decision to move originates, or if there is some higher-level unconscious process going on in other areas of the brain as well. The activation was recorded not only in the primary motor cortex and supplementary motor area, but also in the frontopolar cortext, where predictive information about whether the individual will choose left or right was present up to 7 s before the movement as well as in the parietal cortex. The timing of the decision, on the other hand, could be predicted around 5 s before the motor decision, making the differences between the findings here and in Libet's approach significant.

But the main question remains: what do these results tell us about free will? If we accept the position of nonreductive physicalism, the sole fact that the brain is active and that there is a brain activation related to our mental processes is not a surprise. Only for dualists could the mental events happen without an underlying physical activity. However, with the former it remains to be shown how mental events could be causal in a physical world, and potentially closed under causal closure of physics, and how the consciousness and mental events are not epiphenomenal, since we don't experience them as such.

Among the main points of disagreement concerning Libet's experiments was the nature of the RP and its role in the action generation. Another modified version of Libet's experiment was conducted and described by Prescott et al. (2016). Subjects were asked to choose one of four letters presented to them, at a time of their choice. They noted the moment in which the decision was made by either pressing a button, in 50% of the trials, or not doing anything at all in the other 50%. In both trials the RP was recorded, which led to the conclusion that RP might reflect some decision-related or anticipatory processes rather than preconscious motor planning.

In yet another variation of the experiment, hypnosis was used to investigate if RPs are also observed in actions made under suggestion and thus not completely free (Schlegel et al., 2014). This version of the experiment consisted of five phases, which included two hypnotic inductions. Those inductions were supposed to influence the participant to squeeze a stress ball (in the experiment the participants were asked to squeeze the stress ball in a chosen moment while watching a video clip) with a specific hand. The hypothesis here was that the movements were initiated and then conducted without any conscious feeling of will on the side of the participant. Also in those experiments the RPs were recorded, which led to the conclusion that they might be unrelated to the conscious intention to move or to a conscious decision.

FREE WILL AND NEUROSCIENCE—TSE'S APPROACH

A completely different angle of the problem was investigated by Peter Ulric Tse (2013) who, unlike Libet, treats the problem of free will not as a problem of causes for motor action, but as an instance of the broader mind—body problem. The main focus of the approach thus does not lie in the causes of specific motor actions, but rather in the possibility of the mind's causal influence on the world. Additionally, also unlike Libet, Tse extends his understanding of voluntary action including not only overt but also internal actions requiring no muscular activation (e.g., intentionally shifting your attention from one object to the other).

Contrasting the (mental) conscious decision with the (physical) brain activation as competing causes of an action leading to conclusions about free will in general can only work, if we consider that all free actions are those that require some overt activity. However, as was discussed in previous parts of this chapter, in those by now almost standard approaches, the origins and basis of the conscious decision remain somewhat mysterious. Tse, on the other hand, starts with the idea that free will might not in fact be an event but rather a "durationally extended process" (Tse, 2013) that is constructed from various (I assume both conscious and unconscious in this case) subprocesses, which could leave space for intentions and their role in the conscious decision formation. The will is here defined as "the agent's choice or decision to perform or not to perform an action, either later (via formation of distal intentions) or right now (via proximal intentions). Will is the capacity to so choose or decide to act" (Tse, 2013, p. 20). Unlike Libet's definition, this one emphasizes the importance of intentions in the action generation as a necessary part of exercising the will. Additionally, it allows for incorporating the frequent gaps between an intention and an action that we often experience (e.g., we have an intention to go on vacation to Spain and in 3 months we actually go).

Assuming that for will to exist an account of mental causation is needed Tse focuses on the possibility of free will rather than on the exercise of free will, and the former is in his approach granted on a very basic neuronal level, in the initiation of what will much later become the overt action. This means that mental causation would influence the physical world in the beginning of the whole process, and the mental would be causally efficacious already at the neuronal level, which is explained by the core idea of his approach, *criterial causation* (explained further in the next paragraph). Tse argues that it not only explains how mental causation would be possible but also solves what he considers the core problem, namely the impossibility of self-causation (the idea that our decisions are influenced by our motives, which is intuitively the core of the idea of mental causation if we consider ourselves as the choosing agents, who cause overt actions in the world). The criteria discussed here are a "set of conditions on input that can be met in multiple ways and to differing degrees" (Tse, 2013, p. 22). Neurons assess incoming information based on the criteria and, if met, the neurons fire.

The neuronal model of criterial causation by Tse has three distinct stages: (1) "new physical/informational criteria are set in a neuronal circuit on the basis of preceding physical/mental processing at time $t1$, in part via a mechanism of rapid synaptic resetting that effectively changes the inputs to postsynaptic neuron"; (2) "at time $t2$, inherently variable inputs arrive at the postsynaptic neuron; and (3) at time $t3$, physical/informational criteria are met or not met, leading to postsynaptic neural firing or not" (Tse, 2013, p. 133). The physical realization of informational criteria happens on the neuronal level, where the information from the input triggers the neuron to fire, if such criteria are met. Neurons transform and communicate the information by changing individual action potential spikes into spikes that are being transmitted to other neurons, and place criteria for firing that can be met only by a specific subset of information, so only that subset can trigger the neuron to fire. Mental events can, in this account of causation, cause future mental and physical events by changing the criteria for firing used by neurons for the future inputs.

In simpler terms, neurons react on the basis of certain criteria, which can be described as conditions for firing, and only if those conditions are met the neuron will react and fire. Those conditions are met or not because of the information coded somehow in the input entering the brain. According to Tse, new information entering the brain contained in a new input can change the previous criteria according to which the neurons are firing. After the criteria are changed, the next time a new input arrives, these new criteria will have to be met for the neuron to be activated. For simplicity, I will use the same example as the one discussed by Tse, namely the one referring to Orion. Before we have any knowledge about the pattern of stars in the constellation of Orion, or what Orion is, no neurons fire when that pattern is presented as an input. However, once the criteria (i.e., the input looking like the constellation of Orion) are set, the next time we see that pattern, a specific neuron will fire. The pattern of information is multiple realizable, meaning that the neuron will fire as a response to a certain pattern (in this case the pattern of Orion) regardless of whether it is presented in stars, grains of sand, or fruits. The information is, in fact, the pattern, so the transfer of information avoids the issue of energy transfer, suggesting that this approach to causality is not one of production but rather a version of a difference-making account—what makes the difference in the pattern of neuronal firing is the information in the input, which would make it causal. Tse emphasizes that the informational causal chains are all physical, and also states that, in fact, all the chains are physical with some of them also being informational, which possibly brings his approach close to Davidson's anomalous monism (Davidson, 2001).

Mental events themselves are, for Tse, realized in physical events and they can be causes of both mental as well as physical events through initiating a change in the current state of the neuron, its criteria for firing. However, the notion of realization is in itself not clear and is rarely discussed in both philosophy as well as in neuroscience and Tse does not clarify what he exactly means when he says that "even though mental events are realized in physical events they (i.e., their physical realization) can cause subsequent physical and mental events by preparing new decoders, or changing the criteria for firing on already existing decoders" (Tse, 2013, p. 25). It is unclear, if Tse assumes the events to be causal through their physical realizers or to be causal because of the mental side of the event.

Tse bases his solution of the mental causation problem, and through that the free-will problem, on a specific understanding of the causal exclusion argument and causation itself. He assumes that the main issue for mental causation is the impossibility of self-causation, which I argue is not the case. If the main obstacle was in fact self-causation Tse's argument would have shown how to escape it only on a very basic level, but usually the problem of self-determination is connected to conscious decisions rather than biological unconscious processes. If, on the other hand, we assume, as I do, that the problem lies elsewhere, namely in the causal closure principle (the notion that all physical events have sufficient physical causes), Tse's approach does not seem to solve the issue at hand. If we are not willing to accept dualism nor reduction, we need to accept some form of coexistence of mental and physical, and the most widely used is the relation of supervenience, which puts into question the causality of mental qua mental and not qua physical, as put forward by Jaegwon Kim (1998, 2005). Since the mental events are always also physical (as they have the physical basis) in order to show how mental causation would work, we would have to show that the sole cause of another event is the mental part and not its physical basis. Without a doubt, the physical also enters the causal relation here, and given the causal closure of the physical, which states that all physical events have sufficient physical causes, and the lack of overdetermination (except

from the rare cases of genuine overdetermination), the mental is effectively excluded from the causal chain. However, since the explanations
of how mental and physical are related within his account and why the
input would be causal through its mental properties rather than the
physical are missing, in his argument Tse does not solve the puzzle of
causal closure. It might be the case that Tse's approach is closer to a
reduction in which mental events are seen as identical to physical
events, rather than to any form or nonreductive physicalism.

Also problematic is Tse's usage of terms like deciding and choosing,
which puts into question the role played by neurons in his scenario.
The notions connected to decision-making processes are usually attributed to people, or agents, rather than circuits in the brain (discussed in
Bishop, 2014). We tend to assume that an agent makes a decision, and
the decision itself is a more complex computation, not completely
reducible to the lower level of neuronal computation. On the other
hand, Tse seems to attribute those, usually high-level computations, to
neurons rather than individuals, which could suggest an underlying
reduction of agents to brain circuits, which in turn makes the possibility
of mental causation even more questionable. We don't need to fall into
dualism in order to defend the efficacy of mental events (again, we can
agree on a form of nonreductive physicalism), but reducing those
deeply human abilities to neuronal assessment of the incoming information seems to leave very little room for what we consider necessary
for free will. It is unclear how neurons would even be able to assess the
incoming information and choose or decide beyond a simple mechanistic, biological reaction, which is not what we have in mind neither for
free will nor for mental causation. From the description of how the criterial causation is supposed to work, it seems that neurons only possess
the ability to check if the input information fulfils the conditions that
would make the neuron fire or not, and not much beyond that. This
again is not what we expect from a free action, and we tend to think
that what is needed is more of a high-level property of the system, not
possible on this very basic level. It is very hard to see how neurons
would "decide" and decode meaning of abstract notions. Additionally,
it leaves the processes underlying free will largely performed unconsciously, which opposes an argument that is almost universally
accepted—that free will requires consciousness and our free decisions
or actions are done consciously. If all mental causation was limited to
the unconscious resetting of the neuronal criteria, it would put into
question the kind of control we can have over the process and how it
differs from pure randomness. So although mental causation may work
on that very low computational level, it also has to be causal somewhere later in the decision-making processes, directly influencing our
intentions or decision.

Another problematic angle of Tse's argument lies in the notion of information, which, beyond its connection to the notion of input pattern, is never precisely defined. Patterns that connect to physical entities, like the pattern of Orion, are easily definable. On the other hand, if we, following Habermas (2008), treat the notion of pattern as being applicable to both physical (realized in different ways) as well as abstract cultural patterns transmitted through language, the issue becomes much more complex. The scope of the cultural pattern that influences the resetting of the neuronal firing criteria would be very difficult to define, and specific parts of it would be almost impossible to identify. There is an apparent difference as to what kind of information we assume influences what we mean by deliberation and making rational choices and the type of information transmitted through physical patterns. And again, if the notion of information should be causal, and play a role for free will, the influence the information has should be somehow conscious. The automatic changing of the neurons' firing criteria as the basis for free action seems insufficient and resembles more a computer program executing a task rather than a basis for a free action. It is also unclear what role neurons play in the information assessment, and through the usage of notions like deciding, the complete mappability of agenthood onto the neuronal activity is suggested. This puts the concept of the agent as well as the importance of higher-level computation into question. If everything about the human action would then be just a flow of energy between neurons, the opposite of what Tse would like to argue for (further discussed in Bishop, 2014) would potentially be proven—the mental is reducible to the physical and even if it isn't it's epiphenomenal (plays no causal role in the cause—effect chain). Tse either implicitly allows reduction when he states that everything is just particles interacting with other particles (Tse, 2013), or just leaves the mental as being epiphenomenal.

However, there are great advantages to Tse's account, and they come from philosophical reflections on the nature of free will. First of all, Tse recognizes the extended-in-time nature of our decision-making processes resulting in an overt action. He tries to build an account in which free will is given on such a basic level as to possibly account for all the necessary steps in making a choice. However, this should be connected with an account of how mental causation could additionally be used on higher levels of computation, where it could be connected with consciousness, and not only on the neuronal one. This potentially might entail either adding different levels of computation or accepting the notion of emergence. Then mental causation will potentially happen in two different points of the process leading to action, both fulfilling a very different role.

The other essential notion is the one of information. I have already mentioned some of the problems with how the notion is used in Tse's account, yet the advantages of both information as well as the idea of patterns should not be overlooked. It has been proven by neuroscientists that the human brain remains plastic throughout most of our lives (neuroplasticity), which undermined the previous opinion that it develops rapidly during childhood and then remains static. Changes in the structure of the brain can be caused by physical events but potentially also by all interactions and cultural patterns with which we come into contact. All of them possibly have impact on the physical structure of our brain and through that on the decisions we are making; or at least those are the intuitions we have when we think about what directs our choices. Tse, through the notions of information and pattern, offers a possibility to explain how exactly that happens. The solution is not without its problems as we have seen before, but it might be a first step in a new research direction.

CONCLUSIONS AND FURTHER RESEARCH

In spite of some believing otherwise, the problem of free will is still far from being solved. The issue is, however, more and more pressing, as the topic finds its way to broader discussions in the media (e.g., *Time* magazine) and, potentially, to the court room. Interdisciplinary approaches can further the discussion significantly as we saw in the positions analyzed here, but they need a well-defined scope and definitions. Promising initial results from Tse show the importance of notions such as information and pattern transmission, which should be further investigated. Because of the connection between information exchange and natural language, further analysis could focus on the influence of language and cultural patterns, transmitted through the language, on the brain.

References

Bishop, R. C. (2014, January 31). Review of the neural basis of free will: Criterial causation. *Notre Dame Philosophical Reviews: An Electronic Journal*. University of Notre Dame.
Davidson, D. (2001). *Mental events. Essays on actions and events*. Oxford: Clarendon Press.
Dowe, P. (1995). Causality and conserved quantities: A reply to Salmon. *Philosophy of Science, 62*, 321–333.
Habermas, J. (2008). *Freedom and determinism. Between naturalism and religion: Philosophical essays*. Cambridge: Polity Press.
Haggard, P. (2008). Human volition: Towards a neuroscience of will. *Nature, 9*, 934–945.
Haggard, P. (2011). Does brain science change our view of free will? In R. Swinburne (Ed.), *Free will and modern science*. London: British Academy.
Haggard, P., & Eimer, M. (1999). On the relation between brain potentials and the awareness of voluntary movements. *Experimental Brain Research, 126*, 128–133.

Haggard, P., & Libet, B. (2001). Conscious intention and brain activity. *Journal of Consciousness Studies, 8*(11), 47–63.

Kim, J. (1989). Mechanism, purpose, and explanatory exclusion. *Philosophical Perspectives, 3*, 77–108.

Kim, J. (1998). *Mind in a physical world*. Cambridge, MA: MIT Press.

Kim, J. (2005). *Physicalism, or something near enough*. Princeton, NJ: Princeton University Press.

Libet, B. (1985). Unconscious cerebral initiative and the role of conscious will in voluntary action. *Behavioral and Brain Sciences, 8*, 335–343.

Libet, B., Gleason, C. A., Wright, E. W., & Pearl, D. K. (1983). Time of conscious intention to act in relation to onset of cerebral activity (readiness potential): The unconscious initiation of freely voluntary act. *Brain: A Journal of Neurology, 106*, 623–642.

List, C., & Menzies, P. (in press). My brain made me do it: The exclusion argument against free will, and what's wrong with it. In H. Beebee, C. Hitchcock, & H. Price (Eds.), *Making a difference*. Oxford University Press; Oxford.

Mele, A. (2010). Conscious deciding and the science of free will. In R. F. Baumeister, A. R. Mele, & K. D. Vohs (Eds.), *Free will and consciousness: How might they work?* New York: Oxford University Press.

Mele, A. (2014). *Free: Why science hasn't disproved free will*. Oxford: Oxford University Press.

Menzies, P., & Price, H. (1993). Causation as a secondary quality. *The British Journal for the Philosophy of Science, 44*(2), 187–203.

Prescott, A., Schlegel, A., Sinnott-Armstrong, W., Roskies, A. L., Wheatley, W., & Tse, P. U. (2016). Readiness potentials driven by non-motoric processes. *Consciousness and Cognition, 39*, 38–47.

Salmon, W. (1984). *Scientific explanation and the causal structure of the world*. Princeton, NJ: Princeton University Press.

Schlegel, A., Prescott, A., Sinnott-Armstrong, W., Roskies, A., Tse, P. U., & Wheatley, T. (2014). Hypnotizing Libet: Readiness potentials with non-conscious volition. *Consciousness and Cognition, 33*, 196–203.

Searle, J. R. (2010). Consciousness and the problem of free will. In R. F. Baumeister, A. Mele, & K. D. Vohs (Eds.), *Free will and consciousness: How might they work?* (pp. 121–134). New York: Oxford University Press.

Soon, C. S., Braas, M., Heinze, H.-J., & Haynes, J.-D. (2008). Unconscious determinants of free decisions in the human brain. *Nature Neuroscience, 11*, 543–545.

Tse, P. U. (2013). *The neural basis of free will: Criterial causation*. Cambridge, MA, and London: MIT Press.

Woodward, J. (2003). *Making things happen*. New York and Oxford: Oxford University Press.

Histories of the Brain: Toward a Critical Interaction of the Humanities and Neurosciences

M. Della Rocca

University of Pisa, Pisa, Italy

Readily, historians of science might extend their critical thinking to today's neuro-turn. But there is a formidable obstacle in the way: the neuro-turn itself. **Cooter, 2014, p. 147**

"History and neuroscience make strange bedfellows," Daniel Lord Smail wrote in a 2009 article dedicated to the relationship between brain sciences and the dialectics of history (Smail, 2012a, p. 894). But even as strange as this partnership may appear, after the Decade of the Brain (1990–2000) and at the very beginning of 21st century's "big neuroscience" projects, if we look at the state of the art concerning the interaction between neurosciences and historical disciplines, we can trace no less than two yet consolidated (and still growing) research areas.

Namely, these are history of neurosciences and the brand-new discipline of "neurohistory." Either of these fields of academic expertise has to be considered properly as a consequence of the "neuro-turn" in contemporary culture—even if the first one stemmed from the long and sound tradition of "general" historiography of science, rooted in 18th century philosophical research, while the second one is a new development of the "neuro-revolution" that involved humanities at the end of the last millennium. In this chapter, I will try to compare and connect some features of both these intersections, in order to highlight if (and how, eventually) we can trace a third profitable and heuristic way to consider historical reflection in the "Age of the Brain."

The Human Sciences after the Decade of the Brain.
DOI: http://dx.doi.org/10.1016/B978-0-12-804205-2.00005-7 © 2017 Elsevier Inc. All rights reserved.

ON THE HISTORY OF NEUROSCIENCES, THE OLD-FASHIONED WAY

Today, history of the brain and mind sciences represents an already well-established discipline. Not surprisingly, the topic has encountered significant attention, especially since the beginning of the Decade of the Brain until now. The *Journal of the History of Neurosciences* was founded in 1992, followed by the creation of the International Society for the History of the Neurosciences in 1995 (Haines, 1996), and many other journals devoted to the general history of science and technology have consecrated several issues on topics related to brain research in every period of human history. Following the ascent of the neurosciences both as a consilience-aimed interdisciplinary effort and as an organized academic institution, the history of neurosciences has easily gained its own place among science and technology studies. Notwithstanding more than 20 years of official activity, however, the field of history of neurosciences seems still involved in what has been defined as its "ancillary-heroic" phase. In other words, the field still seems mainly focused on framing neuroscientific triumphs into the chronological collection of historical recording, while at the same time looking for "precursors" of the discipline to justify and celebrate current research in this specialty. While scholars have a long history of criticizing the history of science in its "ancillary way" (For a review on this topic, see Pogliano, 2010) I would like to note the fact that historians of neurosciences seem to have concentrated their academic efforts on the "long past" of the discipline, while some historical reconstructions of contemporary neurosciences—in extension, the history of the interdisciplinary scientific complex as it arose in the second half of 20th century, after the very creation of the word "neuroscience" by Francis Schmitt in the early 1960s and the birth of the International Brain Research Organization (IBRO) in 1961—have been carried on mostly by neuroscientists or scholars in biological and medical research, rather than by historians of neurosciences *per se*. How Hagner and Borck noticed at the dawn of the millennium, "despite its richness, the history of brain research in the twentieth century is, with some remarkable exceptions, still very much a territory to be explored" (Hagner & Borck, 2001).

ON NEUROHISTORY, OR ON MAKING HISTORY AFTER THE NEURO-TURN

On the other side of the relationship between historiography and neurosciences, we have "neurohistory," the new research area inaugurated by Daniel Lord Smail in his brilliant 2008 work, *On Deep History and the Brain* (Smail, 2008). Neurohistory has been defined by Smail as an interdisciplinary approach to human history, aimed at incorporating some among the more recent advances in neurosciences into historiographical theory and methodology, in order to allow a new kind of reconstruction of human past, which could start from our "deep history" (Shryock & Smail, 2011)—i.e., the account of our species' long road since the homination until our present. Such a neuroscience-informed historiography, according to Smail, should be able to elude the conventional limits set in Western culture to the starting point of history—usually corresponding to the invention of writing—to include the life of our hominid ancestors, since around 5–8 million years ago. In this way, theories and models from contemporary neurosciences join those from anthropology, archeology, genetics, linguistics, psychology, and primatology to create the conditions of possibilities of a complex new methodology for historiography—a systemic and global dialectic between the "history from within" and the environmental and global history (Burman, 2012).

But aside from this interesting manifesto, aimed at rethinking the very borders of "deep history," the approach developed by Smail promises to be useful, even to give a different account of other, most recent historical periods, like the Middle Age or early modernity. In his 2008 book (as well as in other shorter essays published since then), Smail rereads some of the most important shifts in human societies, looking for what could have affected the brain states of the people during those times—like psychotropic devices as well as social and political events implied in the activation (or deactivation) of stress-reward neural correlates. Essentially based on the assumption of cerebral plasticity as the main feature of human behavior, neurohistory finds its goal in understanding how "some of the great transformations of history were driven, in part, by their neurophysiological outcomes" (Smail, 2012b).

As Smail explained in the pages of *On Deep History and the Brain*, a historiography informed with contemporary neurosciences should be able to account for how the nervous system (and its ability to change) was confronted by complex and endless dialectics with an ever-changing physical and cultural environment. More specifically, Smail's weapon of choice for this historiographical project is mainly focused on the stress-response system[a]—which neurobiology has explored deeply in the last 20 years, revealing its dramatic capability to be molded by

---

[a] "The stress-response system is one of many candidates for the dialectical model of neurohistory [. . .]. Because chronic stress can have social or political consequences, it is easy to imagine that human behavioral patterns have evolved, albeit unconsciously, in ways that allow powerful or dominant individuals or institutions to exploit the latent quality of chronic stress. The stress-response system, in other words, acts as a niche in which human patterns and institutions take shape. At the same time, the stress-response system has a significant degree of plasticity—not in its structure, but in the way in which it works in given individuals. That plasticity can produce a pattern whereby stress can be an inherited feature of certain class or status groups, which may help explain a pattern that figures in the way that Marx and Gramsci sought to explain through ideas such as false consciousness and hegemony" (2012b, p. ivi).

and adapt to environmental conditions. Neuroadaptive changes in brain stress–reward systems, especially those related to psychotropic substances (or similar neural change-inducing behaviors), have been among the most central topics throughout the Decade of the Brain, leading to a complex (and often harsh) debate regarding the epistemological and ethical issues of this topic (Della Rocca, 2015). Smail's goal is to make a historical "niche" of this brain circuit, through which historians can afford a dialectical tool to rethink historical changes.

Thus neurohistory presents itself as a new methodology for doing historiography, based on the findings of current cognitive and brain sciences—especially plasticity, and from this point of view, it is probably one of the most interesting and fascinating products of the "neuro-turn" in social sciences and humanities. As several authors have pointed out, this research area can offer different insights to scholars, focused on historiography, neurosciences, or philosophy of history (Brukamp, 2012).

Aside from both the "apocalyptic" and "integrated" reception of the neuro-turn in the humanities, we should recognize what neurohistory can offer to contemporary historical inquiry (and to humanities at large): i.e., a powerful and original way to rethink the borders and the relationship between nature and culture through times, linking directly "what makes us humans" with what "we make *as* humans." Furthermore, it offers a critical way to escape from some ideological or conventional schemata that Western researchers still use when thinking of history (like Judeo-Christian tradition, for example, or the eurocentric point of view that often affects historiography). Finally, neurohistory, providing an evidence-based tool to fill in the blanks of remote history, could effectively narrow the gap between "factual" and "conjectural" history, a problematic issue in philosophical reflection on historical knowledge—at least since Francis Bacon, and then in the work of Jean Jacques Rousseau and Immanuel Kant (Iacono, 2015).

On the other hand, neurohistory has received (in this sense, the same as for every "neuro-" discipline emerged by the neuro-turn) a good amount of criticism. These concerned mainly the risks surrounding the rise of neurohistory as a 21st century revival of pop sociobiology—in extension, the revival of a deterministic and presentist approach to history itself. Indeed, defining human beings by the cultural idealization and factualization of contemporary neuroscience—especially now that it has been proclaimed the scientific frontier at the time of the "end of the history"—can appear indeed so ahistorical that the idea of building a new way of doing historiography on it can be considered at least very controversial, if not completely weird.

However, the new research area, notwithstanding its "youth," has proven yet to be deeply conscious of what it means to make historical and biological knowledge meet. Indeed, Smail has pointed out several times that it is not only history that needs brain science, but that even the latter should take some lessons—or advice, at least—from the former. In fact, assuming the plasticity of the nervous system as the cornerstone of neurohistorical discourse, Smail could easily criticize the "presentist orientation" of contemporary cognitive neurosciences. As is known, on the wave of the cultural and epistemic position of the so-called "neuroessentialism," in the past decade neuroscientists have gotten used to a certain kind of communicative bias, making universalistic generalizations from their findings, especially about a presumed deterministic connection between the central nervous system and the expression of complex behaviors. In a special 2014 issue of the journal *Isis*, Smail provided the example of compulsive hoarding—a behavior for which neuroscientists recently affirmed to have found the neural correlates, but that cannot be easily found in precapitalist societies (Smail, 2014). Doing this, he showed how neurosciences have no right to claim for the existence of complex behavioral patterns "just in the brain," since they can be correctly detected (and understood) only through the lens of a close dialog between neurobiological explanation and the historical context in which they take place. And in 2012, at the end of an editorial about neurohistory for the *History News Network*, neurophysiologist and historian of neurosciences Cristopher Smith reminded readers that neuroscience can offer no "magic tricks" to historical disciplines, writing that "Charles Darwin taught us, contrary to 'Whig interpretations,' that there are no crystal balls, the future is not predictable, there is no 'point omega' to which biological and/or human history points." This was clearly a correct cautionary note, along with an introduction to the new and powerful research methodology neurohistory.

Thus the developing neurohistory seems to provide itself at least a first level of critical interaction between history and neurosciences, testing the statements of neuroscientific knowledge in the historical practice—and eventually denouncing or correcting its mistakes. But establishing an apparent two-way channel between neuroscience and the humanitites, in order to break the rule of univocal knowledge-transfer, represents just a first—and still very partial—step toward a very critical interaction between historical and neuroscientific disciplines. The next challenge for scholars from humanities involved in the neuro-turn has to be a meta-reflection on this subject: in other words, the need to *critically historicize* neurosciences, as well as make history "of" and "by" it.

Indeed, historians and neuroscientists (and, more importantly, historians of neuroscience) should be aware even of the risk of "de-historicizing" both what neuroscience is and what it can provide to humanities. Because, after all, neuroscientific knowledge is historically situated knowledge—theories, models, cultural representations of the brain—i.e., *a historical product itself*. It is true that there are no "points omega," but we also have to keep in mind that they also don't exist in the future *nor* in the present—not even in neuro-turn's great promises and hopes.

In the previously cited special issue of *Isis*, Roger Cooter (2014) and Max Stadler (2014) correctly argued that the human image offered by contemporary neurosciences is characterized by some features that are tightly linked to the ideology of neoliberal capitalism and technoscientific society, features that seem unable to fit into a proper historiographical methodology. But they also called on historians of science for a critical historiography of neuroscientific discourse itself—even including the neuro-turn itself—extending what Abi-Rached and Rose, in a 2014 article, explained as the need to "historicize neuroscience" (Abi-Rached & Rose, 2014). In this, the meeting of neurohistory and history of neurosciences can probably develop a third way—i.e., a *historical critique of science*—of the discourse that emerged from the revolution caused by contemporary brain research.

To be clear, let me return to the "ancillary" history of neurosciences for another brief comparison with neurohistory. Despite the obvious differences, these two models of intersection between brain research and historical disciplines share a common trait, for they contemplate little or nothing of conceptual and active feedback from humanities to brain/mind studies at all. The brain research's *episteme*—ideas, keywords, and methods—is supposed to pass from the brain lab to the theoretical and practical field of history, even to be adjusted or limited in some cases as we have seen, but not *vice versa*. The neuro-turn allows history to use neuroscience's concepts for its own goals and to systematize them in a progressive (and sometimes implicitly teleological) narration, but it leaves no space for *contextualization* and *historicization* of the neuroscientific inquiry in itself.

What are the causes of this phenomenon? As I previously argued, while many historians of science have dedicated themselves to the study of the historical precursors of neurosciences, they have seldom dedicated their efforts to the study of 20th century brain research, notwithstanding the awareness of the very post-WWII origins of neuroscientific enterprise as we know it. This occurred mostly in social studies of science, technology, and medicine, as it has been the case for the connection between neurosciences and the "translational research" paradigm, the current model that subordinates basic scientific research to the development of new technologies, or other applications. This feature of translational research can be considered as a symptom of the postmodern character of contemporary science (Forman, 2010), which in turn operates a deep modification in how we conceptualize the historical dimension of science itself. I think we should consider, at least for some aspects, neurohistory and the ancillary history of neurosciences as a particular "degenerated" (in the neurological sense of the word) form of this translational approach. The difference, in this case, is that there is no translation "from bench to bedside," as it usually happens in medicine, but "from bench to desk and archive." What does not differ, here, is the very effect of the neuro-turn as a translational and postmodern phenomenon that leads scholars—especially historians—to "reject the possibility of conceptually characterizing any substantial chunk of the past, reject the hierarchic ordering of causes of any historical happening—indeed, the very idea of causes—and, consequently, take complexification as the only legitimate goal of historical research and exposition" (Forman, 2010, p. 159).

Then, going further in denouncing the "presentist orientation" of contemporary neuroscientific knowledge, we should assume that not even theories and models from brain and mind research are outside of the history (Fuller, 2014; Smith, 2012). Criticisms, then, should be directed not against a history informed by brain facts, but rather to the overall vision of a "neutral" and "absolute" neuroscience—in other words, the last version of the hard-to-kill legacy of a positivistic, progressive, and cumulative vision of science. More than any other scholars, historians of science are among those who can demonstrate that both neuroscientific knowledge (and above all the "neuroscientific discourse," *à la* Canguilhem) is indeed historical knowledge in itself, tightly linked to the historical causal chains of the context in which it develops. Consequently, if we really want to profit from the neuro-turn in historiography and the humanities, avoiding the disappointment of "reinventing the wheel" at any paradigm shift in brain/mind knowledge, we should definitely call for a "historical-turn" on the neuro-turn.

Since we are completely immersed in the very same period of the historical process we have to put under the lens of critique, the task can appear difficult, if not impossible at all. The entanglement of science, the scientific discourse, and the cultural production based on them can seem puzzling, of course: but actually, history of science has recently demonstrated itself able to accomplish a similar goal, and at least a noteworthy example should be able to support this statement. Consider the famous work of American psychologist Julian Jaynes, *The Origins of Consciousness in the Breakdown of the Bicameral Mind*, published in the mid-1970s, one among the most influential and controversial books of the second half of the 20th century, for the history of neuroscience as well as for cultural studies. In this work, as it is well known, Jaynes formulates the hypothesis of a "bicameral" mind model, characterizing human cognition before the development of our conscious Self, which according to the author took place approximately three millennia ago.

The theory expressed in Jaynes' book sought comparison in Homeric and pre-Homeric cultural representations (from literature, religions, and arts), and it mostly relied on the theoretical and experimental acquisitions on hemispheric specialization and lateralization of brain functions, a recent achievement of that time to justify itself. Intriguing and shocking at the time, Jaynes can be considered without a doubt the first author (and still one of the most important, in this sense) of a sort of "neurohistory," at least since the development of contemporary neuroscience.

Today however, everyone minimally acquainted with current models of the brain would probably reject Jayne's hypothesis. And this refusal would be based not only on the rejection of the "hard-wired" hemispherical specialization, which is the core of Jayne's analysis, but even on the awareness of some historical considerations. As noted by Andrea Cavanna and his colleagues in a paper devoted to this subject, the bicameral mind is not fully consistent with 21st century neurophysiological knowledge, and it could unlikely fit into any evolutionary perspective (Cavanna, Trimble, Cinti, & Monaco, 2007). But Cavanna also correctly notes how "Jaynes' speculations appear to be shaped by the striking insights derived from the early studies on "split-brain" patients by Sperry and colleagues" (Cavanna et al., 2007, p. 12). Today, we have no problem dismissing the "hard" version of Sperry's theories about hemispheric specialization, mostly because of the new insight on neural plasticity and on the integration and degeneracy of the central nervous system. Nevertheless, the legacy of Sperry's research on Western culture appears to be still alive—the Internet is still plenty of memes on the "left" and the "right" brain, as everybody can easily notice—and Jaynes' work proved how Sperry's ideas easily influenced even the academic *milieu* of humanities for more than a decade.

Thus a "deep" critical interaction between neurosciences and history (and humanities at large, of course) could be the active assumption of the historical character of neuroscientific knowledge, both in its "turn to humanities" and in itself. Such a position does not necessarily imply adhesion to the Fleckian and Kuhnian models, nor even it should be dismissed as a form of mere sociologism or cultural relativism, rather it is based on the very history of post-WWII brain studies. Historians of neuroscience that would investigate further the context of Sperry's discovery could add new insight to the understanding of the historical dimension surrounding this specific fact and the *ante litteram* neuro-turn that inspired Jaynes' lecture of ancient history. But in a similar way, contemporary scholars in this field should address even contemporary theories on plasticity (and their "translation" in the neuro-turned historiography, of course), looking for consistent contextualization of it in the late-20th century.

TOWARD A CRITICAL HISTORY OF THE NEURO-TURN, OR "PLASTICITY AND ITS DISCONTENTS"

In order to explain the need for a critical approach of humanities to neurosciences then, we should concentrate ourselves, more in detail, on that significant case study plasticity. Indeed, as we have previously seen, Smail's project for a new kind of historiography, able to account for the dialectics between brain mechanisms and their changes through historical ages, relies almost completely on the contemporary idea of brain plasticity.[b] Indeed, the concept of neuroplasticity is today hegemonic in the neurosciences, as well as in the neuroscientific discourse. In fact, it should be considered the very central and *sine qua non* feature of the "neuro-turn" in historiography. After all, historians—in Smail's words, again—"have been disinclined to deal with behavioral or psychological patterns that have the appearance of being universal or hardwired, since the realm of the universal offers little traction for arguments about historical change" (Smail, 2012a, p. 895). It has been only since the exaltation of brain plastic features that a biologically informed historical accounting could develop itself, finally free from the Procrustean bed of sociobiological determinism.

However, the high value of plasticity as catalyst of the neuro-turn has to be generalized to all discipline of the humanities, since a plastic vision of the central nervous system can offer "a new way of thinking about two old ideas: the interaction of biology and culture, and the division of mind from brain" (Littlefield & Johnson, 2012, p. 15), consequently allowing scholars interested in the study of every human activity—from arts to economics—to harmonize their reflections with the contemporary scientific accounts of human biology.

Plasticity is in many ways the very core of the neuro-turn, since it is a perfect metaphor, a powerful model, and a key concept for our times—i.e., the very *trait d'union* between neurobiology and human culture. However, notwithstanding its centrality, plasticity has proven to be a puzzling keyword, both in and of itself and as a full explanation of the "neuro-revolution." The term, in fact, has found its home in almost every publication about neurosciences and the new "neuro-disciplines" in the last decade, but it has revealed to be very "polymorphously perverse," each time associating itself with an apparently different vision (and judgment) of the neuroscientific enterprise, as well of its effects on culture and society.

[b]"A neurohistorical perspective on human history is built around the plasticity of the synapses that link a universal emotion, such a disgust, to a particular object or stimulus, a plasticity that allows culture to embed itself in physiology" (Smail, 2008, p. 115).

Indeed, plasticity seems to embody both the epistemological and cultural triumph of the "brain revolution" of our age and the cautionary note on the limits of neuroscientific explanation about the human dimension at the very same time. In a review of the so-called "neuro-skeptical" literature, published in *The New Yorker* on September 9, 2013, journalist and essayist Adam Gopnik rightly noted how "all of the neuro-skeptics argue for the plasticity of our neural networks" (Gopnik, 2013, p. 87), but plasticity has also been considered (as it should be) the herald of brain centrality, often leading to "neuro-centrism," the conventional view so common in contemporary neurosciences (and sometimes even in neuroscientific culture) that the human brain constitutes the sole substrate needed for a consistent explanation of human mental functions.

This apparent contradiction has struck not only laymen and readers in general, but it seems to be extended even in the academic field. Just to make the case in the neuro-turned humanities, while Walter Glannon challenged and rejected the neuro-essentialist approach to the definition of human Self, he argued that what we call mind emerges from a constant retroactive interaction among the nervous system, the whole body and the environment, and that the product of this interaction is "the plasticity of the brain, the ability of nerve cells to modify their activity in response to change" (Glannon, 2009, p. 324). But at the same time, both Paul Churchland and Joseph LeDoux—both champions of the neuro-turn in Western culture and both targeted by Glannon's critical argumentations—made neural plasticity a cornerstone of their own discourses.

The situation is no better if we try to solve this puzzle looking at the neuroscientific definition of plasticity, or at least at its interpretations as given by researchers working on this topic. At the dawn of our century, there was no sign of a shared general theory of plasticity, and "many neuroscientists use the word neuroplasticity as an umbrella term" meaning "different things to researchers in different subfields" (Shaw & McEachern, 2001, p. 3). In their historical reconstruction of the term, Berlucchi and Buchthel (2009) argued that plasticity seems affected by a multiplicity and variety of meanings and interpretations, as demonstrated by the heterogeneity of many chapters, reviews, and books that still continue to appear in the neuroscientific and psychological literature.

Thus what is plasticity? How should it be considered, especially in relationship with the neuro-turn? In Gopkin's previously cited article, it was stated that "myths depend on balance, on preserving their eternal twoness" (Gopnik, 2013, p. 86)—and indeed, the concept of plasticity seems to be characterized by a strong duality, if not by a plurality of meaning.

Shall we consider plasticity as a part of a larger brain mythology? Contemporary philosophy has addressed this question, studying plasticity as one of the foundational metaphors of our times and culture—with special attention on its connection with the historical dimension of man. In her renowned 2004 essay "What Should We Do With Our Brain?," Catherine Malabou analyzed how "plasticity is the dominant concept of the neurosciences" and how "today it constitutes their common point of interest, their dominant motif, and their privileged operating model, to the extent that it allows them to think about and describe the brain as at once an unprecedented dynamic, structure, and organization" (Malabou, 2008, p. 4). Further, Malabou observed that plasticity (as a popularized myth) has conveyed a sort of "neuronal ideology," tightly linked to that of contemporary capitalism. However, no matter how intriguing Malabou's critique is (and useful to a political reflection on contemporary neurosciences) as noted by Tobias Rees her work lacks an up-to-date historical and epistemological account of brain plasticity (Rees, 2010)—consequently not lifting any veil of mythology from this troubling concept, although revealing an important connection between the scientific and the ideological dimensions behind this term.

Thus there is both a culture and a philosophy of plasticity, both treating the concept as a contemporary myth in some ways. This is the point where we have to turn back to the history of neuroscience. After all, it is history that represents the alternative to mythology when looking at the past. Notwithstanding, some reconstructions of the history of this concept in the vocabulary of neurosciences has been carried out only recently—and only after several complaints on the ambiguous and unclear use of the term had accumulated in the scientific and philosophical literature, at least since the second half of 20th century.

While some historical accounts have declared that it is possible to find the roots of plasticity in Greek antiquity and English empiricist philosophy (Markram, Gerstner, & Sjöström, 2011), the main shared view of the history of brain plasticity has shown that this idea was formulated at the end of 19th century in order to account for a specific range of brain phenomena—i.e., behavioral modifiability, maturation, adaptation to a mutable environment, and learning and compensatory adjustments in response to functional losses (both traumatic and pathological) (Berlucchi & Buchthel, 2009). Thus for at least a hundred years, the term "plasticity" has been used in a variety of circumstances pertaining to the development and functioning of neural circuitry. Although varying in certain conceptual aspects, the fundamental meaning of the term may be found in its numerous appearances across several areas of scientific literature and historical periods. In its most generalized sense, plasticity refers to the capability of the central nervous system to be "molded," "shaped," and modified by learning, experience, and repeated exposition to environmental stimuli. As such, the concept has occupied important positions in psychological and neuropsychological theories since its first formulation by William James in 1890: in 1893 the Italian neuropsychiatrist Eugenio Tanzi (and later, in 1898, his disciple and colleague Ernesto Lugaro) proposed that the repetition of the neural activity could cause a hypertrophy of the neurons involved, creating new activity-dependant associations between neural cells. In those same years, plasticity was included in the definition of the "neuron doctrine" by Santiago Ramon-y-Cajal. Later in the 20th century, plasticity was consecrated in the neurophysiological theory of cell assemblies by Donald Olding Hebb, in which it was postulated that an increase in synaptic efficacy arises from the presynaptic cell's repeated and persistent stimulation of the postsynaptic cell, leading to structural and functional modification of neuronal networks.

Other accounts—less interested in "precursors" and more focused on contemporary neurosciences—underlined how Brenda Milner and Eric Kandel carried out the decisive experimental works that consolidated the plastic brain theory. These historical reconstructions even acknowledged Michael Merzenich and his team for being the first ones to publicly defend the idea that neural plasticity is capable of overcoming its functional dimension, extending the phenomenon of structural reshaping of the central nervous system beyond the embryonic phase, and the role of Alain Prochiantz's findings, which suggested the possibility of morphogenetic processes in adult brains (Rees, 2010). Abi-Rached and Rose pointed out how plasticity as a major pathway in neuroscience developed through the discovery of the nerve growth factor by Rita Levi Montalcini in the 1970s (Rose & Abi-Rached, 2013, p. 49), while Schwartz and Begley cited the experiments on the cortical remapping of monkeys led by Edward Taub in the same decade (Schwartz & Begley, 2002).

What is sure, according to these reconstructions, is that plasticity is not a brand-new feature of 21st century neurosciences. Its history is longer than appears at first sight, and the idea of neuronal plasticity lied at the foundations of neuroscientific research, since its very beginning.

However, historians of neurosciences that confronted plasticity ended their reconstruction in the 1990s—the moment when, as we have seen, plasticity started to act as the springboard both for the neuroscientific revolution and the neuro-turn in humanities.

Then, if we try to historicize the concept of plasticity as the catalyst of neuro-turn, the real questions for a critical historiography of science essentially become: why did plasticity gain such a hegemony in cultural production? Why did this happen in the 1990s and not before?

A possible answer to this question, which is rooted in the most recent history of contemporary neurosciences, should consider the rise of the large-scale brain simulation projects and the connected field of the neuromorphic hardware industry. These two issues have seldom been incorporated in the historical reconstruction of neurosciences,[c] and even more rarely addressed in the philosophical reflection on humanity's neuro-turn. However, it should be noted that, as the start of the neuro-turn coincides with the "scientific factualization" (*sensu* Fleck) of plasticity as the dominant one in neuroscience, the hegemony of plasticity coincides with the discovery of this concept by applied and translational technoscience.

Indeed, it was 1990 when Carver Mead introduced the term "neuromorphic engineering" to describe the interdisciplinary approach to the design of biologically inspired neural information-processing systems, able not only to emulate the nervous system functions but also resemble the brain in structure (Mead, 1990). In what is considered its "manifesto," Mead underlined how synaptic and structural plasticity were the key features of neuromorphic systems, allowing them to simulate *in silico* the mechanisms of adaptation and learning shown by biological systems.

While history of contemporary neurosciences rightly identified the 1990s and the Decade of the Brain for having been the triumph of the "neuromolecular gaze" (Abi-Rached & Rose, 2010), they were also the years when the view of neuroscience as technoscience started to gain its hegemony. And this shift in perspective is tightly linked both to plasticity and the advent of the brain simulation project. In fact, we can consider that it has been the discourse about plasticity, as it emerged in the 1990s, that has directed the neuroscientific agenda until today. As Garis et al. pointed out, plasticity—i.e., the connectivity patterns and the dynamical parameters of brain activity—is one of the two essential conditions of possibility for the new generation of artificial brain projects (De Garis, Shuo, Goertzel, & Ruiting, 2010).

For example, it was Henry Markram, the leader of the Blue Brain Project (and among the heads behind the discussed Human Brain

[c]Some important and noteworthy exceptions are represented by the previously cited article by Max Stadler (2014) and the work of David William Bates, who recently connected the developments of new AI research to those in neurosciences: "It would seem that the recent intensification of interest in the inherent plasticity of the brain—its developmental openness, its always evolving structure in the adult phase, and its often startling ability to reorganize itself after significant trauma—puts considerable pressure on the technological conceptualizations of the brain that assume a complex but definite automaticity of operation. [...] However, it is now the case that the neurophysiological phenomenon of brain plasticity is rapidly becoming assimilated to computational models and digital representations of the nervous system. [...] The processes governing the determination of the plastic brain as it experiences the world are obviously much more complex, but the basic principle still holds. Therefore even the contingent determination of the plastic brain can, it is thought, be rigorously modeled by a virtual computer simulation" (Bates, 2015, p. 195).

Project) who first demonstrated in the 1990s the millisecond relative timing of single pre- and postsynaptic action potentials, describing the highly precise learning mechanism today commonly known as "spike timing-dependent synaptic plasticity" (STDP) (Markram, Lübke, Frotscher, & Sakmann, 1997)—i.e., the process that "adjusts" the strength of connections between the neurons in the brain, and the very cornerstone of the brain simulation projects still running. To cite an example of Markram's awareness of the epistemic value of his findings, in a 2004 study on structural and functional connectivity of pyramidal neurons—published a year later in the *Proceedings of National Academy of Sciences*—he proposed to the public a model of "the neocortical microcircuit as a tabula rasa" (Kalisman, Silberberg, & Markram, 2005), which legitimized both the establishment of a "tabula rasa principle" for the account of plasticity in synaptic connections, as well as the most presentist historical reconstructions of the plasticity concept (Markram et al., 2011).

In the same way, the contemporary neuromorphic hardware industry relied completely on this epistemological tenet. Plasticity represents the very need to consider (and project) intelligent artificial autonomous systems, and therefore researchers have been devoting significant effort the understanding and modeling plasticity mechanisms. In particular, since the 1990s the field of neuromorphic engineering has focused on the development of full-custom hybrid analog/digital electronic systems for the implementation of models of biological computation and learning in hardware, leading to the recent creation, in 2015, of TrueNorth, the first neuromorphic chip, by IBM.

Plasticity, then, was at the core of the neuro-turn as well as at the core of the techno-neuroscientific revolution that is today dictating the rules of mainstream brain research. In order to historicize contemporary neurosciences and to outline a full picture of our scientific and cultural panorama, we simply cannot avoid considering the link between the recent material and epistemological developments in the field.

Then, avoiding any historical determinism, historiography of contemporary science and culture has to incorporate this conjuncture between the advent of plasticity in technological thinking and the hegemony of the same in concept in culture: because if it cannot, it can never fully explain how plasticity made its way into Western culture, leading to so many aspects of the neuro-turn as to a general reframing of brain sciences—i.e., the conceptual circular system of feedback involving humanities, neurosciences, and technology.

Acknowledging this fact has an important consequence on our idea of historiography in and of neurosciences, because if we apply the idea of coevolution to the history of brain sciences, of neuro-turned culture and society in the "age of the brain," we give life to a tight dialog between material, scientific, and cultural history. And this dialog, in turn, can generate a third way of thinking about the relationship between them—a critical one.

The acknowledgment of our being in history should be axiomatic for a discipline that wants to be dialectical—and this is valid for neurohistory as well as for environmental history, in which this kind of reflection has developed in the last years.[d]

While this may appear radical, it should be noted that a critical framing of plasticity in its economical and political context cannot be dismissed as "skeptical" or "relativistic." Plasticity *is a fact* (again, in the sense Ludwik Fleck gave to the term): moreover, it is probably the most important fact of contemporary neurosciences, providing an extremely powerful model to explain the continuous and circular relationship between biological underpinnings, mental functions, and historical and environmental conditions (Morabito, 2015). What is at stake here is that—paraphrasing the Italian physicist and epistemologist Marcello Cini—we need full acknowledgment of plasticity as something essentially "not neutral," even if it is neither an "opinion" (Gagliasso, Della Rocca, & Memoli, 2015).

This means to be aware of the specific historical dimension of neurosciences, linking the need for a critical historiography that has been described by Abi-Rached and Rose to the peculiar features of what Paul Forman has defined as the epistemic context of "postmodern science" (i.e., its strong connections to the necessities and the values of contemporary neoliberal capitalism, as well as to its production processes). This issue should be part of a more general epistemological reflection, both for neurosciences and humanities, but it takes the form of a necessity for scholars in historiography. As Malabou recognized for the hegemony of the brain/computer metaphor in the first decades of the post-WWII Western world, a model that serves at the same time as a central scientific tenet and a cultural keyword is probably very involved in the "material history" of its time, especially in the intersection between political and techno-scientific agendas, which is a common characteristic of contemporary scientific enterprise.[e] Plasticity does not

---

[d]"Clearly, primatology and other animal sciences are the subject of the history of science more than environmental history, but it's through the lens of such science that environmental historians often see the animals they write about, so studies of the social and cultural construction of science are critical aspects of the environmental historian's toolkit" (Walker, 2014, p. 62).

[e]"One of the subsections of Jeannerod's book *The Nature of Mind* is entitled 'The Comparison Between Brain and Computer Is Not Pertinent.' This comparison dates to the 1950s and reigned until the end of the 1980s. It allowed research in Artificial Intelligence to make considerable progress" (Malabou, 2008, p. 34).

seem to be different in this sense, which should lead us to include this level of contextualization in our historical and epistemological analysis.

Walter Benjamin wrote in his *Theses on the Philosophy of History* (1940) that, in order to build a history that can be useful to comprehend our present, we have to think about the questions that arise from our *Jetztzeit*. While the "here and now" of contemporary neurosciences (both as a scientific and a cultural fact) has declared plasticity as the new frontier of knowledge, at which almost every research area should look, historians (especially those involved in science and technology studies) should answer dialectically to this position. That is, historiography has to reclaim the discourse on plasticity—both in neurosciences and in the "neuro-turned" humanities—as a part of its own object of study, in order to highlight the genesis of the concept and the reasons behind its success. Paraphrasing Benjamin again, we should "brush history"—in and of neurosciences—"against the grain" (Benjamin, 1968, p. 257), since only through the accomplishment of this task can we achieve a critical and full engagement between neurosciences and the humanities, without the risk of being involved in a new dogmatic neuro-mythology.

References

Abi-Rached, J. M., & Rose, N. (2010). The birth of the neuromolecular gaze. *History of the Human Sciences, 23*(1), 11−26.

Bates, D. W. (2015). Automaticity, plasticity, and the deviant origins of artificial intelligence. In D. W. Bates, & N. Bassiri (Eds.), *Plasticity and pathology: On the formation of the neural subject*. New York: Fordham University Press.

Benjamin, W. (1968). Über den Begriff der Geschichte [Theses on the philosophy of history] (H. Zohn, Trans.). In H. Arendt (Ed.), *Illuminations* (pp. 253−264). New York: Schocken (Original work published 1940).

Berlucchi, G., & Buchthel, H. (2009). Neuronal plasticity: Historical roots and evolution of meaning. *Experimental Brain Research, 192*, 307−319.

Brukamp, K. (2012). Neurohistory: Being in time. In E. Russell (Ed.), *RCC perspectives: No. 6. Environment, culture, and the brain: New explorations in neurohistory* (pp. 75−78). Munich: Rachel Carson Center for Environment and Society.

Burman, J. T. (2012). History from within? Contextualizing the new neurohistory and seeking its methods. *History of Psychology, 15*(1), 84−99.

Cavanna, A. E., Trimble, M., Cinti, F., & Monaco, F. (2007). The "bicameral mind" 30 years on: A critical reappraisal of Julian Jaynes' hypothesis. *Functional Neurology, 2*(1), 11−15.

Cooter, R. (2014). Neural veils and the will to historical critique: Why historians of science need to take the neuro-turn seriously. *Isis, 105*(1), 145−154.

De Garis, H., Shuo, C., Goertzel, B., & Ruiting, L. (2010). A world survey of artificial brain projects, part I: Large-scale brain simulations. *Neurocomputing, 74*(1−3), 3−29.

Della Rocca, M. (2015). In the Flesh? Appunti critici sul neuroessenzialismo. *Scienza e Società, 21/22*, 35−42.

Forman, P. (2010). (Re)cognizing postmodernity: Helps for historians—Of science especially. *Berichte zur Wissenschaftsgeschichte, 33*(2), 157−175.

Fuller, S. (2014). Neuroscience, neurohistory, and the history of science: A tale of two brain images. *Isis, 105*(1), 100−109.

Gagliasso, E., Della Rocca, M., & Memoli, R. (Eds.), (2015). *Per una scienza critica. Marcello Cini e il presente: filosofia, storia e politiche della ricerca* Pisa: ETS.

Glannon, W. (2009). Our brains are not us. *Bioethics*, 23(6), 321–329.

Glannon, W., "Our brains are not us", in *Bioethics*, **23**(6), 2009, 321-329. Gopnik, A. (2013, September 9). Mindless. *The New Yorker*, pp. 86–88.

Hagner, M., & Borck, C. (2001). Mindful practices: On the neurosciences in the twentieth century. *Science in Context*, 14(4), 507–510.

Haines, D. E. (1996). Announcement: International Society for the History of Neurosciences. *Cerebrovascular Diseases*, (6), 20.

Iacono, A. M. (2015). De l'histoire théorétique ou conjecturelle. In P. Paoletti (Ed.), *Que faire de l'histoire? Philosophie et conscience histoirique au siècle des lumières* (pp. 63–76). Pisa: ETS.

Kalisman, N., Silberberg, G., & Markram, H. (2005). The neocortical microcircuit as a tabula rasa. *Proceedings of the National Academy of Sciences*, 102, 880–885.

Littlefield, M. M., & Johnson, J. M. (Eds.), (2012). Introduction: Theorizing the neuroscientific turn—Critical perspectives on a translational discipline. In *The neuroscientific turn: Transdisciplinarity in the age of the brain*. Ann Arbor: The University of Michigan Press.

Malabou, C. (2008), *Que faire de notre cerveau?* [What should we do with our brain?] (S. Rand Trans.). New York: Fordham University Press (Original work published 2004).

Markram, H., Gerstner, W., & Sjöström, P. J. (2011). A history of spike-timing-dependent plasticity. *Frontiers in Synaptic Neuroscience*, 3(4). Available from http://dx.doi.org/10.3389/fnsyn.2011.00004.

Markram, H., Lübke, J., Frotscher, M., & Sakmann, B. (1997). Regulation of synaptic efficacy by coincidence of postsynaptic APs and EPSPs. *Science*, 275(5297), 213–215.

Mead, C. A. (1990). Neuromorphic electronic systems. *Proceedings of the IEEE*, 78(10), 1629–1639.

Morabito, C. (2015). Neuroscienze cognitive: Plasticità, variabilità, dimensione storica. *Scienza e Società*, 21/22, 13–24.

Pogliano, C. (2010). "A Mere Luxury." Per una storia della storia della scienza. In M. L. Frank, & C. Pogliano (Eds.), *Scorci di storia della scienza*. Pisa: PLUS-Pisa University Press.

Rees, T. (2010). Being neurologically human today: Life and science and adult cerebral plasticity (an ethical analysis). *American Ethnologist*, 37(1), 150–166.

Rose, N., & Abi-Rached, J. (2013). *Neuro: The new brain sciences and the management of mind*. Princeton, NJ: Princeton University Press.

Schwartz, J., & Begley, S. (2002). *The mind and the brain: Neuroplasticity and the power of mental force*. New York: Regan Books.

Shaw, C. A., & McEachern, J. C. (Eds.), (2001). Is there a general theory of neural plasticity?. In *Toward a theory of neuroplasticity*. Philadelphia: Psychology Press.

Shryock, A., & Smail, D. L. (Eds.), (2011). *Deep history: The architecture of past and present*. Berkeley, CA: University of California Press.

Smail, D. L. (2008). *On deep history and the brain*. Berkeley, CA: University of California Press.

Smail, D. L. (2012a). Neuroscience and the dialectics of history. *Análise Social*, 205 (xlvii)(4), 894–909.

Smail, D. L. (2012b). History meets neuroscience. *History News Network*. Retrieved from <http://historynewsnetwork.org/chapter/147619> Accessed 15.09.15.

Smail, D. L. (2014). Neurohistory in action: Hoarding and the human past. *Isis*, 105(1), 110–122.

Smith, C. U. M. (2012). What can historians learn from neuroscience? *History News Network*. Retrieved from <http://historynewsnetwork.org/chapter/147627> Accessed 15.09.15.

Stadler, M. (2014). Neurohistory is bunk? The not-so-deep history of the postclassical mind. *Isis*, 105(1), 133–144.

Walker, B. L. (2014). Animals and the intimacy of history. In A. C. Isenberg (Ed.), *The Oxford handbook of environmental history*. Oxford and New York: Oxford University Press.

PART II

THE NEUROSCIENCES OF SOCIAL SCIENCES AND ETHICS

6

The Theory of Brain-Sign: A New Model of Brain Operation

P. Clapson

Birkbeck, University of London, London, United Kingdom

> The unity of consciousness—our sense of self—is the greatest remaining mystery of the brain. As a philosophical concept, consciousness continues to defy consensus. *Nobel Laureate, Eric Kandel (2013, p. 546)*

INTRODUCTION

This chapter presents a theory of brain function that I have developed in recent years, called *the theory of brain-sign*. The relevance to this volume is that the theory is a scientific construct that reconciles our supposed mental life with neural function and, by the same means, reveals conceptually the neural mechanisms of human social behavior. The theory reconstructs the brain phenomenon hitherto supposed as consciousness, thus demonstrating how one organism's cooperative action with others is entirely physical. This eliminates the mind—body problem and dissolves the question of how we can know other minds, so to act in concert with them. There are no minds.

Many disciplines are directly concerned with brain function. Neuroscience of course, but also psychology and the social sciences, medicine, and psychiatry. Indirectly many human activities depend on

The Human Sciences after the Decade of the Brain.
DOI: http://dx.doi.org/10.1016/B978-0-12-804205-2.00006-9

© 2017 Elsevier Inc. All rights reserved.

an interpretation of brain function, explicitly or not, e.g., law, ethics, education, politics, international relations, music.

However, there is a profound schism in the approaches of the disciplines. On the one hand, human beings are characterized as transcendent to physical nature by possessing a mind with faculties of perception, feeling, and thought. This demands a different explanatory framework than determinist physicality. On the other, humans are seen as entirely biological, which is more allied to the rigors of a physicalist account. The schism is apparent in the lack of a common vocabulary that would derive from a common view of the human being. It may also be seen in the debate about whether humans have free will, or whether their behavior fits into the causal structure of the physical universe. The debate goes back through history as a clash between the divine view of human beings (the mind being a gift of God, or of other transcendent origin) and the material view.

From the scientific stance the universe and everything in it, including organisms, are wholly physical and, in principle, can be described by science. But we are immediately faced with an apparent paradox. Science is a construct of the human organism. Is the statement that humans are physical and describable by science not itself dependent on the very thing that enables the description...the human organism? How can the organism describe itself when it is the source of description?

The paradox is apparently solved in the phenomenological approach, as in this passage by Maurice Merleau-Ponty:

> The whole universe of science is built upon the world as directly experienced, and if we want to subject science itself to rigorous scrutiny and arrive at a precise assessment of its meaning and scope, we must begin by reawakening the basic experience of the world of which science is the second-order expression. (1962, p. viii)

Experience, as consciousness, appears to have priority in establishing what the world is. Of course, we must investigate how consciousness gives us the world by exploring consciousness itself; however, consciousness is the source of knowledge, including scientific knowledge. For phenomenology this is obvious. But it is a supposition. The obvious has often been overturned, not least in science itself. As a supposition, the brain phenomenon could result from what we really are, which is not what phenomenology claims.

A scientific theory about what we really are and how we operate could clarify the situation. Its effectiveness would result from two factors. First, it would dissolve the problem about how to reconcile our supposed mind with the physical universe. Second, it would offer a framework that is neurobiologically plausible, indeed necessary. It could then be legitimized from hypothesis to validation by how it satisfies behavioral data and experimentation—the normal scientific process.[a] That is what brain-sign theory sets out to do.

Another introductory point should be made. Between the disciplines there is not only a schism of approach but a fundamental incompatibility. Neuroscience is concerned with brain function in the structure and electro-chemistry of its operation. Psychology is concerned with a generalized view of mental states and behavior across individuals. The social sciences address the dynamics of social behavior, and the impact on the individual of the beliefs and behavior of others. But if their subject matter is the biophysical universe, there should be common ground in these disciplines. They could be seen as subsets of an overarching or underpinning biophysical discipline. In practice, practitioners are protective of their own concepts and vocabulary. If those are gone, it is feared, so is the discipline. Though psychology departments now engage in neuroscience and social science discussions, the issues between the disciplines cannot be resolved because their *modus operandi* are different.

In the flyer for the vast volume on cognitive neuroscience, Gazzaniga and Mangun note:

> The fifth edition of *The Cognitive Neurosciences* [2014] continues to chart new directions in the study of the biological underpinnings of complex cognition—the relationship between the structural and physiological mechanisms of the nervous system and the psychological reality of the mind.

No demonstration is given that the mind exists, nor does it have a scientific definition, yet it is said to be sourced by the (brain and) nervous system. Since, as Eric Kandel says, there is no consensus among practitioners on what the conscious mind is, the claim that the mind is a psychological reality *and* is underpinned by the nervous system is inevitably suspect. But all brain-related disciplines, even those that are materialist, derive their viewpoint from philosophical history, which is dominated by the concept of mind.

Brain-sign theory eliminates the conscious mind. The scientific model does not accept that we are mental subjects who have immediate access to, or direct knowledge of, ourselves or our condition in the world.[b] Therefore all disciplines must change. There will be no triumphant discipline to which all others conform. The key criterion is that the model should satisfy the organism's biophysical operation.

---

[a]In other words, in proposing that there is a state we "really are," we are not proposing that we can know what we really are, but rather hypothesizing a scientific explanation that works better under certain conditions (in this case that the universe is wholly physical; Cf. Thomas Kuhn's *The Structure of Scientific Revolutions* (1996)).

[b]That we do have such access was Descartes' supposition, the approach eventually spawning phenomenology.

BACKGROUND ON BRAIN-SIGN THEORY

We might suppose an organism with a brain and nervous system requires nothing more to survive in the world. The organism has evolved; its brain will develop adaptively toward the world during its lifetime, and it will be able behaviorally to accomplish all necessary tasks until it eventually dies. What manifestly is missing from this account, however, is the state assumed as consciousness.

But the apposite question here is not why consciousness came into being but rather how organisms, operating purely with a brain and nervous system, could exist on their own. They could not. Nor could they come into existence without other like organisms, not to mention prior evolutionary history. But if being with other organisms is their fundamental condition, how do they cooperate in the physical world? There must be a cooperative link between them in the situations of their interaction. That link is the sign, a condition of biology, that enhances the survival of the individual and/or group. Signs are intrinsically physical and biologically ubiquitous.

When one organism controls the operation of another, that link can be an explicit sign as with the dance of signer bees who, *via* the physical transmission of electromagnetic radiation, alter the brain and therefore the behavior of receiver bees (if it works). The receiver bees go off to collect nectar.[c] But a problem arises when a task carried out by more than one organism does not develop in a precise or predictable way moment by moment. This uncertain situation applies to most of human life and to the lives of other specific organisms.

In our so-called experience, it is obvious that when we make a purchase in a shop and hand over the money, we do so by looking at the sales lady, seeing her held out hand, and placing the money from ours into hers. What could be simpler? But if the control in our heads were causal assemblies of neurons, neurotransmitters, glia, etc., how could the transaction be achieved? How did our brains gain the ability to proceed through the transaction successfully? For the process in space must be judged moment by moment and negotiated (Is her hand there? Is it moving?) and, while in our 'experience' every movement is apparently seen and understood, we have no sense that any such causal negotiating activity is taking place in our head beyond seeing the movements and acting appropriately.

On the other hand, why should we not suppose our neural brains could achieve the money-handling transaction? Surely it is the causal brain of cells and chemistry that engineers the result. The answer is that

---

[c]Even more rudimentary are the shapes of a flat fish or stick insects that have evolved in such a way as to protect them from predators whose brains fail to distinguish them from their surroundings—defense not cooperation.

there appears to be no signification involved between the brains of the sales-lady and ourselves. If all interorganism cooperation requires signification, where has it happened? Is signification simply where each of our hands are placed? Do our neural brains track the other in this process? If so, what is the relevance of our seeing anything?

We seem to have an explanatory impasse:

1. Our brains should be doing the work of the transaction because they contain the physical material that engineers causality.
2. But we appear to see, and we suppose our seeing is relevant to the causal process.
3. In seeing, however, the operational mechanics of our brain's activity that causes the accurate movement of our hands is absent. All that is apparent to us are unfolding states of the world. We act, as it were, by invisible forces. So how can seeing be the cause of our acting when no causal mechanics are evident in the seeing?
4. Moreover, neuro-scientifically, how our operational brain can generate sight as a mental cause is completely inexplicable.
5. And, if our neural brains do the work *and* our supposed seeing also does the work, we have too many states doing the same work— causal overdetermination.

Other points could be raised. One might say that handing the money over is a practice we have learned from our earliest years, and all that is happening is a habit. In neuroscience terms this could correspond to brain assemblies having adapted to generate the money-exchange process, and there is nothing new in this particular exchange (what Edelman and Tononi (2000) dubbed "the remembered present"). One might add that the explicit signification took place years ago when a parent said "hand over the money," which has been forgotten in the habit.

But none of this actually addresses the situation as we suppose we find it. For in the transaction we seem to be in this present where the activity is taking place. To say, for example, that automated motor functions occur unconsciously (as is often said) merely leaves us with the question of why there is any consciousness at all, particularly when there is no actual account of how it could work.

AN OVERVIEW OF BRAIN-SIGN THEORY

The previous section may have initiated the question: Does this mystifying complexity imply that consciousness is a failed theory? Surely the situation has to be reconceived.

Brain-sign theory does assume that the neural brain is the means by which the actions of the organism take place because the brain has the physical conditions for causality. These are not unconscious actions because there is no consciousness and therefore no unconscious. The brain is a causal organ, not a knowledge organ. In addition, however, interorganism cooperative action requires signification to facilitate it. These two foundational tenets, causation and signification, generate the new account.

The key interorganism element in the money transaction is the money itself.[d] However, the mechanisms for the exchange are the hands and arms appropriately positioned; these too are a causal focus. For the sales-lady, our bodily position is crucially relevant, as is hers for us. Then there is the distance between us over the counter and the whole environment of the transaction (the shop we are in, the street where it is positioned), all of which are relevant to varying degrees to the causal operations car-ried out by the neurobiology of our individual brains.

But the causal story by itself is inadequate because what has to be signified between our brains are the ongoing world targets of that joint activity. That is why our brains signify the money transacted, the hands and arms, the bodily presences, and so on. They interpret their causal activity at each moment into the sign of each world target. They can do this because their causal orientation toward each of the elements in the transaction is adequate. The represented money, the hands and arms are, as signs, drawn (as it were) by the brain from the causal orientation it has toward them, which could allow physical interaction with them.

By removing consciousness from the process we acknowledge that a neural brain has no means to know what is happening—it is just neu-rons, neurotransmitters, glia, etc. What has changed is that the brain phenomenon is what 'we' are as brain-sign, and 'we' perform a crucial neurobiological function.[e] With great speed and complexity, the brain creates, representationally, the relevant and changing world of its causal orientation, which functions as a common target between

[d]In his best-selling book *Sapiens: A Brief History of Humankind* (2014, p. 36), Yuval Noah Harari states that, "Ever since the Cognitive Revolution, Sapiens have been living in a dual reality. On the one hand, the objective reality of rivers, trees and lions; on the other hand, the imagined reality of gods, nations and corporations." A key imagined category is money, which, Harari proposes, is developmentally crucial for Homo sapiens and also does not exist. The power to imagine things that do not exist, together with human collectivity, has, according to Harari, facilitated human control of the world and other species. Yet Harari neither explains what imagination is (consciousness, of course) nor how it can actually operate in the physical world. Brain-sign theory states that dualism (including a "dual reality") has to be dispelled for humans to place themselves wholly in the physical world.

[e]In the following text, emphasis (e.g., 'we') indicates that a mentalist term is being used for a brain-sign state, which is defined in a quite different way. Further development of the theory will obsolete this.

organisms, enabling their collective causal functioning in the world in fact, the world the brain cannot see. This biophysical construct has developed as part of human genetic inheritance, as it has for many organisms. The notion that we do see, and indeed it is we who see, is a scientific mistake.

The Beneficial Results of Substituting Brain-Sign for Consciousness: (i) Physical Communication; (ii) Rejection of Psychological States; and (iii) Reconstruction of the I or Self

It might seem incongruous that a causal brain can effect monetary exchange yet be dependent on interneural signification to enable the process. But this results from a misconstrual of the brain's causal powers. The supposition of mind already takes for granted that each individual is isolated in the physical world. Humans are not isolated entities. Their brains and nervous systems are structurally linked together by the intervening physical intermediaries of molecular transmission, electromagnetic radiation, and compression waves. This embeddedness in the physical world is also our genetic inheritance.

Electromagnetic radiation streaming to our eyes from the monetary exchange impacts the processes of our brains that concurrently are enacting the monetary exchange program. Electromagnetic radiation between ourselves and the sales-lady impacts the causal operation of our brains, enabling our joint activity. In that transaction we are not separate causal entities, but one cooperative causal unit, bound together by the physical intermediaries of that action. (Brain-sign theory terms this *the unit of the communicated*.) The mechanics are invisible, not because we cannot see the physical intermediaries of communication (electromagnetic radiation), but because 'we' do not see anything. In generating the environment of our brain's causal orientation as brain-sign, there is no requirement to portray the brain/body's method of causality, for the biophysical role of our brain and body is to effect the causality, ultimately for survival and reproduction.

However, that we are convinced we do see is biofunctionally effective. The 'conviction' is an element of brain-sign itself. If we were not convinced, biological communication would fail. This is a decisive example of 'our' inability to know about 'our' operative condition as brain-sign. The magic of mentalism has been replaced by a wholly physical account.

If we are blind, we cannot interact with the sales-lady as can the 'sighted', not because our bodies cannot move, but because our brains do not receive signals from our eyes and, therefore, the world is not accessible for our actions in that way.

In the state of *blindsight*, however, as a result of brain damage, individuals can act in the world but say they do not know how they act because they cannot see what they are doing. The supposition has been that their brains, in this regard, do not generate consciousness, yet they can act unconsciously. But the lack of consciousness is not their problem: rather it is their brain's incapacity to communicate their ability to act. Brain damage has destroyed their communicative capacity as brain-sign (see the work of Milner and Goodale (2006) and further analysis in Clapson (2012)).

Brain-sign theory also allows a different characterization of the structure of the brain phenomenon. Rather than assuming the psychologically vital role of *attention*—our attention is at one moment fixed on getting our hands in synchronization, at another the money is passed over, and so on—it is the brain's causal activity that is primary. Its identification at each moment of what aspects of the world it is controlling generates brain-sign content. Thus redundant psychological explanation, as attention, can be discarded because what is causing our bodies' movement and the context of our actions has already been established.

But there is a further reconstruction to address. In the money-exchange example reference has been made to 'our', as in 'our bodies'. Consider this quotation from a recent paper by Giulio Tononi and Christof Koch published by the Royal Society in May 2015:

> "That I am conscious, here and now, is the one fact I am absolutely certain of—all the rest is conjecture." Which is then followed by these words. "The past two centuries of clinical and laboratory studies have revealed an intimate relationship between the conscious mind and the brain, but the exact nature of this relationship remains elusive."

The authors rightly refer to Descartes as the modern originator of this dualist position. However, to claim it is a legitimate scientific position (which nonetheless and revealingly remains "elusive") returns us to the quotation from Merleau-Ponty. Long before these positions, Kant stated that:

> It must be possible for the "I think" to accompany all my representations; for otherwise something would be represented in me which could not be thought at all, and that is equivalent to saying that the representation would be impossible, or at least would be nothing to me. (1933, pp. 152–153)

For Kant, the I, self, or subject was not a representation but the foundation of the very possibility of consciousness, as representation to a subject. Since brain-sign theory claims that consciousness does not exist, the subject of that supposed seeing also does not exist. Indeed, the I that is 'certain' puts science in an impossible (dualist) position. But the

I has *not* been a universally accepted philosophical assumption (an early example was, of course, David Hume). Therefore it cannot claim to be a scientifically accepted foundation under Thomas Kuhn's definition of "normal science" (1996). Having restructured the brain phenomenon as (monist) brain-sign, the 'sense' of the 'I' *not* as a self can be identified.

For the biofunctional structure of brain-sign, the causal actor in the world is the human body. The relation between the 'I' and what is supposedly seen is the relation between the body in its causal orientation as established by the brain, and what in the world it is causally orientated toward. The 'sense' of being 'I', therefore, is a neural representation signifying the body's causality. It 'seems' to remain a constant because the 'I' is the same represented causal organism. Indeed, it accounts for why Kant supposed the 'I' could know nothing about itself beyond existing as the foundation of consciousness. Nothing could be known by the representation of the body's causal orientation, and it is not a transcendental subject. And of course there is no 'sense' that 'I am' the body's represented causality. Yet 'I' serves as neurobiological communication between brains, which know nothing of the world.

Thus the mysteriousness of consciousness, which Tonini and Koch even project out to panpsychism, has been repositioned by brain-sign theory as physical biology.

Now the question might be raised: How is it that I am aware that I see the money changing hands if I am not the subject of my mental states? This is addressable without mental awareness. At a first level the causal body is represented in the brain (the 'sense' of 'I') conjoined with that in the world toward which the causality is directed. The 'I' has a nonprominent character, but is rather the 'sensed' locus of 'seeing' *per se*. However, another level of representation can take place. It is the status of communication in which 'I', my body's causality, asserts its orientation. It is often expressed in language. "I can see the money in my hand." (Language will be addressed shortly.) This is not a mental I aware of its own states. It is an added level of delivery for interneural communication.[f]

---

[f]By contrast, Sartre terms it a reflective consciousness of prereflective consciousness, in which the ego emerges. However, Sartre was a dualist—i.e., consciousness is not a physical state. Therefore his account does not, nor sets out to offer a scientific model. His ego is an nonphysically defined philosophical "object" (1958, p. xxvi).

II. THE NEUROSCIENCES OF SOCIAL SCIENCES AND ETHICS

The Three Categories of Brain-Sign: (i) Categories-of-the-World; (ii) Categories-of-Interaction; and (iii) Language

Before proceeding further on brain-sign structure and content it is important to be clear about a specific aspect of brain operation according to brain-sign theory. The brain is a causally effective physical object. Our bodies act *via* the adapting brain in their interactions with the world.[g] We can never know how the brain operates, or how the world is in fact because 'we', as brain-sign, are derived from the brain's causal orientations. So Kant was correct in a sense. He proposed the world was inaccessible to us (or the mind) as *things in themselves*. But Kant was not developing a scientific theory of the brain, or working out its relation to the mind. The modern debate is about how the mind can be the brain, and there are many conflicting attempts to square this circle.[h]

However, it is not sufficient to agree that the mind is in the head and cannot interact directly with the world, and its knowledge is consequently not absolute. What must be established, as far as possible, is what the brain phenomenon is and does as physicality, and its relation to the brain of which it is an aspect. Thus there is no concurrence here with the modern program to identify the conscious mind in the brain, as do Tononi and Koch. The conscious mind is a failed theory.

(i) Objects, events, or situations in the world (such as handing over the money to the sales lady), which the brain selects for brain-sign generation, are termed categories of the brain-sign world, or simply *categories-of-the-world*—because the only world 'we are' is brain-sign. We may appear to see things differently from our fellows, but this is not because we have different minds. To suppose that differences result from personality, attitudes, or preferences is already to employ mentalist-based categories that create the dilemma from which brain-sign theory offers an escape. The aim is to generate a scientific theory that exposes the complexities that mentalism obscures.

For example, if you and I look at a car we may disagree about its color. The difference has often been termed *subjective* precisely because (supposedly) color belongs to a subject's mental states, though precipitated by the object, rather than the color existing in the world as seen.[i] We probably do not disagree it is a car we see. Yet the problem remains: what is the subject's experience as generated by the brain that is establishing the difference? "I see it as green, while he sees it as gray;

---

[g]Of course, there are other physical mechanisms in the body; the distinction here is for the elimination of mind.

[h]Steven Pinker, for example, says straightforwardly that the mind is what the brain does (1997). Unfortunately this is said without apparent concern for how this could be so, or how we could know it is so, or what knowledge would be (see further Clapson, 2012, 2014).

[i]Historically this derives from John Locke's approach (see further Clapson, 2012).

but then he's color blind." That is, we are both aware of the physical deficiency of one of our pairs of eyes.

The brain-sign account clarifies. There are two organisms, call them A and B. A's brain, from electromagnetic radiation to the eyes, generates brain-signs of an object (a car), one facet of which is a differential identification termed color, specifically green. B generates gray. Their brains' causal orientation to the object *per se* serves their bodies' ability to function in relation to the object. Their brain-signs serve their brains' ability to act collectively in relation to the object. In principle, the 'apparent' same color would add to that ability. That the color differential is not the same could initiate A's attempt to alter B's brain-sign account (or vice versa) because one or the other might serve in a more biofunctionally efficient way, e.g., distinguishing the green one from an actual gray car. Thus brain-sign not only supports collective action, but also disagreement resolution, and/or the triumph of one over the other. (This will be discussed further.) Consider how impossible this would be if brain-sign did not exist, and the complexity and difference of causal interneural states could not establish a common 'world' (though not an identical one).

To us it seems we see, and thereby argue over color, but this is a scientific inaccuracy. Neurobiology does not supply direct knowledge to 'us' as brain-sign. Color, so-called, is a category-of-the-world.

Electromagnetic radiation is not the only interneural medium. Compression waves serve the brain-sign states that mentalism refers to as hearing and molecular transmission that of smell. However, space precludes an extended account and the principles are similar.[j]

(ii) Let us now move to categories of organismic interaction, or simply *categories-of-interaction*. The principle here is that categories-of-the-world always have accompanying categories-of-interaction. The latter signify the body's response to that in the world to which it is causally reacting. It is vitally important for the collective action of human beings, and other appropriate organisms, because it signifies the physical state of each organism's engagement. This account reconstructs what mentalism has referred to as feelings, senses, and emotions.

It is incomprehensible that brains, which are physical objects, are distressed by pain,[k] sense danger, or love or hate. Organisms are more

---

[j]We have also not addressed touch, purely for space reasons. Further analysis reveals major reconceptions of what has been taken as our experience (see further Clapson, 2012, 2014).

[k]Pain itself being a category-of-the-world because it signifies an organism's physical condition. This can be demonstrated by a quotation from Frith. "The mental experience of pain is associated with activity in the anterior cingulate cortex. People who suffer from severe chronic pain were sometimes treated by cutting out this area of the brain (cingulotomy). After surgery these people still felt the pain, but they no longer had an emotional response to it" (2007, p. 150f). Of course, feeling pain is confused with emotional response, which results from ambiguous and inaccurate mentalist ontology and language.

complex than lumps of rock or trees, but they cannot rise beyond physicality. However, as a means of communication, categories-of-interaction are effective because apposite brains generally generate these states and therefore there can be a communicative equivalence or identification, just as there are for categories-of-the-world.

Thus while mentalism proposes that a mental subject has a sense of foreboding, is happily warm or distressingly cold (warm and cold being categories-of-the-world), or can be in a towering rage, categories-of-interaction characterize the responsive state of the physical organism in space and time as a physical sign.[1] The degree to which responsiveness occurs will depend on the biological significance of the event, which will depend substantially on already established conditions of the brain in its history.

A category-of-the-world might be a dark gothic building before which the organism stands, generating a 'sense' of anxious foreboding, a category-of-interaction. The two categories are interwoven—the object *per se* and the response of the organism. The category-of-interaction derives directly from the causal orientation of the brain toward the object. The building 'appears' threatening because, for example, in the physical processing of the brain it has an unusual and strangely complex structure, demanding elaborate organizational analysis. Moreover, to enter it, hypothetically (the brain processes by causal orientation, and buildings are to be entered), is to be in a dark and unrevealing space—'darkness' being the brain's signifying representation of the absence of electromagnetic radiation by which it can operate in the world—which may well contain hidden threats to survival. The brain does not know any of this, but the history of its interactions in the world has established in general its causal orientations and responses, and even more so if innumerable films have influenced the structure and orientation of the brain. The category-of-interaction, therefore, can be located directly as physical states (cf. analogously the chameleon's changing skin patterns).[m]

Thus the 'sense' of foreboding is decomposable to physical description as the impact on the organism of its causal orientation toward the object. It does not occur as decomposed, but as a realm of communicative immediacy ('fear', 'hope', 'awe', etc.), which the brain generates directly, i.e., not from the associated category-of-the-world.

---

[1]Feeling, sense and emotion have a confused and uncertain usage because there is no physical ground for their classification.

[m]The account here is not, of course, inevitable. Brains with different structures will react differently, perhaps with 'intrigue' or 'excitement'. Decomposition can be similarly obtained.

Thus whereas mentalist foreboding cannot be in the brain, the brain-sign 'sense' of foreboding will be identifiable in the neural fabric.

In contrast, Bennett and Hacker state that "seeing an ordinary table or chair does not evoke *any* emotional or attitudinal response whatsoever in normal circumstances" (2003, p. 40). It is worth dwelling on the issue here.

The approach of these writers does not reconcile with neuroscience (they are critics of much neuroscientific theorizing) because their supposition is that there are independent mental states as seeing, feeling, sensing, thought—faculty psychology. Each faculty plays a particular role in the knowledge and causality of the organism. But how are they then integrated? Herein, hypothetically, is neuroscience's *binding problem*, the apparently unified nature of consciousness.[n] (Kant termed it "the transcendental unity of apperception.")

For a neuroscience that does suppose there are separate psychological faculties, the problem of binding exists. But brain-sign theory proposes that this results from the false supposition of psychological faculties. Brain-sign theory proposes that the brain receives input from sensory organs and other sources (e.g., internal organs), but processes all as physical states that do not maintain the separateness of mentalist faculties. The brain's function is to generate causality, which indeed results from the impact of organ input, but it derives mostly from historically established brain structures (assemblies). Brain-sign is generated by the brain from the amalgam of its physical processing differentiated into the *appearance* of the input of faculties or organs (i.e., our seeing, hearing, etc.), which do coexist for interneural communication, but have not been integrated *as* seeing, hearing, etc.—i.e., not apperception. 'Apparently' differentiated faculty inputs are a communicative representation of the body world of sensory organs (eyes, ears) because the sensory organs are part of the actual world as represented by brain-sign.[o] Put another way, while seeing and hearing 'appear' to coexist, 'we' do not see or hear because brain-signs do not function that way. Therefore there is no neuroscientific requirement to demonstrate how faculties are integrated.

Here is a supportive illustration. For a healthy adult, the hill we are looking at appears easy to walk up. But if we have incurred a problem of the heart or lungs, the hill looks much steeper. Yet the hill has not changed. What has happened is that the causal orientation of the brain

[n]This occurs in the heading quotation from Eric Kandel. Kant addresses it in the first *Critique* (1933).

[o]Michael Anderson (2014, Chapter 4) addresses the question of whether the brain processes by faculties in designated areas of the brain and, from empirical evidence, rejects it. However, he does not address consciousness.

II. THE NEUROSCIENCES OF SOCIAL SCIENCES AND ETHICS

has altered in characterizing the hill. Our physical problem alters the 'appearance' of the hill (category-of-the-world), and our associated response (category-of-interaction) is one of 'concern' or 'fear', states which are explanatorily decomposable. Both categories result from the brain's causal orientation. So we might say to our companion, "It looks mighty steep. I'm not sure I can make it."[P]

There is no objectivity for the organism about the world by which objects or events can be judged as knowledge, nor are categories-of-interaction subjective because there is no subject (no I) for them.

The biophysical link between the 'I', the represented world and bodily response, is the brain-sign 'we are'. There is no escape from this neurally generated condition until death. At each moment 'we are' the multidimensional story the brain is telling about its moment by moment causal orientation and response, which allows the brain (in principle) continually to communicate. It is not telling that story to us both because there is no mental I, and because we are already a biophysical condition over which, as brain-sign, 'we' have no control. Free will, so-called, is a (neural) misrepresentation of our (its) condition.

(iii) The third category of brain-sign is language, but not as mentalism supposes it, with unresolved complications of ontology, organization, and function. In keeping with brain-sign theory's physicalist approach, language has no meaning and humans do not understand it. There is no thinking by language. Nevertheless brain-sign language is a crucial component of interneural communication.

The problem with *meaning* is revealed in the explanatory procedure. To explain the meaning of a word, one produces other words for which further explanation has to take place. The situation continues without end. Eventually one stops trying to arrive at the fundamentals of meaning, as for any other word. However, as brain-sign, the occurrence of a word gives the 'sense' that it is understood. What is the ontological foundation of that 'sense'?

In 'reading' words (i.e., electromagnetic radiation from the page impacts the retina of the organism and is transduced to the brain) the brain's causal orientation is altered. The reader takes for granted they understand the words, but brains do not understand words. From the causal orientation, the 'reader's' brain as brain-sign (i.e., the interpretation by the brain as the 'images' of the words on the page), is brought into physical correspondence with the 'writer's' brain. Because of a

---

[P]This is a well-known phenomenon. Anderson refers to it as follows (2014, p. 178):
"'Ability'... has shown to have systematic effects on perception and estimation....
Perception of the slant of a hill systematically varies with the expected effort of the walk up it; a hill is perceived as being steeper when one wears a heavy backpack" (see Bhalla and Proffitt (1999) for further examples).

time lapse between 'writing' and 'reading', this is not usually concurrent. However, with speech, approximate concurrence is the case. The 'sense' of understanding *is* brain-sign signifying a currently available causal orientation correspondence—a neurobiological condition.

How does language in the individual brain come about? The infant's brain develops causal orientations to objects in the world. Language development is directly related to the causal orientations of the infant's brain. Compression waves ('speech') from parents and others who are 'teaching' the child language, associate in the infant's brain with the infant's neural causal orientations (chair, window, doggie). Thus the brain develops language structures that drive the ability to 'name out loud' objects in the world (that are not actually seen).

However, while, ontologically, the language brain-sign of the infant is in the brain, this is not a mental world. The structure of brain-sign language in this case comprises the category-of-the-world (as 'apparent' chair, window, dog) associated with the word ('chair', 'window', 'doggie'). The coincidence of the infant's and parent's brain-signs signifies a shared neural causal orientation.

The infant's brain will also express categories-of-interaction to the object with (perhaps) what mentalism terms pleasure: "nice pussy cat." Thus a comprehensive interneural exchange between parent and child takes place.

However, in the adaptive modification of the brain, brain-sign language develops more than object words (nouns), since categories-of-the-world also entail events and situations. Thus, as mentalism has it, verbs designate actions, adverbs qualify verbs, and adjectives qualify nouns, etc. Sentences organize the elements into complex constructs. Indeed, brain-sign language becomes so efficient that associated categories-of-the-world become hardly or not at all 'apparent' 'to' the infant 'speaker', as with the adult. (And it is well-established in mentalist terms that words arise in the speaker before awareness of what they are going to be.)

Brain-sign language is a complex discriminatory neural method of modifying the causal orientation of another brain (see further Clapson, 2012, 2014).[q]

However, since the brain operates by linking the production of language to the causal orientation of the brain, words are both arbitrary (as Saussure proposed, and as indicated by the numerous world languages), and inevitably sustain ambiguity in relation to the brain's causal orientation. While 'we', as brain-sign, 'sense' both an understanding of words and, generally, accuracy in what 'we' convey, in practice 'we' are locked

[q]Anderson (2014, Chapter 7) does emphasize the communicative function of language but fails to account for his sustained use of mentalism.

in the limitations of the neural ability to translate the brain's causality into a communicative projection. Brains will signify different causal orientations (i.e., actual neural assemblies) by the same word. Hence, the possibility of mentalism's misinterpretations and misunderstandings. And, in language evolution, words become related to quite different causal orientations. "Gay" is a topical example.

Ontologically, the brain-sign of 'language' is exactly as previously stated about brain-sign generally: it is a neural sign of the boundary or limit of causal orientation, not, as consciousness, knowledge about the world. Indeed, how often do we 'sense' or 'say' "I don't know how to express myself" or "I cannot describe this," not because we, as a mental subject, do not have the appropriate vocabulary, but because the brain does not, or cannot, adequately interpret its causality into its method of communication as brain-sign.

The elements of language expand into what mentalism terms ideas and concepts, theories and beliefs. These derive from elaborate structures of neural causality toward the world, and enable complex forms of group behavior that have been extraordinarily successful in human manipulation and control of the world. Science and law, medicine and political activity are obvious examples. However, by the same means, it has rendered the species excessively destructive in competitive and dominant conditions.

SOME NOTES ON NEURAL FUNCTIONING

Before considering in the final section the state of the human sciences after brain-sign theory, a few points should be made about neural functioning.

Obviously brain-sign must be located in the brain, but no account of it is yet given. To do so requires direct neurobiological inquiry. It is likely, however, that fMRI scans of the brain relating to an individual's activity identify causal conditions rather than brain-sign, not least because brain-sign is a more comprehensive representation of the brain's operation than a specifically identified activity. Moreover, since brain-sign is derived from the causal orientation of the brain, brain-sign will occur later in time, which fits with the well-known findings of Libet, Gleason, Wright, and Pearl (1983). What fMRI scans do not identify is the location of mental states, despite that assumption by some theorists and the popular press.

Brain-sign theory does not endorse the rationale of cognitive neuroscience. The brain operates by neural assemblies that can cause action, actual or potential. Brain-sign is a derivative neural communication method, and not, as cognitive neuroscience presupposes, mental states to be identified in the brain as additional causal entities.

All the categories of brain-sign have the same kind of function. They do not impact each other. Much of brain-sign content is not of the world that is directly before the organism. These occur as *repetitions* of former brain states ('memory') and *reformulations* of brain states ('imagination'). However, the brain-sign account obviously differs from mentalism. For example, the mentalist distinction between semantic and autobiographical memory erroneously presupposes the function of mind.

Neural science is not a discipline carried out by conscious human subjects. It is an account of the brain (and body) by itself. Scientific discourse is generally expressed as if its vantage point is outside the world being explored. Indeed, so is consciousness for, supposedly, the world is the object for consciousness (i.e., mentalist aboutness). Brain-sign theory alters this fundamentally because brain-sign is generated by the brain as part of the world for a requisite physical function. However, the expression of brain-sign theory (as shown here) inevitably retains characteristics of scientific expression precisely because 'we', as brain-sign, cannot be the causal brain that is generating the 'discourse'. The 'reader' of brain-sign theory, therefore, is positioned by their brain as brain-sign.

That 'we', as brain-sign are a derivative result rather than a cause has precedent in the literature, though not of course as brain-sign.[r] What this means is that (for example) the 'sense' that we can act to change our supposed mental states by mental states themselves, as if the will is within consciousness and has causal power, is a scientific error. We cannot decide to make ourselves happy. That 'we feel unhappy' is already a brain-sign of a neural causal orientation expressing a state of biologically determined inadequacy. The brain may then alter its causal orientations, including perhaps actions.

THE HUMAN SCIENCES AFTER BRAIN-SIGN THEORY

The human sciences are distinguished from the physical sciences. As Roger Smith says

> One simplifying response...[is] to take the distinction in German between *Naturwissenschaft* (natural science) and *Geisteswissenschaft* (the science of spirit) and erect it into a basic distinction between different kinds of knowledge appropriate for different subject matters: the former for nature which just 'is', and the latter for culture which displays 'values'. Some modern writers do this in order to differentiate between the natural and the human sciences. (2013, p. 91, and further below)

[r] At least as early as Schopenhauer's volume, *On the Freedom of the Will* (1985), and later Wegner's *The Illusion of Conscious Will* (2002).

Rather than subjecting humans and their social ways to the rigors of a determinist science, where cause and effect are dependent on specifiable physical properties, the human sciences provide accounts by descriptive vocabulary. Neural science is in principle entirely physical, for brain elements and operations are the substance and condition of matter. Yet, according to mathematician Robert Crease:

> Newton's strange new world was found in our world—but it is not our world, either, nor one we could live in. We humans, even the scientists among us, inhabit what philosophers call the 'lived world', amid designs, desires and purposes: we live in an Aristotelian world. (2008, p. 64)

Science's success lay in changing the Aristotelian physical world that was, in aspects, founded on designs, desires, and purposes. Yet Crease supposes we live (whatever he means exactly) in that world, not the Newtonian world. His characterization of our "life" certainly fits with the human sciences. How has brain-sign theory altered the situation?

Instead of supposing that consciousness exists (as presumably does Crease), in which designs, desires, and purposes are deemed to exist, we have inserted into human biology (and other specific creatures) a neural communicative mechanism that is physical. Brain-sign theory demonstrates that human beings do not "live" in an Aristotelian world.

By reclassifying the brain phenomenon not as consciousness, we show how Aristotelian categories enter the frame. They are examples of the brain's self-interpretation of its causal orientations in a way that allows human neural communication. (Other ways involve gods, fate, immortality, and paradise.) The reality of neural collective activity, by contrast, is in vast assemblies of causal neural processes that could, in principle, be identified in the brain (though not known). But they could never function as effective communication, both because of their vastness and the intrinsic differences between them. Thus, in evolutionary development, the brain has generated a simplified account of the world (categories-of-the-world), its own responses to the world (categories-of-interaction), and the means to communicate about the world by mutually altering their causal orientations, brain-sign language.

Paul Churchland also rejects folk psychological terms as causal categories, most recently developing an account of consciousness based on intertheoretic reduction: e.g., light is electromagnetic radiation. Thus consciousness is brain states. He locates this reduction in recurrent neural networks. "*Creating* consciousness in the first place [as a recurrent neural network]...was a firmly *neurobiological* thing" (2007, p. 17). So, for example, a concept is "a proprietary *volume* within [the network's] activation space, a volume that confines all of the possible points (i.e., neuronal activation vectors) that count as determinate cases of the abstract determinable that is the concept proper" (2007, p. 144).

Eliminating dualism is desirable and his account of brain function, as network processing, is instructive. But he offers no biofunctional framework for network states as *consciousness*. For example, we know of light, hypothetically, by seeing it, but seen light is not reducible to electromagnetic radiation (see Clapson, 2012, pp. 113–114).

Brain-sign theory removes the problem by identifying the brain phenomenon as interneural communication. The world of our actions is shared between brains as brain-sign, but not the causal mechanisms in the brain and body for those actions.

Folk psychology is not a failure of scientific insight or method: its bioneural function is a neural self-explanation of its causal orientation, couched in simple mentalist terms. It has served humans for millennia.

But crucially, the brain has no direct insight into the reality of the world. It operates by genetically inherited means including trial and error and imitation. Brain-sign theory proposes that the brain can communicate about itself in a more scientifically efficient way than the mentalist terms of folk psychology by 'positing itself' as purely physical. The result is a more complex and lengthy explanation, but more scientifically likely. Indeed, Churchland states that "The *contents* of consciousness—especially in our intellectual, political, artistic, scientific, and technological elites—have been changed dramatically" (2007, p. 17). Yes, but not as consciousness.

Science legitimizes itself by experimental success. This is the path for brain-sign theory.

The implications are straightforward. Designs, desires, and purposes may be a way the brain self-describes, and humans may continue to use such terminology—or rather, brains may continue to communicate, as brain-sign, in these terms (as we still say the sun rises in the morning, though the scientific explanation of the planetary system does not). But brain science need not employ such language. A new vocabulary will be developed from the tenets of brain-sign theory. Examples are brain-sign itself, and the redefinition of the I or self in section "The Beneficial Results of Substituting Brain-Sign for Consciousness" point (iii).

Brain-sign theory, and the associated vocabulary, can unify the current approaches of neuroscience, psychology, and the social sciences, for the theory provides the underpinning scientific ground and the overarching vocabulary for humans as physical constructs.[s] It can be employed in a beneficial reformulation of the human sciences—the law, ethics, international relations, music, and so on—by its clarification of the physical foundations of human life.

---

[s]Perhaps this is what Gazzaniga hypothesized in his book *The Mind's Past* (1998), with the sentences: "Psychology is dead" (p. xi) and "The odd thing is that everyone but its practitioners knows about the death of psychology" (p. xii). However, Gazzaniga, as already mentioned, has not overcome the category of mind despite his emphasis on neuroscience and other brain-related disciplines (Gazzaniga, 1998).

II. THE NEUROSCIENCES OF SOCIAL SCIENCES AND ETHICS

It may also be conjectured that brain-sign theory will help remove approaches that generate human conflict expressed as desires, purposes, beliefs, and indeed values.

References

Anderson, M. L. (2014). *After phrenology: Neural reuse and the interactive brain*. Cambridge, MA: MIT Press.

Bhalla, M., & Proffitt, D. R. (1999). Visual-motor recalibration in geographical slant perception. *Journal of Experimental Psychology, 25*, 1076–1096.

Bennett, M., & Hacker, P. (2003). *The philosophical foundations of neuroscience*. Malden, MA and Oxford: Blackwell Publishing.

Churchland, P. M. (2007). *Neurophilosophy at work*. Cambridge: Cambridge University Press.

Clapson, P. (2012). *The world without knowledge: The Theory of Brain-Sign* (Doctoral thesis, Durham University). Retrieved from <http://etheses.dur.ac.uk/3560/>.

Clapson, P. (2014). Knowledge, science and death: The theory of brain-sign. *Activitas Nervosa Superior, 56*(4), 105–120. <http://www.activitas.org/index.php/nervosa/article/view/184/205>.

Crease, R. (2008). *The great equations: Breakthroughs in science from Pythagoras to Heisenberg*. New York: W. W. Norton & Co.

Edelman, G., & Tononi, G. (2000). *Consciousness: How matter becomes imagination*. London: Penguin.

Frith, C. (2007). *Making up the mind: How the brain creates our mental world*. Oxford: Blackwell.

Gazzaniga, M. S. (1998). *The mind's past*. Berkeley, CA: University of California Press.

Gazzaniga, M. S., & Mangun, G. R. (2014). *The cognitive neurosciences* (5th edition). Cambridge, MA: MIT Press.

Harari, Y. N. (2014). *Sapiens: A brief history of humankind*. New York: Vintage Books.

Kandel, E. (2013). The new science of mind and the future of knowledge. *Neuron, 80*(3), 546–560.

Kant, I. (1933). *Critique of pure reason* (2nd ed.; N. K. Smith, Trans.). London: Macmillan Press (Original work published 1781; second edition published 1787).

Kuhn, T. (1996). *The structure of scientific revolutions* (3rd edition). Chicago, IL: University of Chicago Press.

Libet, B., Gleason, C. A., Wright, E. W., & Pearl, D. K. (1983). Time of conscious intention to act in relation to onset of neural activity (readiness potential): The unconscious initiation of a freely voluntary act. *Brain, 106*, 623–642.

Merleau-Ponty, M. (1962). *Phenomenology of perception* (C. Smith, Trans.). London: Routledge (Original work published 1945).

Milner, A. D., & Goodale, M. A. (2006). *The visual brain in action* (2nd edition). Oxford: Oxford University Press.

Pinker, S. (1997). *How the mind works*. London: Allen Lane and The Penguin Press.

Sartre, J.-P. (1958). *Being and nothingness* (H. E. Barnes, Trans.). London: Methuen (Original work published 1943).

Schopenhauer, A. (1985). *On the freedom of the will* (K. Kolenda, Trans.). Oxford: Basil Blackwell (Original work published 1839).

Smith, R. (2013). *Between mind and nature: A history of psychology*. London: Reaktion Books.

Tononi, G., & Koch, C. (2015). Consciousness: Here, there and everywhere? *Royal Society, Philosophical Transactions B*, Section 2: Here. Retrieved from <http://rstb.royalsocietypublishing.org/content/370/1668/20140167>.

Wegner, D. M. (2002). *The illusion of conscious will*. Cambridge, MA: MIT Press.

7

On the Redundancies of "Social Agency"

A. Tillas

University of Düsseldorf, Düsseldorf, Germany

INTRODUCTION

What determines social behavior? Do we have the capacity to act independently of social influences or does our sociocultural setting drive our actions? In this chapter, I present a philosophical argument relative to the "structure vs agency" debate (S vs A). The S vs A debate concerns the primacy of structure (S) or agency (A) as the main determinants of behavior and is one of the most central debates in current social sciences (e.g., Barker, 2005; O'Donnell, 2010) to the extent that it cuts across many of the most fundamental aspects of human cultural life.

A refers to individuals and particularly to the capacities of individuals to act on their own goals, to exercise their free will and engage in instrumental action both as individuals and as part of larger social groups. Furthermore, A might refer to the capacity of individuals to consciously and willfully change their surrounding environment, to make choices in society that are not guided by anything else other than their own desires, goals, wishes, and agendas. In this sense, the key aspect of A is the notion of individual choice and our capacity to act at will and independently of social pressures and influences. In this sense, the notion of A is closely related to that of free will. (Further elaborating on this issue extends beyond the scope of this chapter.)

The Human Sciences after the Decade of the Brain.
DOI: http://dx.doi.org/10.1016/B978-0-12-804205-2.00007-0

© 2017 Elsevier Inc. All rights reserved.

S, on the other hand, refers to the social arrangements and scaffoldings that are believed to influence choices of individuals. In this sense, S is what constrains us from acting as autonomous agents, and refers to either material or immaterial/cultural factors. Often S refers to the process of socialization, i.e., the process of internalizing existing social norms, principles, inferential patterns, and so forth (cf. Searle, 1995), while structural factors include social class, education, religion, gender, customs, etc. Furthermore, S also refers to biological and genetic factors.

Finally, it is worth clarifying that S does not refer to something static but rather concerns a dynamic framework. Consider, for instance, the differences in the ways in which products are produced, distributed, and consumed in feudal and in capitalist societies, and the changes required to bring about this transition.

Importantly the S vs A debate points to a deep contradiction in human nature. Namely, the conundrum is that while S is necessary for social life (e.g., conventions like driving on the left side of the road while in the UK are crucial for commuting safely), at the same time it constrains us from doing what we would like to do (e.g., driving in the middle of the road to use a caricature of the previous example). In this sense, S is necessary and fundamental yet restrictive. Furthermore, S often appears as natural, even though it is actually conventional, e.g., the ways in which young children behave seems to us natural even though there are significant systematic discrepancies in the ways children from different cultural backgrounds behave.

There are various strands in the S. vs A. debate. First, Peter Berger and Thomas Luckmann (1966) construe the relation between the two as dialectic in nature and promote neither S nor A as the primary determinants of behavior. Second, there are those opting for the primacy of S over agency. Consider, for instance, structuralists (e.g., Claude Lévi-Strauss), functionalists (e.g., Auguste Comte, Herbert Spencer, and Talcott Parsons), and marxists (e.g., Louis Althusser). Third, there are those arguing for the primacy of A. For instance, methodological individualists like Max Weber roughly argue that social agents act on rational choices, while interactionsists (e.g., George Herbert Mead, Herbert Blumer) argue for the subjective nature of social reality and that social processes reduce to interactions of individuals.

It is worth clarifying that different approaches to the S vs A debate most often correlate with different approaches to several other issues in social sciences such as those concerning social ontology and the nature of the social realm and what qualifies as a cause and what as an effect in the social world. In this chapter, I focus solely on the S vs A debate without considering such related issues and without arguing for the primacy of either S or A as the main determinants of human behavior. Instead, given that the aspects of behavior in which social scientists are interested concern conscious actions based on rational decision-making, I construe A as being strongly related to higher cognitive processes.[a] In turn, I treat agents as thinking subjects. Against this background, I appeal to cognitive science and neuroscience and suggest an empirically supported view of thinking. Continuing, I use this view to shed light on the notion of A as used by social scientists and in turn on the S vs A debate. On the basis of the suggestions made in the following pages, the S vs A debate is rendered redundant.[b]

THERE IS NO SUCH THING AS *SOCIAL* AGENCY

Participants in the S vs A debate most often take into account processes of internalizing social norms and principles (socialization). Interestingly however, they assume that agents can nevertheless think and ultimately act autonomously. In contrast to this view, and without advocating holism, I argue that agents cannot act independently once they have internalized existing social norms and principles. For only behavioral patterns based on thinking processes can qualify as (genuine) cases of (social) action. In turn, thinking is heavily influenced by our experiences with aspects of S, e.g., cultural environment, language and customs. In this sense, A is saturated with information concerning structural features. On these grounds, I argue that A is heavily influenced by S or that A is "structured." My starting point is that concepts are built out of perceptual representations (e.g., Prinz, 2002) and become individuated in terms of these representations. That is, we activate a given concept in terms of activating a/the perceptual representation(s) that make up that concept. For example, we think of a tree in virtue of reactivating a perceptual representation of one (cf. Barsalou, 1999). Second, I appeal to the widely accepted hypothesis according to which concepts are the building blocks of thoughts in the sense that we form a thought in virtue of activating the appropriate concept. For example, we form the thought "black cat" in virtue of activating our concepts BLACK and CAT. Given (a) the close relation between A and thinking, (b) the relation between thinking and concepts, and (c) concepts and S (in terms of our experiences with aspects of S), A is saturated with structural features or succinctly is "structured."

[a]In this chapter I do not consider issues such as sense of agency, feeling of agency and sense of ownership, which feature prominently in the agenda of philosophers of mind and cognitive science. See Synofzic, Vosgerau, and Newen (2008) for a discussion.

[b]See also Parsons (1937), who calls this debate a pseudo-debate, and Wittgenstein (1976) for a critical analysis of rule-following and private language (skeptical paradox).

Assuming a light-hearted reductionism, thinking is neuronally implemented. In turn, the ways in which thinking processes occur is highly contingent on the weightings of the synaptic connections between neuronal groups underpinning or grounding thinking. In this sense, the process of internalizing social norms (socialization) is precisely a process of adjusting the weightings of the synaptic connections in question (conditioning). Inevitably, socialization processes, and in this sense S, has a decisive impact on agency. In turn, even though we often have the impression that thinking occurs freely or spontaneously, this is not the case (cf. Christoff, Ream, & Gabrieli, 2004).

Consequently, given that A is itself structured, in the above sense of the term, there is little value in arguing over the primacy of S or A. More concisely:

1. Any A-related claim in the social sciences is a claim about willful action.
2. Any willful action is the result of a free choice.
3. Any choice is the result of reasoning.
4. Reasoning is thinking.
5. Thinking is structured.
6. If thinking is structured (5), then A is itself structured.

Conclusion: The S vs A debate is redundant.

Next, I clarify a number of potentially controversial issues in the above syllogism, and continuing I focus on premise 5, which clearly requires further elaboration.

Clarifications

First, consider the hypothesis that thinking is the result of rational processes (from premises 3 and 4 of the above syllogism). Clearly, there are instances of thinking that do not stem from rational processes. The importance of reflection in decision-making has been extensively studied in the literature, and it has often been pointed out that its role is overestimated (e.g., Dijksterhuis & van Olden, 2006). Furthermore, Bortolotti (2011) argues that not all reasoning leads to "wise" actions and that not all "wise" choices that experts make are the result of reasoning.

Consider also the classic study of Nisbett and Wilson (1977) on confabulation. Famously, they asked subjects in a bargain store to judge which one of four nylon-stocking pantyhose was of the best quality. The stockings, which were in fact identical, were presented on racks spaced equal distances apart. As situation would have it, the position of the stockings had a significant effect on the subjects' choice. In fact, 40%

of the subjects chose the far right—and most recently viewed—pair. This is a clear-cut case of decision-making that does not reply on reasoning, but rather on perceptual input. Famously, when subjects were asked to explain their judgments, most of them attributed their decision to different characteristics such as the knit, weave, and elasticity of the stockings that they chose to be of the best quality. Confabulation could thus be used as a toy model of our phenomenology that thinking and decision-making derive from rational processes, even though this is not always the case (see also Bortolotti & Cox, 2009; Haidt, 2001).

Furthermore, our inferences about unknown features of the environment are systematically error and biased prone. For they are not based on laws of probability but on "quick and dirty" heuristics (e.g., Kahneman, Slovic, & Tversky, 1982). Similarly, Gigerenzer and Todd (1999) consider intuitions to be "fast and frugal" heuristics. Such heuristics provide computationally cheap solutions to complex problems, either theoretical or practical for which we cannot clearly distinguish between two given alternatives. For instance, we decide whether Hamburg or Liverpool has the larger population in terms of which one is most popular, assuming that bigger cities are often more popular. Finally, there are numerous studies focusing on the role of intuitions in decision-making (cf. Dreyfus, 1997; Osman, 2013).

In light of the above, conscious reasoning is not our only guide to decision-making. Nevertheless, aspects of S do influence decision-making processes. For intuitions, in the sense of unconscious stored information influencing thinking, are themselves the output of perceptual encounters with our sociocultural (and physical) environment as much as reasoning is. Next, I elaborate on the nature of reasoning and the contribution of unconscious stored information in higher cognitive processes.

Thinking Is "Structured"

My starting point is the aforementioned and ubiquitously accepted view that concepts are the building blocks of thoughts. In turn, concepts are built out of perceptual representations formed during experiences with our sociophysical environment and stored in long-term memory. I only present this view here briefly as I discuss this in detail elsewhere (Tillas, 2010, in press). The key point here is that all concepts are linked back to experience temporarily, but they do so *indirectly*.

Concept Learning

On perception of a given object, a representation is formed and stored in long-term memory. The precise locus for storing a given representation is influenced and often determined by top-down effects from stored representations in the subject's mind. On encounter with a subsequent instance that is recognized as a subsequent instance of a given kind (e.g., "this thing is similar to the one I saw earlier"), a representation is also formed. At this point, a scanning process is initiated and a match is sought for in the subject's memory. The same scanning process is also initiated during encounter with the first instance, but it does not yield any matching stored representations (cf. Barsalou, 1999; Demarais & Cohen, 1998).

The scanning process begins with a highly demanding similarity level, which drops if a match is not found after a scan. If the similarity level drops below a certain point (after a certain number of scans), a new "mental file" (cf. Perry, 2001) is created for the category at hand. Once a match between the currently formed and a stored representation is found, then the existing representation is activated and drives perception in a top-down manner. For instance, if on perception of the first instance of a tree, a subject has formed a representation of the tree's trunk, and a match with a stored representation is found, then selective attention will be driven to the rest of the tree parts (e.g., branches and leaves) that have been attended during the encounter with the initial instance (cf. Barsalou, 1999).

Second, activation of matching stored representations leads to storage of the currently formed representation at the same locus or in any case "closer" to the stored matching representations. In this way mental files become informationally enriched. A useful way to construe the claim that representations are stored "closer" together is in terms of representations with stronger positive memory effects in cognitive tasks. This hypothesis builds on evidence showing that unlike bottom-up attention, i.e., attention captured by salient features, top-down attention enhances formation of representations of attended features (e.g., Corbetta, Miezin, Dobmeyer, Shulman, & Petersen, 1990; Noudoost, Chang, Steinmetz, & Moore, 2010) as well as on evidence showing that information attended through top-down attention will later on be relevant for memory formation (Uncapher, Hutchinson, & Wagner, 2011) or later remembering (Craik, Govoni, Naveh-Benjamin, & Anderson, 1996).

Despite the aforementioned significant role of top-down effects, learning does not merely rely on such effects of stored representations on perceptual processes. Had that been the case, the perceptual process would have been heavily influenced by accidents of the first encounter with a certain instance and potentially miss out on certain statistical regularities. To this end, top-down influences have to be construed more liberally, in the sense that some nonoverlapping information is allowed as a byproduct of not being too skewed by the bias of the first encounter.

Concepts are built in virtue of an abstraction process selecting similarities across representations formed during encounters with instances of a given category and stored in a given locus in memory—think of this locus in terms of Perry's (2001) aforementioned mental file metaphor. Similarities are understood here in terms of frequencies of occurrences to the extent that the more frequently occurring features across members of a given set would naturally capture the greater similarities among them. In line with Hebb's (1949) widely accepted rule of learning according to which "neurons that fire together, wire together" (associative long-term potentiation), the more frequently a certain feature occurs across instances of a given kind, the stronger the connections between neurons that ground perception of that feature will grow. Thus abstraction selects only representations carried by stronger connections, i.e., representations of properties that most, if not all, members of a given category bear. In turn, these representations are used in concept formation (cf. Tillas, 2014).

Conceptual Characteristics and Thinking

A crucial characteristic of concepts with regards to their role in thinking is that they are associationistic in their causal patterns. That is, every concept is associated with other concepts (cognitive content), and their connections influence the ways in which they become activated. In line with Hebbian associationism, the more frequently a subject is confronted with uses of certain concepts in her linguistic community, the stronger the connections between them will become.

Once a given concept becomes activated, the concepts strongly associated to it become subactivated or primed. Subactivated representations are representations that are not fully blown in consciousness. More specifically, the notion of subactivation refers not to the equivalents of discrete values—"0," neuron not firing and "1," neuron firing—but to values within the range 0−1. In this sense, subactivated connections between different concepts are largely unconscious and are what we phenomenologically experience as intuitions or gut feelings (cf. Trafford & Tillas, 2015).

On the hypothesis that concepts are the building blocks of thoughts, subactivation of certain concepts (or the neuronal ensembles that ground them) constrains or primes activation in a specific way, i.e., toward specific thoughts associated with the subactivated links of the appropriate conceptual network. These unconscious cognitive processes are influenced by the relative weights between their associations, and are projected into conscious cognitive processes (cf. Trafford & Tillas, 2015). As a result, the stronger the connection holding between a given set of concepts, the more intuitive it will seem to us that a certain given thought will entail (or at least be followed by) a certain other thought.

Evidence in support of the suggested associationistic view of thinking can be found in the work of Elman et al. (1996), even though I am not committing to a connectionist view of cognitive architecture. Briefly, according to Elman et al., artificial neural networks can be highly constrained by the network's current weight assignment. That is to say, the pattern of activation set by a connectionist network is determined by the weights or strength of connections between the units. Given that these weights model the effects of the synapses between different neurons in the human brain, different levels of activation of synapses that connect one neuron to another place a significant constraint on what new ideas the mind can explore next. In this sense, thinking is heavily conditioned by the associationistic patterns of concepts, i.e., by the associations between representations formed during experiences with our sociophysical environment, which ultimately comprise all of our sociologically nontrivial conceptual repertoire.

Finally, the associationistic nature of concepts is also related to their rich cognitive role properties. Specifically, concepts are endogenously controllable or can be activated in the absence of their referents (e.g., Barsalou, 1999; Prinz, 2002).[c] In this sense, concepts, and in turn thoughts, can be formed not only on the basis of processes of perceiving (bottom-up) but also of processes of thinking (top-down). Here, I assume that endogenously controlled thinking is a form of associative thinking, in the sense of current thinking caused by earlier thinking. Given connections between concepts, activation of a given concept primes other associated concepts. For instance, consider someone uttering the word "trip" and another agent mistakenly hearing the word "grip," who may resultantly start to think about friction and laws of physics, rather than traveling. This is a case where an agent is forming a thought in the absence of an appropriate stimulus, seemingly spontaneously but actually in an associative manner.

Against this background, I argue that thinking—the very medium of decision-making—is itself "structured," in the above sense of the term (saturated with our previous experiences of aspects of S). For thinking occurs in virtue of deploying concepts that are built out of representations formed during experiences with our sociophysical environment. Thus thinking is molded by socialization—a process that precisely captures these repetitive situations to which subjects are more than often exposed. Consider, for instance, the case of gender socialization during

---

[c]Endogenous control over a given concept is acquired by associating the set of perceptual representations comprising a given concept to a perceptual representation of a word or goal-directed action over which we already have endogenous control. See Tillas (2015) for a detailed analysis.

which parents handle, speak to, and dress their sons and daughters differently (cf. Oakley, 1974).

On the hypothesis that thinking is structured, there is probably little value in arguing over the primacy of A or S as determinants of social behavior. For free-willing agents or agents that can act independently and make their own free choices, simply do not exist. Thus we could possibly acquire a better understanding of social behavior by focusing on the ways in which S is internalized and how it influences A.

A different way to approach the issue of how S influences A is by looking at the relation between language and thinking. For instance, there is a growing consensus that natural language plays a significant role in our cognitive lives. Building on this claim various thinkers have gone on to suggest that thinking is secondary to language. For instance, Davidson (1975) argues that thoughts are only attributable to creatures that are interpretable. For Davidson, a creature that we cannot interpret as capable of meaningful speech is a creature that we cannot interpret as capable of possessing contentful attitudes. Furthermore, Brandom (1994) argues that thought does not take place in language but thought can only be attributed to linguistic agents. For Brandom, thought and language acquire content through their mutual interrelations. Nevertheless, Brandom promotes the significance of language over that of thought since he argues that the objectivity of conceptual norms derives from public linguistic practice. Finally, Carruthers (1998, 2005, 2008) argues that natural language, e.g., English, becomes a language of thought in the sense that natural language is constitutively involved in conscious thinking and that conscious thinking occurs in a form of inner speech.

Even though I have argued against a constitutive view of the role of language in thinking elsewhere (cf. Tillas, 2015), language does play a crucial role in thinking and heavily influences reasoning.[d] To the extent that natural language is one of the most clear-cut cases of structural features, the claims made in the previous pages gather pace.

POTENTIAL CRITICISMS

Structural Changes and Agency

As explained above S is dynamic. Consider, for instance, the changes entailing the transition from feudalism to capitalism to use the

---

[d]If it is assumed that a constitutive relation holds between thinking and language, in the sense that one cannot exist without the other, it is hard to explain how it is that human subjects at early developmental stages or nonlinguistic animals are capable of entertaining conceptual thoughts—an intuitive and widely accepted view.

aforementioned example. On the hypothesis that A is structurally saturated, it seems that it is not agents who bring about changes in S, like the aforementioned transition from feudalism to capitalism. But how does S change, if not by A? In reply, a common way for changes to occur is by the presence of tensions stemming from differences between interacting individuals. As shown presently, there are key differences between interacting social forces obtained at various levels from the subliminal self—see above for the claims about intuitions being saturated by our experiences with S—to the level of social institutions. In this sense, it is *S itself* that allows for changes, while a crucial role in this process is bestowed on A.

Furthermore, S does not determine agents in the same manner or to the same extent. In this sense, structurally determined agents are not mere automata whose behavior is fixed. For we only have experiences with fragmented aspects of a dynamic, indeterminate, and at times locally specific S. In addition, there is not a one-to-one relation occurring between thoughts and actions, and there are several reasons no such relation is obtained. For instance, sociologically nontrivial actions are the outcome of sophisticated cognitive processes taking into account a large array of information, beliefs, desires, goals, agendas, and contextual features, which in turn is impossible to (perfectly) overlap across different agents. These differences between agents become even more pronounced if we also consider the role of intuitions, which, as explained, reflect our past histories and contribute greatly to decision-making processes.

An additional factor that contributes to structural changes concerns communication—the sine qua non of social life—which is most often imperfect. Communication is imperfect since there are crucial differences at the level of conceptual networks that different interlocutors deploy at a given point to form and communicate their thoughts. And even though successful communication only requires a partial overlap between the conceptual networks that interlocutors deploy, certain details might not get across, allowing, in this way space, for further differences and tensions to occur (cf. Tillas & Trafford, 2015).

Naturally, more changes occur more rapidly today. For technological advances allow us to travel and communicate easier, faster, and more often than we have ever been able to. As agents we are more exposed to different structural features, we become influenced by them, and try to adjust (aspects of) them to fit our own sociocultural microenvironment. Tensions occur in the process.

This widely varying assortment of subjective experiences, conceptions, and interpretations of structural features that are ultimately biased, as well as the largely differing goals, beliefs, and desires of different agents, leads to interactions between diverging forces. Without necessarily implying a clear-cut Heraclitan dialectic, existence of key differences among interacting individuals creates tension, which in turn provides a fertile ground for changes to occur.

Diverging characteristics, which are important for changes, are not limited at the level of individuals, and can also be found at a more macroscopic level. Consider, for instance, Ogburn's (1922) cultural lag theory, where changes are understood as the result of different societal subsystems (e.g., institutions) being out of sync with one another. For example, technological innovations in industrial businesses occur much more rapidly than changes in government bureaucracies. Thus certain institutions set a pace with which other institutions struggle to keep up. As a result, this brings about tensions between these subsystems and in this way paves the way to social changes.

In a nutshell, S does not determine agents uniformly, which gives rise to tensions between interacting parties (e.g., individuals and institutions) entailing, in turn, changes. In this sense, structural determinism is construed here rather liberally.

Who Made Who?

A further potential criticism concerns the "origins of S." Specifically, on the hypothesis that there are no agents, how is S—an artificial "framework"—brought about? In reply, despite being a hugely complex and sophisticated arrangement, S is reducible to simple bits of information, linguistic conventions, behavioral patterns, actions, and ultimately movements of coexisting and interacting parties. To this extent, I am not denying that there are free-willing agents with regards to simple movements or simple goal-directed actions such as drinking to quench one's thirst and cooperating with someone else to carry a heavy load. However, with regards to actions in which social scientists are interested, there are no agents, as explained below. In an attempt to shed light on social behavior, I look into the origins of A (sections "How Does Agency Come About?" and "Volition and Consciousness Awareness"), while I provide evidence for the claim that social agents are structurally determined in sections "Structure and Top-Down Effects in Perception" and "Reactivation of Stored Representations and Top-Down Effects."

EMPIRICAL EVIDENCE FOR AGENCY AND AGAINST SOCIAL AGENCY

How Does Agency Come About?

Taken at face value, the picture described in the previous pages might seem bleak to the extent that A is described as saturated by S or as so heavily influenced by experiences with aspects of S that they are unable to act autonomously. However, as hinted on in section "Who Made Who?," this is not the case with regards to simple actions. In support of this claim, I present neuroscientific evidence about how A is established. This evidence is also intended as a reply to the aforementioned "who made who" criticism.

Volition is most often construed as a process at the end of which an action is decided on and executed. A crucial but largely neglected aspect of these volitional processes involves inhibition of antagonistic motor programs (voluntary and involuntary movements; e.g., Filevich, Kuhn, & Haggard, 2012). According to Ganos et al. (2014) volition may strongly depend on these inhibitory processes. In this sense, volition emerges through a clear developmental trajectory both in anatomical and functional terms (e.g., Leisman, Machado, Melillo, & Mualem, 2012).

Consider, for instance, the seemingly involuntary and uncoordinated movements of subjects at early developmental stages and how they transform to perfectly orchestrated and context-appropriated movements in adults, contributing to achieving set goals. Most likely, we learn to control unintentional movements, because of unwanted outcomes, negative rewards, and so forth. Inhibition of involuntary movements is what grounds A. For "intentional inhibition"—as Ganos et al. (2014) call it—allows subjects insight about how they can control their bodies and in turn change their environments.

In shedding light on the processes contributing to A, I follow Ganos et al.'s work (2014) on neurodevelopmental disorders, such as Gilles de la Tourette syndrome (GTS), during which not all elements of motor development undergo the aforementioned developmental process.

GTS is a striking example of a neurodevelopmental disorder that is characterized by the presence of motor (and phonic) tics. Tics characteristically bear great kinematic similarities with normal movements, and also share most neurophysiological properties with them, while complex tics may resemble voluntary actions (Paszek et al., 2010, in Ganos et al., 2014). However, the main difference between tics and normal movements is that the former lack the flexibility and context-appropriateness of human voluntary action control, while their phenomenology is that of involuntary, repetitive movements.

In order to achieve useful goal-directed action GTS patients need to recruit additional control mechanisms. In particular, they might need to strategically compensate for involuntary movements by enhanced cognitive control. Consistent with this view, GTS subjects showed normal or even supra-normal motor performance in manual and saccadic control tasks (where participants must visually fixate on a central cue and then look at a target; e.g., Baym, Corbett, Wright, & Bunge, 2008; Buse et al., 2012 in Ganos et al., 2014) suggesting that they learn to efficiently control their actions.

Specifically, Ganos et al. examined brain activation of GTS patients under two conditions. Under the first, subjects were allowed to tic freely, while under the second subjects were asked to voluntarily inhibit their tics. The results showed a significant reduction in tic rate under the inhibition condition. Focusing on the neural differences between the two conditions, they recorded an increase of activation (regional homogeneity (ReHo) in the left inferior frontal gyrus (IFG) under tic inhibition in comparison to the free tic condition).[e] Furthermore, they found strong positive correlations between inhibition-related increases in IFG activation and the behavioral capacity to inhibit tics.

Volition and Consciousness Awareness

Haggard, Clark, and Kalogeras (2002) focus on voluntary actions and conscious awareness and show that the central nervous system deploys a specific neural mechanism that produces intentional binding of actions and their effects in conscious awareness.

In studying consciousness of action, they focused on the perceived time of both intentional actions and their sensory consequences. Specifically, they studied the links between representations of actions and their effects, by first comparing the perceived times of voluntary actions (baseline) with the perceived times of involuntary movements induced by transcranial magnetic stimulation (TMS). Continuing, they examined how the perceived times of these events shifted when such events triggered an auditory stimulus. In particular, they asked subjects to watch a conventional clock face and to judge the onset times of four events. In the voluntary action condition, subjects pressed a key at the time of their choice. In the involuntary condition subjects noted the time of a muscle twitch produced by stimulation of the motor cortex. In the third condition subjects noted the time of an audible click made by the TMS applied to the parietal cortex—without inducing motor activation (sham condition). Finally, in the auditory condition, subjects noted the time of a tone.

---

[e]Given that tic inhibition lacks a behavioral marker to serve as an event for event-related designs, brain processes underlying voluntary tic inhibition are often studied using resting state fMRI (RS-fMRI), which records brain activations in more or less natural conditions. This is thought to reflect the "intrinsic" functional organization of the brain. In this experiment, Ganos et al. used ReHo—a specific method of analyzing RS-fMRI, which captures the synchrony of resting-state brain activity in neighboring parts of brain areas (voxels).

In the operant conditions, voluntary action and motor cortical TMS were followed 250 ms later by a tone. Crucially, the presence of an additional event (tone) in the operant context caused large perceptual shifts. In particular, subjects' awareness of the voluntary key press action was shifted later in time, closer to the following tone. At the same time, awareness of the tone was shifted earlier in time, toward the key press. In contrast, involuntary, TMS-induced movements produced perceptual shifts in the opposite direction. The results for the sham and auditory conditions show minimal perceptual shifts, suggesting that no binding occurs for arbitrary unrelated events.

In a second experiment, Haggard et al. (2002) investigated the effects of temporal interval on intentional binding. In particular, 12 different subjects were asked to voluntarily press a key, while key presses were followed by a tone at intervals of 250, 450, or 650 ms. Subjects judged the time of tone onset in separate fixed blocks where all trials involved a single interval as well as in three additional blocks containing a randomized combination of all intervals. They tested fixed and randomized blocks in counterbalanced halves of the experiment. As in the aforementioned experiment, a single-event, baseline block of tone-only trials was measured in each half and used to calculate perceptual shifts.

An ANOVA measurement on shifts in judgment revealed significant effects of schedule and lag and a significant interaction between the two. An ANOVA on the changes in variability across trials showed no significant effect of schedule, a trend toward an effect of lag due to increased variability at 450 ms only, and no significant interaction between them. On the basis of these results, Haggard et al. argue that there exists a binding effect that correlates with temporal contiguity and temporal predictability. It is worth noting that this binding does not seem to depend simply on improved allocation of attention at the time of the effect.

In this sense, conscious representations of sensorimotor events surrounding voluntary actions are bound by a specific cognitive function. The second of the conducted experiments suggests that this cognitive function obeys two key general principles of association. The binding effect is regulated by temporal contiguity and temporal predictability. In turn, the obtained results suggest that the recorded perceptual shifts may be a conscious aspect of a general linkage through time between representations of actions and their effects. In light of the above results, Haggard et al. argue that our brain lies to us in order to show us how to bind stimuli beyond strict perceptual experiences and to construct a coherent conscious experience of our own A.

It is worth noting that schizophrenic patients with hallucinations and delusions often attribute external events to their own A or may attribute their own actions to external sources (e.g., Frith, 1992, as reported in

Haggard et al., 2002). Haggard et al. speculate that these misattributions could reflect excessive or impoverished intentional binding, respectively. In this way, the suggested binding process predicts the notoriously compromised A of schizophrenic subjects.

The above results suggest that A comes about in a way that occasionally disregards strict perceptual experiences. However, this does not mean that we can also disregard information stored in our memory in light of reoccurring experiences with our natural environment or S. I examine the influence from stored information on perception as well as cognition next.

Structure and Top-Down Effects in Perception

Top-down effects in perception are ubiquitously accepted. Nevertheless, there are also skeptics like Fodor (1983) and Pylyshyn (1999) who argue that cognition does not affect vision (or perception in general) directly. Rather, argues Pylyshyn, cognition only produces top-down effects indirectly through attention and decision-making.

Starting from a modular view of the mind, Pylyshyn (1999) assumes that early vision is cognitive impenetrable (qua a module). In arguing that bottom-up information is resilient to (contradicting) top-down information, Pylyshyn appeals to evidence from work done on perception of optical illusions. Consider, for instance, the Müller-Lyer (M-L) illusion, where the observer perceives the two lines as different in length, even though one knows that they are of equal length. For Pylyshyn this is clear evidence that perception is cognitively impenetrable.

In contrast, Gregory (1970) offers an alternative interpretation of the same effect, which precisely builds on top-down influences. Gregory explains that when confronted with the M-L illusion we perceive one line to be longer than the other regardless of our knowledge that they are both of equal length due to the fact that the brain "interprets" the inward-pointing arrowheads as distance cues similar to the ones read off when looking at the upper corner of a rectangular room, where walls and ceiling meet. In turn, the data convey information about one of the lines "standing out" while the other one is "standing back." Given that both lines subtend the same angle on the retina, the line that is construed to be farther away, is perceived as larger. The brain makes this correction in light of our geometrical knowledge. (see also Barnes, Bloor, & Henry, 1996).

For Gregory, subjects are susceptible to this illusion because they live in highly "carpentered" environments in which rectangular shapes, straight lines, and square corners abound. This hypothesis is confirmed by Segall, Campbell, and Herskovitz (1966) who studied susceptibility to optical illusions, among other groups, of Zulus who interestingly live in round huts and plough their land and fields in circles rather than in rows, as is the norm in western countries. The obtained results show that Zulus are significantly less susceptible to the effect. Contra Pylyshyn, information stored in the mind of Zulus does influence perception in a top-down manner. However, given that Zulus do not live in carpentered environments, their brains do not correct the perceived stimulus in the way this is done in the case of westerners.

Further evidence in support of the claim that there are top-down effects in perception can be found in the literature on phoneme-restoration. In a typical phoneme restoration experiment, a subject would hear a word like "table" with the "b" sound obscured by noise. The results obtained during similar experiments show that subjects in their majority hear "table" with "b" in place even though this has not been present in the auditory stimulus. Even though this evidence is interpreted in various, and often competing, ways, Elman and McClelland's "trace model" (1985) offers the most convincing interpretation of results from phoneme-restoration experiments (see Norris, McQueen, and Cutler, 2000, and Samuel, 1997, among others). Briefly, they argue that a perceptual input activates stored information, with various competing and similar to each other representations influencing perception of individual phonemes in the inputting signal. These results highlight the influence of stored representations on perception. Given that concepts are built out of perceptual representations, my claim is that our previous experiences (including experiences with S) play a key role in thinking and of course in action. It is on these grounds that I argue that A is saturated with S.

Reactivation of Stored Representations and Top-Down Effects

Previous experiences influence not only perception but also cognition, as suggested by evidence showing that stored representations become activated during imagery. For instance, Demarais and Cohen (1998) examined whether visual imagery (required by a task) evokes saccadic eye movements. Also, they examined whether visual imagery determines the spatial pattern of the saccades. In testing these hypotheses, they asked subjects to solve transitive inference (or syllogistic) problems with the relational terms "left | right" and "above | below" while the horizontal and vertical eye movements were recorded by electrooculography.

Subjects were given transitive tasks using ordinary household objects, which can be easily imagined side by side on a kitchen counter, or arranged on a vertical axis, e.g., on the shelves of a kitchen cabinet. Subjects listened to prerecorded instructions such as "a jar of pickles is below a box of tea bags, the jar of pickles is above a can of coffee, where is the can of coffee?."

The obtained results show that subjects made more horizontal and fewer vertical saccades while solving problems with the "left│right" terms than while solving identical problems with "above│below." Furthermore, subjects made more vertical saccades while dealing with problems using "above│below" relational terms than when using "left│right" terms. Thus tasks that evoke spatially extended imagery trigger eye movements that reflect the spatial orientation of the image. The strong similarities between recorded saccadic eye movements and auditory stimuli show that stored representations drive selective attention (via influencing saccadic eye movements) in a top-down manner.

In a similar experiment, Spivey and Geng (2001) showed that eye movements coordinate elements of a mental model of imaged counterfactual scenarios. Specifically, they asked subjects to imagine or recall objects that were no(t) (longer) present and recorded their eye movements using an eye tracker. The center of the pupil and the corneal reflection were tracked and also a scene camera, yoked with the view of the tracked eye, provided the examiners with an image of the subject's visual field.

While looking at a white projector screen, subjects listened to five prerecorded scene descriptions with upward, downward, leftward, rightward, and nondirectional spatiotemporal dynamics in random order. In the majority of cases, scene descriptions avoided explicit directional terms such as "left," "up," and "above." All subjects (two groups) spontaneously and systematically looked at particular blank regions of space as if they were manipulating and organizing spatial relations between mental images.

Thus, argue Spivey and Geng, interpreting a linguistic description of a visual scene requires activation of a spatial mental representation, and position markers allocated in a visual field manipulate visual attention. Searching for stored mental representations is accompanied by an oculomotor search of external space. Against this background, I argue that stored representations become reactivated even in the absence of the appropriate stimulus and drive selective attention in a top-down manner (see also Chao, Haxby, & Martin, 1999).

The above evidence shows that activated stored representations drive attention in a top-down manner. Building on that, I argue that previous exposure to our sociocultural, political, and physical environment influence the way in which we perceive, conceptualize, and understand the world. Specifically, both conscious (concepts) and unconscious (intuitions) determinants of sociologically nontrivial actions derive from perceptual encounters with our sociophysical environment. In turn, sociologically nontrivial actions are saturated with our experiences with S.

CONCLUSION

In the previous pages, I put forth a view according to which thinking is structured. Against this background, I showed that there is little value in arguing over the primacy of A or S as the main determinants of social behavior. For free-willing agents—in the sense explained above—simply do not exist, since A is saturated with our experiences with S. Even though these structural influences are rather liberal in nature, as explained, they determine social behavior. Finally, I have argued that focusing on the process of socialization, and in turn on the ways in which S is internalized and influences thinking, can shed light on the nature of social behavior.

Acknowledgments

I am grateful to James Trafford, Patrice Soom, and Gottfried Vosgerau for their helpful comments on earlier drafts of this chapter.

References

Barker, C. (2005). *Cultural studies: Theory and practice*. London: SAGE.
Barnes, B., Bloor, D., & Henry, J. (1996). *Scientific knowledge—A sociological analysis*. London: Athlone.
Barsalou, L. W. (1999). Perceptual symbol systems. *Behavioral and Brain Sciences, 22,* 577–609.
Baym, C. L., Corbett, B. A., Wright, S. B., & Bunge, S. A. (2008). Neural correlates of tic severity and cognitive control in children with Tourette syndrome. *Brain: A Journal of Neurology, 131,* 165–179.
Berger, P. L., & Luckmann, T. (1966). *The social construction of reality: A treatise in the sociology of knowledge*. Garden City, NY: Anchor Books.
Bortolotti, L., & Cox, R. (2009). "Faultless" ignorance: Strengths and limitations of epistemic definitions of confabulation. *Consciousness and Cognition, 18*(4), 952–965.
Bortolotti, L. (2011). Does reflection lead to wise choices? *Philosophical Explorations, 14*(3), 297–313.
Brandom, R. (1994). *Making it explicit: Reasoning, representing, and discursive commitment*. Cambridge, MA: Harvard University Press.
Buse, J., August, J., Bock, N., Dorfel, D., Rothenberger, A., & Roessner, V. (2012). Fine motor skills and interhemispheric transfer in treatment–naïve male children with Tourette syndrome. *Developmental Medicine and Child Neurology, 54,* 629–635.
Carruthers, P. (1998). Conscious thinking: Language or elimination? *Mind and Language, 13*(4), 457–476.
Carruthers, P. (2005). *Consciousness: Essays from a higher order perspective*. Oxford: Clarendon Press.
Carruthers, P. (2008). Language in cognition. In E. Margolis, R. Samuels, & S. Stich (Eds.), *The Oxford handbook of philosophy of cognitive science* (pp. 382–401). New York: Oxford University Press.
Chao, L. L., Haxby, J. V., & Martin, A. (1999). Attribute-based neural substrates in temporal cortex for perceiving and knowing about objects. *Nature Neuroscience, 2,* 913–919.

Christoff, K., Ream, J. M., & Gabrieli, J. D. (2004). Neural basis of spontaneous thought processes. *Cortex*, *40*(4-5), 623−630.

Corbetta, M., Miezin, F. M., Dobmeyer, S., Shulman, G. L., & Petersen, S. E. (1990). Attentional modulation of neural processing of shape, colour, and velocity in humans. *Science*, *248*, 1556−1559.

Craik, F. I. M., Govoni, R., Naveh-Benjamin, M., & Anderson, N. D. (1996). The effects of divided attention on encoding and retrieval processes in human memory. *Journal of Experimental Psychology General*, *125*, 159−180.

Davidson, D. (1975). *Thought and talk. Inquiries into truth and interpretation* (pp. 155−170). Oxford: Oxford University Press.

Demarais, A. M., & Cohen, B. H. (1998). Evidence for image-scanning eye movements during transitive inference. *Biological Psychology*, *49*(3), 229−247.

Dijksterhuis, A., & van Olden, Z. (2006). On the benefits of thinking unconsciously: Unconscious thought can increase post-choice satisfaction. *Journal of Experimental Social Psychology*, *42*, 627−631.

Dreyfus, H. L. (1997). Intuitive, deliberative, and calculative models of expert performance. In C. E. Zsambok, & G. Klein (Eds.), *Naturalistic decision making* (pp. 17−28). Mahwah, NJ: Lawrence Erlbaum.

Elman, J. L., Bates, E. A., Johnson, M. H., Karmiloff-Smith, A., Parisi, D., & Plunkett, K. (1996). *Rethinking innateness: A connectionist perspective on development*. Cambridge, MA: MIT Press.

Elman, J. L., & McClelland, J. L. (1985). An architecture for parallel processing in speech recognition: The TRACE model. In M. R. Schroeder (Ed.), *Speech recognition* (pp. 6−35). Basel: S. Krager AG.

Filevich, E., Kuhn, S., & Haggard, P. (2012). Intentional inhibition in human action: The power of "no." *Neuroscience and Biobehavioral Reviews*, *36*, 1107−1118.

Fodor, J. (1983).) *The modularity of mind: An essay in faculty psychology*. Cambridge, MA: MIT Press.

Frith, C. D. (1992). *The cognitive neuropsychology of schizophrenia*. Hove, UK: LEA.

Ganos, C., Kahl, U., Brandt, V., Schunke, O., Bäumer, T., Thomalla, G., et al. (2014). The neural correlates of tic inhibition in Gilles de la Tourette syndrome. *Neuropsychologia*, *65*, 297−301.

Gigerenzer, G., Todd, P., & the ABC Research Group (1999). *Simple heuristics that make us smart*. New York: Oxford University Press.

Gregory, R. L. (1970). *The intelligent eye*. New York: McGraw-Hill.

Haggard, P., Clark, S., & Kalogeras, J. (2002). Voluntary action and conscious awareness. *Nature Neuroscience*, *5*, 382−385.

Haidt, J. (2001). The emotional dog and its rational tail: A social intuitionist approach to moral judgment. *Psychological Review*, *108*, 814−834.

Hebb, D. O. (1949).) *The organisation of behaviour*. New York: John Wiley & Sons, Inc.

Kahneman, D., Slovic, P., & Tversky, A. (1982). *Judgements under uncertainty: Heuristics and biases*. Cambridge: Cambridge University Press.

Leisman, G., Machado, C., Melillo, R., & Mualem, R. (2012). Intentionality and "free-will" from a neurodevelopmental perspective. *Frontiers in Integrative Neuroscience*, *6*(36), 1−12.

Nisbett, R. E., & Wilson, T. D. (1977). Telling more than we can know: Verbal reports on mental processes. *Psychological Review*, *84*(3), 231−259.

Norris, D., McQueen, J. M., & Cutler, A. (2000). Merging information in speech recognition: Feedback is never necessary. *Behavioral and Brain Sciences*, *23*, 299−325.

Noudoost, B., Chang, M. H., Steinmetz, N. A., & Moore, T. (2010). Top-down control of visual attention. *Current Opinion in Neurobiology*, *20*, 183−190.

Oakley, A. (1974). *The sociology of housework*. London: Martin Robertson.

O'Donnell, M. (2010). *Structure and agency*. London: SAGE Publications Ltd.

Ogburn, W. F. (1922). *Social change with respect to culture and original nature*. New York: B. W. Huebsch.

Osman, M. (2013). A case study: Dual-process theories of higher cognition—Commentary on Evans & Stanovich 2013. *Perspectives on Psychological Science*, 8(3), 248–252.

Parsons, T. (1937). *The structure of social action: A study in social theory with special reference to a group of European writers*. New York: McGraw-Hill. (Reprinted from Glencoe IL: The Free Press, 1949)

Paszek, J., Pollok, B., Biermann-Ruben, K., Muller-Vahl, K., Roessner, V., Thomalla, G., et al. (2010). Is it a tic?—Twenty seconds to make a diagnosis. *Movement Disorders: Official Journal of the Movement Disorder Society*, 25, 1106–1108.

Perry, J. (2001). *Knowledge, possibility, and consciousness*. Cambridge, MA: MIT Press.

Prinz, J. (2002). *Furnishing the mind: Concepts and their perceptual basis*. Cambridge, MA: MIT Press.

Pylyshyn, Z. W. (1999). Is vision continuous with cognition? The case for cognitive impenetrability of visual perception. *Behavioral and Brain Sciences*, 22(3), 341–423.

Samuel, A. G. (1997). Lexical activation produces potent phonemic percepts. *Cognitive Psychology*, 32, 97–127.

Searle, J. (1995). *The construction of social reality*. New York: The Free Press.

Segall, M., Campbell, D., & Herskovitz, M. J. (1966). *The influence of culture on visual perception*. New York: Bobs-Merrill.

Spivey, M. J., & Geng, J. J. (2001). Oculomotor mechanisms activated by imagery and memory: Eye movements to absent objects. *Psychological Research/Psychologische Forschung*, 65(4), 235–241.

Synofzic, M., Vosgerau, G., & Newen, A. (2008). Beyond the comparator model: A multifactorial two-step account of agency. *Consciousness and Cognition*, 17, 219–239.

Tillas, A. (2010). Back to our senses: An empiricist on concept acquisition (Doctoral thesis). University of Bristol, Bristol, UK.

Tillas, A. (2014). How do ideas become general in their signification? In E. Machery, J. Prinz, & J. Skilters (Eds.), The Baltic international yearbook of cognition, logic and communication (Vol. 9). Manhattan, KS: New Prairie Press. <http://newprairiepress. org/biyclc/vol9/iss1/12/>.

Tillas, A. (2015). Language as grist to the mill of cognition. *Cognitive Processing*, 16(3), 219–243.

Tillas, A. (in press). On the origins of concepts. In: C. Kann, D. Hommen, & T. Osswald (Eds.), *Concepts and categorisation*. Paderborn: Mentis.

Tillas, A., & Trafford, J. (2015). Communicating content. *Language and Communication*, 40, 1–13. <http://dx.doi.org/10.1016/j.langcom.2014.10.011>.

Trafford, J., & Tillas, A. (2015). Intuition & reason: Re-assessing dual-process theories with representational subactivation. *Teorema*, XXXIV(3), 197–219.

Uncapher, M. R., Hutchinson, B. J., & Wagner, A. D. (2011). Dissociable effects of top-down and bottom-up attention during episodic encoding. *The Journal of Neuroscience*, 31(35), 12613–12628.

Wittgenstein, L. (1976). *Philosophical investigations* (G. Anscombe, Trans.). Oxford: Blackwell.

8

Two Kinds of Reverse Inference in Cognitive Neuroscience

G. Del Pinal[1] and M.J. Nathan[2]

[1]Zentrum für Allgemeine Sprachwissenschaft (ZAS), Berlin, Germany
[2]University of Denver, Denver, CO, United States

INTRODUCTION

Cognitive neuroscience is based on the plausible idea that our understanding of the mind has much to gain from investigations into the workings of the brain. Over the last few decades, this steadily growing field has substantially advanced our studies of relatively modular systems, like vision and touch, and promises to seriously contribute to the resolution of longstanding disputes in various domains of higher-cognition. To achieve this ambitious aim, one of the most commonly used techniques is *reverse inference*, i.e., the practice of inferring, in certain tasks, the engagement of cognitive processes from patterns or locations of neural activation. Since different psychological theories often make incompatible assumptions about the processes underlying a specific cognitive task, reverse inference can, in principle, be used to discriminate between competing hypotheses.

Scientists and philosophers often talk about reverse inference tout court. However, this chapter shows that it is crucial to distinguish between two different types of reverse inference. In the first kind, cognitive processes are inferred from the particular locations of neural activation observed in particular tasks. We examine these location-based inferences through a case study on the nature of mind-reading. Some

The Human Sciences after the Decade of the Brain.
DOI: http://dx.doi.org/10.1016/B978-0-12-804205-2.00008-2

121

© 2017 Elsevier Inc. All rights reserved.

prominent scientists have argued that mirror neurons provide decisive evidence for embodied theories of mind-reading. Their argument is based on a paradigmatic location-based reverse inference (LRI). In the first part of this chapter, we show that this argument fails (see section "Location-Based Reverse Inference" of this chapter) and, in doing so, we highlight some inherent problems with this kind of inference (see section "LRI: General Problems and Limitations" of this chapter).

Critiques of location-based inference are widespread. Indeed, prominent researchers have gone as far as suggesting that reverse inference should be removed from the toolkit of cognitive neuroscience. In the second part of this essay, we maintain that this more radical step should be resisted. Drawing on a recent case study of recognition memory, we argue that a second kind of inference, based on pattern-decoding techniques, overcomes the problems faced by location-based inferences. In particular, we show that pattern-based inference does not presuppose any problematic "neo-phrenological" assumptions about functional localization in the brain. As a result, pattern-decoding techniques overcome some of the oldest and most resilient objections that have been raised against the methodology of cognitive neuroscience (see section "Pattern-Based Reverse Inference" of this chapter). Although pattern-based inferences are quickly gaining popularity among cognitive neuroscientists, they are still largely ignored in most methodological discussions.

LOCATION-BASED REVERSE INFERENCE

To introduce LRI, consider some neuroscientific studies of "mind-reading," the capacity to identify and predict the mental states and behavior of others. The two main competing cognitive explanations of this capacity are the *theory-theory* (TT) and the *simulation theory* (ST).

> (TT) According to the theory-theory, mind-reading is based on a science-like folk theory of mind, which includes law-like generalizations over symbolic representations of the following categories: (a) observable inputs and mental states; (b) mental states and other mental states; (c) mental states and observable outputs. (Fodor, 1992; Gopnik & Wellman, 1992)
>
> (ST) According to the simulation theory, mind-reading is based on the capacity to take other agents' perspective by simulating their mental states and actions as if they were one's own. This allows us to access directly the intentional states or actions that, in our own case, would cause and result from the simulated states, and attribute them to others. (Gallese & Goldman, 1998)

In emphasizing the symbolic and abstract nature of higher cognition, TT follows the cognitivist tradition of Descartes and Kant and, more recently, Chomsky and Fodor. In contrast, ST's assumption that the vehicles of cognition have a sensory-motor format the empiricist tradition of Locke and Hume.

It should be clear that TT and ST provide distinct, intuitively plausible, and yet mutually incompatible accounts of mind-reading. How can we choose one over the other? Some authors have argued that *mirror neurons*—brain cells that fire both when an organism acts and when the organism observes the same type of action performed by someone else—provide decisive evidence in favor of ST over TT (Gallese & Goldman, 1998; Iacoboni, 2009; Rizzolatti & Sinigaglia, 2010).[a] The key data was obtained using experimental variations of two tasks (Kilner & Lemon, 2013; Rizzolatti, Fogassi, & Gallese, 2009). The first task involves *executing* a basic-level motor act, such as grabbing a cup, or more complex tasks, such as grabbing a cup to drink from it; the second task involves *observing* another agent performing the same type of motor acts. To see whether these discoveries provide evidence in favor of ST or TT, let us begin by spelling out the predictions, at both the cognitive and neural levels, of each hypothesis.

According to TT, subjects possess concepts for simple motor actions, such as GRAB. These simple concepts, conceived as a-modal symbolic representations, can be recursively combined with other concepts to represent more complex motor acts, such as GRAB A CUP TO DRINK, and invoked in folk-psychological generalizations expressing regularities such as "subjects who grab cups to drink tend to be thirsty."[b] The crucial point is that, according to TT, the same concept, GRAB, is tokened both when an act is categorized as a "grabbing act" and when it is intentionally executed, i.e., both when we categorize (or imagine) the grabbing actions of others and when we form an intention to grab. TT's prediction that both *execution* and *observation* tasks involve tokens of

---

[a]There are various reasons why studies of motor acts and mirror neurons are especially apt to clarify the structure of LRI. First, the relevant neural-level data is quite clear and has been extensively replicated. Second, such data was obtained by using single-cell recordings, which allows us to focus on reverse inference *per se*, without addressing tangential problems regarding more controversial data-gathering tools, such as fMRI.

[b]Although these concepts are a-modal, they interface with sensory and motor processes. On the sensory side, these concepts can be applied based on perceptual cues; on the motor side, they can be used to form intentions to act so that, when we form an intention to *grab that red cup*, one usually executes the corresponding motor act and, indeed, grabs the red cup as opposed to, say, biting it. These interface conditions are sometimes called "legibility constraints" since, to be usable, a conceptual system (even an a-modal one) must be legible at its input-output interfaces, so that perceptual inputs can lead to the tokening of concepts, and action concepts can be translated into the appropriate motor commands.

II. THE NEUROSCIENCES OF SOCIAL SCIENCES AND ETHICS

GRAB (or, in more complex cases, tokens of GRAB CUP ∨ DRINK) has implications at the neural level. If TT is correct, then the neural implementation—and, consequently, the neural activation pattern that codes for tokens of GRAB—should be instantiated when grabbing actions are both *executed* and *observed*. Of course, TT does not presuppose that such neural patterns are identical; all it requires is that both representations be tokens of the same type, i.e., tokens of the concept GRAB.

Next, consider ST. Recall that, in this view, simulation is the default mind-reading process. When subjects perceive others as grabbing a cup, they simulate this basic-level motor action as if they were executing it themselves. This simulation allows a subject to retrodictively determine the intention that she would have when performing that same act herself, in similar conditions.[c] The key point is that the process tokened in *observation* is a subcomponent (or is structurally analogous to a subcomponent) of the corresponding process tokened in *execution*. Hence, at the cognitive level, ST predicts that a subset of the execution process is also tokened in observation processes. At the neural level, this entails that the implementation of the part of the simulation that is shared with the execution of the motor process should be instantiated in both *execution* and *observation*.

At this point, it should be clear that deciding whether TT or ST provides the correct explanation of mind-reading, based on the experimental contrast between *execution* and *observation*, is not as straightforward as it is sometimes assumed, as both theories predict that there is a key cognitive component present in both experimental conditions. According to TT, this is the tokening of an action concept; ST takes this to be the tokening of subsets of motor processes. Yet, if one could distinguish between tokens of motor-action concepts and tokens of motor-action simulations at the neural level, this data could be used to adjudicate between TT and ST. To see whether this can be done, we now consider the neural data.

Recall that the neural data was gathered in experiments that compared two basic conditions: in *execution*, subjects perform a basic-level motor act; in *observation*, subjects observe full or partial evidence that others are executing the same type of basic-level motor act. The neural data collected in these studies was obtained using single-cell recordings

---

[c]The details of the simulation process can be spelled out in various ways. On one view, the process is rather direct, in the sense that the observed motor action is directly matched by a corresponding representation, which initiates the understanding of that action "from the inside" (Rizzolatti, Fogassi, & Gallese, 2001). In more complex models, subjects generate a candidate goal for the observed action and then simulate such action as if it were their own. If the action matches the goal, then that is the goal assigned to the perceived action; in cases of mismatch, the process can be repeated (Gallese & Goldman, 1998). For our purposes, we can remain neutral between these two models.

of macaques (Kilner & Lemon, 2013; Rizzolatti et al., 2009), yet fMRI studies suggest that the basic findings also apply to humans (Rizzolatti & Craighero, 2004). Neural-level analyses revealed both an *activation pattern* and a *location pattern*. A group of mirror neurons, call it *MN*, is selectively activated in both execution and observation, and *MN* is localized in the premotor cortex.

Two clarifications are now in order. First, to say that some neurons are "selectively activated in some task" implies that these same neurons are not differentially activated in other relevant tasks. This has three implications. (1) Mirror neurons are not differentially activated by acts that look similar to (but are not) motor acts, e.g., motions of a hand that are not instances of a grabbing action, but look relevantly similar (Rizzolatti, Fadiga, Gallese, & Fogassi, 1996). (2) Mirror neurons are activated by instances of the same type of basic-level motor act even if the cues vary in modality. To illustrate, mirror neurons are activated not only when a subject sees the full action of grabbing, but also when she sees only part of the action, or even if she hears evidence of a grabbing action (Umiltà et al., 2001). (3) Finally, the mirror neurons activated in the *execution* and the *observation* of a basic-level motor act, such as *grabbing a cup*, are also activated when the act is embedded in a more complex one, such as *grabbing a cup to drink*. Interestingly, if you compare a set of *execution vs observation* conditions involving a complex act (grabbing a cup to drink) with a set involving a different complex act (grabbing a cup to clean), some mirror neurons fire in both sets of conditions, since both sets involve grabbing acts, whereas others fire in only one of the two (drinking vs cleaning) sets (Fogassi et al., 2005; Iacoboni et al., 2005). Second, while we identified two different components of the result—the pattern across tasks and the location of the mirror neurons involved—ST theorists rarely separate these two aspects. However, keeping patterns and locations distinct will allow us to determine the relative inferential weight carried by the specific location of mirror neurons. Importantly, both kinds of inferences—from patterns of neural activation to cognitive processes and from locations of activation to cognitive processes—are common in neuroscience. However, neither of these kinds of reverse inference should be confused with the *pattern-decoding techniques* discussed in section "LRI: General Problems and Limitations" of this chapter.

With all of this in mind, we can now examine the reverse inference that allegedly supports ST over TT. Recall that, in reverse inference, the engagement of a cognitive process (in a set of tasks) is inferred from neural data, usually consisting of specific patterns or locations of neural activation. The conditional probability that a particular cognitive process is engaged, given a set of tasks and neural data, depends on the probability of the neural activation in the task, given the hypothesized cognitive process. Hence, in order to determine whether *MN* provides any reason to select TT over ST, or vice versa, we need to determine the likelihood of the neural results, given TT and ST, respectively.

Consider, first, the mirror neuron *pattern*, i.e., the result that a particular set of neurons selectively fires at the same rate in both *execution* and *observation* tasks. As noted, TT assumes that both conditions engage tokens of the same motor concept. Consequently, TT predicts some neural overlap in both conditions, namely, the pattern that codes for these tokens of the same concept type (in our example, the uniform and selective firing rate of *MN*). ST makes essentially the same prediction but, in this case, the uniformity is due to partially overlapping motor processes being engaged in both conditions, as opposed to tokenings of the same concept.[d] In short, TT and ST both predict an overlap in part of the neural pattern observed in both conditions (*execution* and *observation*) but neither makes specific predictions regarding the fine-grained structure of this pattern.

Next, consider *location*, the result that *MN* (the set of mirror neurons that selectively fire at the same rate in both *execution* and *observation*) are located in the premotor cortex. These considerations seem especially relevant because of a key difference between TT and ST. According to TT, what *execution* and *observation* share is the tokening of the same concept, which is taken to be an abstract representation that cannot be reduced to sensory or motor processes. In contrast, ST claims that what both tasks share is a partial overlap of the same type of motor process; in one case as part of a motor act and, in the other, as part of a simulation. At this level of informal description, the location of mirror neurons might seem to provide decisive evidence in favor of ST over TT for, as ST theorists note, most neurons in the premotor cortex are known to be involved in motor acts.[e] This suggests that mirror neurons implement subsets of motor plans rather than a-modal motor-action concepts.

[d]Strictly speaking, TT and ST only make this prediction on the assumption that tokens of the same cognitive process have uniform neural implementations. Although, in principle, this uniformity assumption could be challenged (e.g., based on radical forms of multiple realizability), it is a fundamental presupposition of cognitive neuroscience that we shall take for granted in our discussion.

[e]Indeed, ST theorists sometime insist that it is the location of mirror neurons in the premotor cortex that provides decisive evidence in favor of ST over TT (Rizzolatti & Sinigaglia, 2010). For instance, given that mirror neurons fire across modalities, one might be tempted to conclude that these brain cells are a-modal. ST theorists, however, resist this move and propose instead that, since mirror neurons are located in the premotor cortex, they should be conceived as translating various modalities to the motor modality, as opposed to translating modalities into a nonmodal abstract representation. A further consideration often used to defend ST appeals to deficit patterns caused by neurodegenerative diseases that affect the motor system. However, once we admit that motor concepts could be tokened in and interface with premotor areas, much of this evidence becomes irrelevant to the debate between TT and ST (cf. Hickok, 2014).

Despite the intuitive appeal of these considerations, we argue that ST does not predict or explain the location of mirror neurons better than TT; strictly speaking, both theories are equally compatible with the neural results. The problem with the above reasoning has two sources. First, it misconstrues the predictions of TT regarding the neural encoding of motor-act concepts; second, it overestimates ST's prediction regarding the neural implementation of simulations. We now elaborate both points, in turn. Beginning with TT, as noted, this hypothesis entails that, when planning a motor act such as grabbing a cup, agents form an intention that tokens the concept GRAB CUP. This intention then interfaces with and instructs motor regions that carry out the computations required to execute the action. In the observation case, the corresponding act is understood as an instance of GRAB CUP together with a representation of a different agent. While, strictly speaking, TT does not predict that tokens of motor-act concepts are encoded in premotor areas, such a discovery is perfectly compatible with TT. Given that premotor areas are involved in motor actions, this location is a very plausible candidate for the interface between tokens of motor intentions (involving motor-act concepts) and the operations involved in motor executions. In short, although, strictly speaking, TT does not predict the location of mirror neurons, it is certainly not undermined by it either.[f]

This conclusion might be viewed as good news for ST. However, to conclude that ST is supported by the neural data, it is not sufficient to just show that ST is compatible with the MN location. In addition, we must also show that ST predicts that data in a stronger sense than TT. Does ST's prediction fit the neural data more naturally or accurately than the analogous prediction of TT? Unfortunately, the answer seems negative. ST is surely *compatible* with motor simulations being implemented in the premotor cortex, but the theory does not entail this result

[f]This has interesting implications for the debate between "a-modal" and "embodied" theories of concepts. Nothing currently known about the nature of neural computation and representation prevents one from holding that a-modal concepts about, say, tactile, visual, or auditory domains are encoded in neural locations that are topologically close to the areas that process tactile, visual, or auditory stimuli. Thus, TT can assume, as a reasonable working hypothesis, that concepts for basic-level motor acts are encoded in areas topologically close to the motor areas involved in action execution. To be sure, this closeness between locations of concept tokenings and their corresponding input-output interfaces is not quite predicted by TT. This is because one cannot a priori dismiss other "hardware" solutions to processes such as extracting conceptual categories from sensory and motor modalities, applying concepts to sensory inputs, and using concepts to form motor intentions that can interface with motor processes to execute actions. Still, TT does not presuppose any distance between sites where symbolic concepts are encoded (in long-term and working memory tasks) and the corresponding input-output processing regions with which they can interface.

any more than TT does. To illustrate, suppose that *execution* and *observation* showed instead that the MN pattern was localized in premotor areas in one task but in the prefrontal cortex in the other. Would this undermine ST? Clearly not. The neural location that implements a simulation process could be different from the location that implements the simulated process for various reasons. For instance, this could be a hardware solution to getting and keeping the simulation processes off-line. Of course, there would have to be a way to determine that the neural pattern is the implementation of a simulation of the target process, but this information could be revealed by the details of the actual firing pattern, independently of the location. Note that this sort of reasoning is not unusual. This is the way in which experimenters often try to show that, although the perirhinal cortex is not the locus of spatial processing, it carries spatial information, and although the hippocampus is not the locus of item processing, it carries item-related information (more on this below).

Admittedly, questions concerning which areas of the brain *could* implement motor-action concepts and simulation processes are somewhat puzzling. The reason for this is that we do not know enough about how cognitive representations and operations are extracted, encoded, and tokened at the neural level to be able to substantially narrow the range of neural locations that could perform the categorization operations of TT or the simulation processes of ST. The conclusion of the above discussion is that TT and ST are both compatible with location data provided by mirror neuron studies. To be clear, we are not denying that we know quite a bit about the function of premotor neurons, and by extension, of premotor mirror neurons, e.g., that they are "involved" in action planning, execution and recognition. However, in debates between TT and ST what we have on the table are two competing accounts of the precise computational form of the processes involved in action planning, execution, and recognition. Pointing out the premotor location in which those operations are implemented does not, by itself, help us select which of the competing operations are actually used.

LRI: GENERAL PROBLEMS AND LIMITATIONS

In the previous section, we have seen that *MN location*, i.e., the discovery that the relevant pattern of mirror neurons is localized in the premotor cortex, does not provide conclusive evidence to select ST over TT (or vice versa) as an explanation of mind-reading. In this section, we extend the discussion into some general limitations of LRI. Specifically, we begin by identifying a key assumption for the proper use of reverse inference, which we call the "linking condition." Next, we argue that this linking condition is extremely hard to fulfill for any study that infers cognitive process from the location of neural activation. In the second part of the chapter, we present an influential technique, *multivariate pattern analysis* (MVPA), supporting a different type of reverse inference, which overcomes the problems faced by LRI.

To spell out the assumption that we maintain any properly conducted reverse inference should meet, suppose that C and D are two competing hypotheses of the cognitive processes underlying some task t. Further assume that C posits the engagement of cognitive process c, D posits the engagement of cognitive process d, $c \neq d$ (c and d are not the same process), and let n stand for a differential pattern of neural activation in some specific location. A reverse inference from the presence of n in t to the engagement of c in t requires the existence of independent studies that establish a link between n and c. The reason for this should be obvious: in order for the observation of n to support C over D, there must be a corroborated connection between n and c, i.e., it must be shown that n is evidence for c and that n is not evidence for d. Let us call these background studies "linking studies." Provided that there is a robust linking study connecting n with c, experimenters can use the observation that n is engaged in t to infer that c (and not d) is engaged in t, providing evidence in favor of C over D.

Important as they are, linking studies are hard to obtain. In particular, there are (at least) two problems that must be avoided. First, the connection between c or d with n cannot be obtained a priori, but must be discovered though painstaking experimental work. This experimental work typically requires determining whether n is connected to c or d, in the context of some task t^* that, to avoid circularity, must be different from task t.[8] However, if C and D differ not only in their predictions for task t, but also in their cognitive-level interpretation of t^* then the linking studies that associate n to, say, c are problematic, as they ignore one of the competing alternative theories, namely, D. Second, even if the linking studies of c and n are properly conducted, in the sense that they are not biased toward any alternative, the region of the

---

[8]To minimize the possibility of violating the linking condition, properly conducted reverse inferences invoke, as part of their background studies, tasks that are relevantly different from those subsequently used to discriminate among the competing cognitive hypotheses. Furthermore, the links to neural data should be established in tasks in which experimenters can control, with reasonable confidence and without ignoring any of the theories that will be subsequently tested, the engagement of the relata on the cognitive side. Of course, in the tasks then used to evaluate the competing hypotheses, the engagement of the cognitive process of interest is at issue, and the probability of its presence is reverse inferred from the resulting location of neural activation.

brain where *n* is located could be multifunctional, i.e., it might also be known to implement other cognitive processes. Obviously, the two problems are not independent: the less selective the brain region of interest, the stronger the chance that linking studies ignored that *n* could also implement *d*.

Let us now apply this reasoning to the debate between TT and ST. The appeal to *MN location* in support of ST over TT falls prey to precisely these shortcomings. Specifically, to give a functionalist-level interpretation of the *MN location*, ST theorists do not appeal to any tasks in which it is reasonably uncontroversial that something like simulation processes (or some key subcomponent) is engaged. Furthermore, in the tasks that are considered—the variations of *execution* that generate mirror neuron activation—what is under dispute is precisely their fine-grained functional interpretation. This is a clear instance of the first problem isolated above: mere activation in premotor areas does not have a cognitive-level interpretation that is relevant to adjudicate between ST and TT. This is because ST and TT provide different cognitive-level accounts of the interface between intentions for and execution of basic-level motor acts. Consequently, they provide different accounts of what is going on in premotor areas, in tasks such as *execution*. As a result, obtaining the same activation in premotor areas in *observation* is compatible with both cognitive-level accounts: TT and ST.

It might seem reasonable to suppose that the mind-reading debate is particularly susceptible to these problems. Perhaps the differences between TT and ST, in the *observation vs execution* experimental context, are especially subtle, or maybe the fine-grained computational diversity of the premotor cortex is unique. These conjectures, however, do not withstand serious scrutiny. More or less overt violations of the "linking condition" can be found in a number of reverse inferences used in studies of higher cognition (Coltheart, 2006a, 2013). This is not accidental. The most common technique employed in studies of higher cognition to make neural data bear on competing cognitive-level hypotheses is LRI, where the engagement of a cognitive process in a task is inferred from a particular location of neural activation (Shallice & Cooper, 2011). LRI faces a systematic difficulty in satisfying the linking condition—i.e., avoiding the two problems identified above—because the brain regions of interest are seldom selective. Most neural areas implement several cognitive processes, ranging from completely distinct and independent to subtly different and interconnected ones. It is this functional diversity, we surmise, that undermines the plausibility and robustness of many LRIs.

At this point, one could argue that the general lack of selectivity of brain regions does not undermine the plausibility of LRI. Rather, it is a methodological "caveat" that might also serve the positive function of a general warning against explaining serious scientific claims in terms of "just so stories." Indeed, recent discussions of reverse inference have promptly noted and addressed the problem that activation in a region could also signal the engagement of cognitive processes other than the ones posited by the hypothesis under scrutiny. Whether expressed in Bayesian (Del Pinal & Nathan, 2013; Hutzler, 2014) or likelihoodist terms (Machery, 2014), if the experimental settings are designed appropriately and the linking studies are reliable, the general solution consists of recognizing that experimenters can ignore cognitive processes that are not part of at least one of the competing theories. In short, while LRI is, in principle, a sound inference, its usefulness is much more limited than enthusiastic supporters often recognize. This is a problem since, as illustrated in the mind-reading case, substantial debates often involve disputes about the fine-grained cognitive functions carried out in particular tasks and neural locations, and we need a way to discriminate between such hypotheses.

The tension between lack of selectivity and LRI occurs at various levels of analysis. At a relatively coarse level, the difficulty arises from the controversial assumption that brain regions are relatively selective for coarsely defined processes. At more fine-grained levels, the difficulty arises from the computational diversity associated with single neurons or groups of neurons in a given region, as revealed by the discussion of mind-reading. Consider the fine-grained computational diversity of premotor areas such as macaque area F5. Some neurons in F5 are selective for action perspective, manner of approach, or final execution strategy. Some neurons are selective for type of goal. Some are selective for particular modalities and others are cross-modal. In particular, some neurons are active *only during observation* and others only during execution (Kilner & Lemon, 2013). Take a neural network with basic units with that much computational diversity, and consider how many different processes—some more simulation-like, others more categorization-like, and all mimicking the MN pattern—you could build out of those basic units. In particular, note that some neurons in premotor areas fire only during observation. Hence, it remains an open question how exactly we should conceive of the representational format of the ones that are mirror neurons, i.e., that satisfy the MN pattern. TT and ST provide two different hypotheses, equally compatible with the data, and both implementable in a location with this sort of fine-grained computational diversity.

In conclusion, although LRI is a sound inference, its correct application is too limited to be of general use in cognitive neuroscience. In the following section, we shall examine a different kind of reverse inference that, we argue, provides a more promising and widely applicable technique to discriminate between competing cognitive-level hypotheses.

PATTERN-BASED REVERSE INFERENCE

Consider, once again, the case of mind-reading. In order to properly assess the relative plausibility of TT and ST, one needs to set up experimental tasks in which the cognitive processes posited by ST (motor act simulation) are uncontroversially engaged, tasks in which the cognitive processes posited by TT (categorization) are uncontroversially engaged and, in each case, map the neural results onto the corresponding cognitive-level process. In a word, one needs the appropriate linking studies. One must then compare those neural-level results with the results obtained in *observation*, the task in which the identity of the underlying cognitive process is under dispute. By doing so, one can determine whether understanding the basic-level motor acts of others is more like a simulation process or more like a straightforward categorization process. In the previous section, we argued that LRI has substantial problems fitting this bill. The root of the trouble is the computational diversity of brain regions, which implies that evidence provided by rough activation in a specific area is of limited value. In this section, we present and discuss a different kind of reverse inference that, we maintain, fares better on this score.

Mappings between cognitive functions and patterns of activity in particular brain regions, even those based on single-cell recordings, are not sufficiently selective. This point was illustrated *via* canonical studies of mirror neurons, which provide a neural pattern, the *MN location pattern*, consisting of a set of neurons that fire at the same rate in *observation* and *execution*. As noted, this firing pattern has not been decoded, in the sense that we do not yet know what sort of process or content is encoded in such firing sequence: that simple pattern could implement a token concept for a motor act, a subset of motor processes, or various other states. However, there are other kinds of mappings that are much more selective. For instance, vector patterns of neural activity, which contain detailed information about fine-grained cognitive-level representations and processes, can be decoded with techniques such as MVPA. MVPA uses tools from machine learning to create statistical machines—called "classifiers"—which can decode the cognitive states or processes encoded in particular neural data sets, such as multivoxel patterns obtained using fMRI. These decoded patterns can then be used to reverse infer the engagement of specific cognitive processes (Poldrack, 2008, 2011; Poldrack, Mumford, & Nichols, 2011; Tong & Pratte, 2012).

To illustrate MVPA, consider a study in episodic memory research about the cognitive processes underlying our capacity to classify items as "old" or "new." We formulate this recognition capacity as follows, where *s* ranges over "normal" adults. A set *E* contains some items that are new to *s* and others that *s* has previously encountered. If *s* is randomly presented from an item *e*∈*E* and has to determine whether or not she has already encountered *e*, *s* can reliably distinguish between "old" and "new" items. Most researchers now accept some version of a dual-process theory of recognition. Two prominent competing explanations are the following (note the interesting parallels with the mind-reading debate):

According to *RR*, recognition decisions are based on two processes that draw on two distinct sources of information: *recollection* of specific details and nonspecific feelings of *familiarity*. Recollection is used by default but, when such information is unavailable, subjects employ familiarity.

According to *RF*, recognition decisions are based on two processes that draw on two distinct sources of information: *recollection* of specific details and nonspecific feelings of *familiarity*. However, neither is the default process: the source of information employed depends on *specific contextual cues*.

RR and *RF* posit the same components to explain recognition decisions; what distinguishes them are the different interactions among those constituents. According to *RR*, recollection information is used by default to determine whether or not an item is "old," and familiarity is only reverted to when such information is unavailable. In contrast, *RF* implies that certain contextual cues will sometimes induce subjects to make familiarity-based recognition decisions even when recollection information is available.

To test these hypotheses, "pattern classifiers" are trained to determine the specific multivoxel patterns associated with recollection and familiarity processes. More precisely, classifiers are trained in tasks where experimenters are able to control which cognitive process is engaged, thereby explicitly meeting one of the linking conditions for reverse inference. For instance, in one experiment, which will serve as our main example, subjects were exposed to singular and plural words, such as "shoe" and "shoes" (Norman, Quamme, & Newman, 2009). These subjects were then scanned while performing recognition tasks involving previously examined items (e.g., a shoe) and unrelated lures (e.g., a bicycle). The recognition tasks were divided into two disjoint sets: *recollection blocks* and *familiarity blocks*. In recollection blocks, subjects were explicitly instructed to recall specific details of the mental image formed during the study phase, and to only answer "yes" if they were successful in that recollection. In contrast, in familiarity blocks subjects were instructed to only answer "yes" if they found the word familiar and to ignore any details they might recollect from the study phase. After a training phase, classifiers were able to reliably determine whether some multivoxel pattern of neural activation is an instance of recollection or familiarity.

What gives MVPA-based inferences a substantial advantage over traditional LRIs is that the reliability of the classifiers can be established within the experiment itself. In our example, this can be done by saving a subset of the recollection and familiarity blocks for later testing (so that they are not used at the "training stage") and then determining the rate at which the classifier correctly categorizes the corresponding neural patterns. This phase of the study, where experimenters can control which process is engaged provides the links between recollection, familiarity, and their corresponding multivoxel patterns that can then be used in reverse inference. Once these links are established, one can test competing hypotheses RR and RF in cases where the engagement of the subprocesses cannot be directly controlled.[h]

In a second phase of the study, subjects were scanned while being exposed to a mixture of previously observed items ("shoe" and "ball"), unrelated lures ("horse" and "box"), and previously unobserved switch-plurality lures ("balls"). The subjects' task was to determine whether the words they encountered were "old" or "new." To test the competing cognitive-level hypotheses, experimenters examined the subset of tested items for which subjects made correct recognition decisions. Note that, since these are cases where both recollection and familiarity information was available to subjects, RR and RF make different predictions about what is going to happen. According to RR, the classifier should categorize all the corresponding voxel patterns as recollections patterns, since this is the default. RF, in contrast, predicts a more variable classification, involving at least some instances of familiarity patterns, given that neither pattern should be used by default.

[h]The question arises whether we can extend the reliability of classifiers obtained from the testing phase to cases in which the experiments cannot determine the engagement of the psychological variables, since the latter inevitably involve some variation on the task. There are various studies which suggest that classifiers perform well under task variations. For example, in one study pattern classifiers were used to predict phonemes. The classifiers were still successful when presented with data from voices that were not presented in the learning phase (Formisano, De Martino, Bonte, & Goebel, 2008). Hence, at least this much variation in the task does not affect performance. In a study of visual working memory, classifiers were trained on data elicited by unattended gratings and then tested on whether they could also predict which of two orientations was maintained on working memory when subjects were viewing a blank screen. Again, their reliability was maintained despite the substantial difference in stimuli and tasks (Harrison & Tong, 2009). Indeed, testing for this kind of robustness relative to stimuli/task variation is usually taken as evidence that the brain region from which the data was obtained really does provide information about the function of interest (Tong & Pratte, 2012).

The MVPA experimental results support *RF* over *RR* (Norman et al., 2009). When both types of information are available, various contextual cues determine whether familiarity or recollection is used as the basis of a subject's recognition decision. In other words, contextual cues determine whether, according to the classifier, the multivoxel patterns underlying recognition decisions resemble more unambiguous familiarity or unambiguous recollection patterns.

Let us call a reverse inference based on pattern-decoding techniques such as MVPA, a *pattern-based reverse inference* (PRI). We conclude our discussion by emphasizing three substantial advantages that PRIs have over LRIs, regarding both experimental practice and philosophical implications. First, PRI allows for the reliability of classifiers to be determined within a phase of the same experiment in which they are employed. This feature of PRI fulfills the linking condition and allows experimenters to quantify their confidence in particular reverse inferences, thus providing a substantial advantage over LRI. As discussed in section "LRI: General Problems and Limitations," successful reverse inference presupposes the existence and availability of robust and accurate links between levels. The linking studies required to establish these bridge laws confronted LRI with several difficulties stemming, for the most part, from the lack of selectivity of most brain regions of interest. Consequently, when the differences between competing cognitive hypotheses are subtle, the linking studies often ignore or underestimate the resources available to one (or both) of these hypotheses. This problem becomes evident in the TT vs ST debate, where the functional interpretation of premotor areas (especially in *execution* tasks) according to TT is basically ignored. To make things worse, even if the linking studies presupposed in a specific LRI are, in themselves, nonproblematic, we still have to consider other tasks that might activate the brain regions of interest and ask, for each task, whether it might be relevant for the hypothesis under scrutiny. Readers familiar with meta-analyses of function-structure mappings know how frustrating and confusing these studies can be (Lindquist, Wager, Kober, Bliss-Moreau, & Barrett, 2012; Poldrack, 2011). It is important to realize that the problem is not primarily due to the resolution of the neuroscientific tools commonly employed in LRI, such as fMRI. In this respect, the example of mind-reading discussed in section "Location-Based Reverse Inference," is particularly instructive. The single-cell recording technique used in mirror neuron studies of macaques reveals quite starkly the fine-grained computational diversity of neurons in the premotor cortex. Hence, the main problem has to do with computational diversity, which goes all the way down to the function of single neurons, making it often difficult—or even impossible—to determine the precise degree to which an LRI should affect our confidence in a given hypothesis.

In contrast, MVPA and other techniques employed in PRI, decode cognitive processes from multidimensional vector patterns—as opposed to regions—of neural activation. As noted, when using these techniques, the reliability of the classifier underlying a particular reverse inference can be established within the experiment. To be sure, different kinds classifiers will vary in their success depending, among other things, on the type of task, the number and specificity of learning trials, the brain regions used for analysis (if restricted), and the nature of the machine learning algorithms (Poldrack et al., 2011). In the above recognition example, classifier accuracy was around 60%. However, in other tasks, such as those involving categorization of basic-level objects, the accuracy of classifiers can be much higher (Haxby, Connolly, & Swaroop Guntupalli, 2014).

A second advantage of PRIs follows from the fact that classifiers do not "assume" the functional localization of cognitive states or processes of interest. The multivoxel patterns used by classifiers can be distributed across the brain. Of course, experimenters can restrict the analysis to particular brain regions, especially if there are independent reasons to believe that a specific brain area is involved in some task, or if what is under scrutiny is the degree to which a region is responsible for executing some task. However, classifiers can also take nonlocalized, widely distributed multivoxel patterns. This is especially useful when comparing complex multistep processes that likely involve several brain regions. All the examples considered in this essay are of this complex kind, as are most models of higher cognitive capacities in general. The philosophical significance of this observation is that PRI does not fall prey to one of the oldest and most resilient objections against the traditional approach of cognitive neuroscience (including LRI), namely, that it assumes a strong and objectionable form of *functional locationism* (Nathan & Del Pinal, 2015). Speaking directly against this misconception, studies applying PRI can shed light on the degree of modularity, specialization, and functional localization of various cognitive processes and brain areas. For instance, MVPA studies consistently show that the ventral temporal cortex carries sufficient information for classifiers to accurately distinguish between animate and inanimate objects (Kriegeskorte, 2011). A similar approach may also help in studies previously explored via LRI. To illustrate, in our discussion of mind-reading, we have seen that the pattern and location of mirror neurons provide too little information to discriminate between competing cognitive hypotheses TT and ST. As some authors have pointed out, this might be because the neural locus of cross-modal action understanding is more widely distributed and likely includes nonmotor areas (Spaulding, 2012). This hypothesis can be explored with MVPA. Indeed, MVPA studies in this area show that most of the voxels used to reliably classify actions of the same type across modalities are distributed in areas that are *not* canonical human motor areas or homologs of the macaque F5 (Oosterhof, Tipper, & Downing, 2013).

A third advantage of PRI is that the information decoded from multivariate vector patterns does not depend on previous assumptions about the coarse function of a given brain region of interest. As discussed above, when using an LRI, assuming that some brain region has a specific function, coarsely identified, often does not help in adjudicating between subtly distinct hypotheses. The TT vs ST case provided a clear example, where the crucial difference lies between a motor-simulation process and an abstract representation of motor goals or execution processes. The debate between *RR* and *RF* illustrated an even subtler case, where the key distinction turns on the dynamics (as opposed to the individual components) of recognition decision processes. Once again, single-cell recordings, such as those recording the firing of mirror neurons in macaques, are extremely instructive since they highlight the fine-grained computational diversity of neurons and neural networks within a given brain region such as the premotor cortex. If one were to build neural networks composed of units with that much computational diversity, one would be able to develop processes that resemble the simulations of ST or the categorization processes of TT. Consequently, even fine-grained activation of these areas in tasks of *execution* and *observation* cannot, by itself, be used to determine which of these processes is actually implemented at the neural level. As shown by the recognition memory example, PRI is especially well-suited to solve these kinds of problems. By training a classifier to "learn" which multivoxel patterns are associated, say, with paradigmatic simulation and abstract categorization processes, a PRI can detect whether these diverse computational micro-units are working together in a way that is more like simulation or more like categorization in the context of a specific task (e.g., *observation*).

CONCLUSION

Cognitive neuroscience has recently faced theoretical, scientific, and popular backlash. Many philosophers and scientists have been consistently critical of the impact of neuroscience on the study of higher cognition (Coltheart, 2006a, 2006b, 2013; Fodor, 1974, 1997, 1999; Miller, 2008; Tressoldi, Sella, Coltheart, & Umiltà, 2012). Similar concerns have also been raised in various popular publications (Satel & Lilienfeld, 2013). This reaction is not completely unjustified. For every article providing neuroimaging evidence against deontological ethics or classical cognitivism, there is another one arguing that neuroscience has taught us nothing of relevance about higher capacities such as decision-making and language processing. Our discussion of reverse inference supports a measured optimism. Careful analysis of the role of mirror neurons in debates about the nature of mind-reading supports the positions of the "neuroskeptics." Indeed, respecting the linking condition of reverse inference is a tough challenge for virtually any LRI. Nonetheless, we tried to show how PRI provides theoretical tools for overcoming some of these problems. To be sure, this and other recent neuroscientific techniques raise a new host of challenges. Still, old adagios such as "lack of selectivity," "excessive functional locationism," and charges of implementing a "new phrenology" are not among them. As the field of neuroscience progresses, it is crucial that philosophers and theorists in general do not merely focus on traditional technologies and familiar objections, but also turn their attention to newer and potentially more powerful pattern-decoding techniques, such as PRI. In an attempt to fuel further discussion and constructive debate we suggested, in this chapter, a significant distinction between "two kinds of reverse inference."

References

Coltheart, M. (2006a). Perhaps functional neuroimaging has not told us anything about the mind (so far). *Cortex, 42*, 422−427.

Coltheart, M. (2006b). What has functional neuroimaging told us about the mind (so far). *Cortex, 42*, 323−331.

Coltheart, M. (2013). How can functional neuroimaging inform cognitive theories? *Perspectives on Psychological Science, 8*(1), 98−103.

Del Pinal, G., & Nathan, M. J. (2013). There and up again: On the uses and misuses of neuroimaging in psychology. *Cognitive Neuropsychology, 30*(4), 233−252.

Fodor, J. (1974). Special sciences (or: The disunity of science as a working hypothesis). *Synthese, 28*, 97−115.

Fodor, J. (1992). A theory of the child's theory of mind. *Cognition, 44*, 283−296.

Fodor, J. (1997). Special sciences: Still autonomous after all these years. *Nôus, 31*, 149−163.

Fodor, J. (1999). Let your brain alone. *London Review of Books, 21*, 68−69.

Fogassi, L., Ferrari, P. F., Gesierich, B., Rozzi, S., Chersi, F., & Rizzolatti, G. (2005). Parietal lobe: From action organization to intention understanding. *Science, 308*, 662−667.

Formisano, E., De Martino, F., Bonte, M., & Goebel, R. (2008). "Who" is saying "what"? Brain-based decoding of human voice and speech. *Science, 322*, 970−973.

Gallese, V., & Goldman, A. (1998). Mirror neurons and the simulation theory of mind-reading. *Trends in Cognitive Sciences, 2*(12), 493−501.

Gopnik, A., & Wellman, H. M. (1992). Why the child's theory of mind really is a theory. *Mind and Language, 7*(1−2), 145−171.

Harrison, S. A., & Tong, F. (2009). Decoding reveals the contents of visual working memory in early visual areas. *Nature, 458*, 632−635.

Haxby, J. V., Connolly, A. C., & Swaroop Guntupalli, J. (2014). Decoding representational spaces using multivariate pattern analysis. *Annual Review of Neuroscience, 37*, 435−456.

Hickok, G. (2014). *The myth of mirror neurons.* New York: Norton.

Hutzler, F. (2014). Reverse inference is not a fallacy per se: Cognitive processes can be inferred from functional imaging data. *Neuroimage, 84*, 1061−1069.

Iacoboni, M. (2009). The problem of other minds is not a problem: Mirror neurons and intersubjectivity. In J. A. Pineda (Ed.), *Mirror neuron systems* (pp. 121–133). New York: Humana Press.

Iacoboni, M., Molnar-Szakacs, I., Gallese, V., Buccino, G., Mazziotta, J. C., & Rizzolatti, G. (2005). Grasping the intentions of others with one's own mirror neuron system. *PLOS Biology, 3*, E79.

Kilner, J. M., & Lemon, R. N. (2013). What we know currently about mirror neurons. *Current Biology, 23*, R1057–R1062.

Kriegeskorte, N. (2011). Pattern-information analysis: From stimulus decoding to computational-model testing. *Neuroimage, 56*, 411–421.

Lindquist, K. A., Wager, T. D., Kober, H., Bliss-Moreau, E., & Barrett, L. F. (2012). The brain basis of emotion: A meta-analytic review. *Behavioral and Brain Sciences, 35*, 121–202.

Machery, E. (2014). In defense of reverse inference. *British Journal for the Philosophy of Science, 65*(2), 251–267.

Miller, G. (2008). Growing pains for fMRI. *Science, 320*, 1412–1414.

Nathan, M. J., & Del Pinal, G. (2015). Mapping the mind: Bridge laws and the psycho-neural interface. *Synthese*, (Online).

Norman, K., Quamme, J., & Newman, E. (2009). Multivariate methods for tracking cognitive states. In K. Rosler, C. Ranganath, B. Roder, & R. Kluwe (Eds.), *Neuroimaging of human memory: Linking cognitive processes to neural systems*. New York: Oxford University Press.

Oosterhof, N. N., Tipper, S. P., & Downing, P. E. (2013). Crossmodal and action-specific: Neuroimaging the human mirror neuron system. *Trends in Cognitive Sciences, 17*(7), 311–318.

Poldrack, R. A. (2008). The role of fMRI in cognitive neuroscience: Where do we stand? *Current Opinion in Neurobiology, 18*, 223–227.

Poldrack, R. A. (2011). Inferring mental states from neuroimaging data: From reverse inferences to large-scale decoding. *Neuron, 72*, 692–697.

Poldrack, R. A., Mumford, J. A., & Nichols, T. E. (2011). *Handbook of functional MRI data analysis*. Cambridge: Cambridge University Press.

Rizzolatti, G., & Craighero, L. (2004). The mirror-neuron system. *Annual Review of Neuroscience, 27*, 169–192.

Rizzolatti, G., Fadiga, L., Gallese, V., & Fogassi, L. (1996). Premotor cortex and the recognition of motor actions. *Cognitive Brain Research, 3*, 131–141.

Rizzolatti, G., Fogassi, L., & Gallese, V. (2001). Neurophysiological mechanisms underlying the understanding and imitation of action. *Nature Reviews Neuroscience, 2*, 661–670.

Rizzolatti, G., Fogassi, L., & Gallese, V. (2009). The mirror neuron system: A motor-based mechanism for action and intention understanding. In M. Gazzaniga (Ed.), *The cognitive neurosciences* (Vol. IV, pp. 625–640). Cambridge, MA: MIT Press, Chapter 43.

Rizzolatti, G., & Sinigaglia, C. (2010). The functional role of the parieto-frontal mirror circuit: Interpretations and misinterpretations. *Nature Reviews Neuroscience, 11*, 125–146.

Satel, S., & Lilienfeld, S. O. (2013). *Brainwashed: The seductive appeal of mindless neuroscience*. New York: Basic Books.

Shallice, T., & Cooper, R. (2011). *The organization of mind*. Oxford: Oxford University Press.

Spaulding, S. (2012). Mirror neurons are not evidence for simulation theory. *Synthese, 189*, 515–534.

Tong, F., & Pratte, M. S. (2012). Decoding patterns of human brain activity. *Annual Review of Psychology, 63*, 438–509.

Tressoldi, P. E., Sella, F., Coltheart, M., & Umiltà, C. (2012). Using neuroimaging to test theories of cognition: A selective survey of studies from 2007 to 2011 as a contribution to the decade of the mind initiative. *Cortex, 48*, 1247–1250.

Umiltà, M. A., Kohler, E., Gallese, V., Fogassi, L., Fadiga, L., & Keyser, C. (2001). I know what you are doing: A neurophysiological study. *Neuron, 31*, 155–165.

9

The Neuroscience of Ethics Beyond the Is-Ought Orthodoxy: The Example of the Dual Process Theory of Moral Judgment

N. El Eter

Paul Valéry University of Montpellier, Montpellier, France

INTRODUCTION

Since the declaration of the 1990s as the Decade of the Brain, neuroscience has become more popular and has started to deal with all phenomena and manifestations of human cognition. It has also started to deal with issues that were for a long time a privileged space of traditional disciplines, especially philosophy. Being one of the last themes that philosophy has retained, ethics became an object of study in cognitive neuroscience. This cognitive approach guided and optimized by the advancement in neuroimaging—notably fMRI—inaugurated a "neuroapproach" that competes with traditional methods of disciplines. In this direction, neuroscience as a new way to understand human morality will face many philosophical implications. One of these problems is the Is-Ought problem. Ethics is supposed to study and to deal with what we "ought to do," based on normative reflection that must

The Human Sciences after the Decade of the Brain.
DOI: http://dx.doi.org/10.1016/B978-0-12-804205-2.00009-4
© 2017 Elsevier Inc. All rights reserved.

give us a rule or a line to conduct notably to resolve a moral dilemma. While neuroscience deals with what "is" happening in the brain to understand the function of our "cognitive machine," so it deals with facts by describing it.

The Is-Ought problem became one of the classic philosophical problems in the 20th century. Before developing into a classic or conventional problem, every philosophical problem is nourished by the cultural context of its time. Before going into the contextual details, I will describe what the Is-Ought problem is.

Put simply, we can say that the Is-Ought problem relies on a dichotomy that highlights the difference between describing things that occur, take place, can be experienced, or exist, and judging how the things, mostly in a moral sense, ought to be. What makes these two statements problematic is that we cannot infer an "Ought" statement from an "Is" statement. For example, aggression is a part of human nature, so humans ought to be aggressive.

The Is-Ought-distinction goes back to the philosopher David Hume (1711–76) and to his book *Treatise of Human Nature*, more specifically book III, where he develops his moral philosophy (Hume, 1739–40/2014). Since Hume, the Is-Ought problem was subject to many transformations and adaptations before taking its actual form in contemporary discussions in neuroethics.

Actually, this problem is usually cited together with the "naturalistic fallacy" introduced by G.E. Moore (1873–1958) in his book *Principia Ethica* (1903/1982). Moore is concerned with the definition of the good as a moral concept with natural properties, whereas Hume's problem is about the problem of inferring an "Ought" statement from an "Is" statement. Even if the two problems accept a fact/value dichotomy they still have many differences and implications. In this perspective, for Ruwen Ogien, Hume's problem is logical while Moore's problem is conceptual (Ogien, 2004).

But what is the relation between this classic problem and neuroethics?

The Is-Ought problem is part of neuroethics, more specifically its subdiscipline neuroscience of ethics, since the emergence of the research field. As agreed in a large part of the relevant bibliography, one can distinguish two subdisciplines of neuroethics. These two domains are the ethics of neuroscience and the neuroscience of ethics (Roskies, 2002), or simultaneously applied neuroethics and fundamental neuroethics (Evers, 2009). While the first one is a specialized branch of bioethics concerning neurosciences, its uses and its social implications, the second one is concerned with the identification of the neural basis of human morality, more specifically human moral judgment. As Adina Roskies anticipates, the ethics of neuroscience will progress faster than the neuroscience of ethics because the latter is facing more epistemic and philosophical problems, and as she said, "will be the area with truly profound implications for the way ethics, writ large, is approached in the 21st century" (Roskies, 2002).

For reasons of clarity, the term "neuroscience of ethics" is chosen in this chapter. Actually, we can have a variety of alternative terminologies to refer to studies that use cognitive neuroscience to explain the function of moral judgment. The issue of terminology seems crucial because each of the terms "neuroscience" and "ethics" could have several alternative meanings. Take for example, the variety of headlines of articles related to the subject: "The Neurobiology of Moral Behavior" (Mendez, 2009), "The Neural Basis of Human Moral Cognition" (Moll, Zahn, de Oliveira-Souza, Krueger, & Grafman, 2005), "The Neural Bases of Cognitive Conflict and Control in Moral Judgment" (Greene, Nystrom, Engell, Darley, & Cohen, 2004), and "Neuropsychology of Moral Judgment and Risk Seeking: What in Common" (Balconi & Terenzi, 2012). In other words, we can find under several terms a large corpus of studies that treat the same object. On the other hand, choosing the term "neuroscience of ethics" refers directly to conventions of the "legitimate"[a] discipline "neuroethics," which is a central reason to adopt it in this chapter.

The Is-Ought problem applied to neuroscience of ethics concerns the research field's central idea of understanding the neural basis of our moralities. The simplified philosophical application of the problem is the following: how can neuroscience explain what "Is" happening in our brain while we are trying to judge morally what we "Ought to do"?

Proponents of the Is-Ought distinction rely on an argument of incompatibility and a contradiction between the method and the subject studied. This contradiction exists, because neuroscience is considered a scientific instrument able to describe the world as a fact, while ethics and morality rely on the world of normativity where the central question is "what we ought to do?" with regard to ourselves and others. It is about the distinction between the world as it is and the world as we would like it to be.

In this chapter I will try to demonstrate how Hume's Is-Ought problem and Moore's "naturalistic fallacy" were taken out of their original context, before becoming an orthodoxy. The first part of this chapter is devoted to clarifying the Is-Ought problem and the naturalistic fallacy and to demonstrating how they are used in different contexts by their authors. For an example I will turn to Richard Joyce's book *The Evolution of Morality* (Joyce, 2006) and his critique of the "naturalistic

[a]In the sense that we can find a large bibliography, research centers, and academic degrees (e.g., M.A. in Neuroethics at George Mason University in the United States).

fallacy" and show that some philosophers have made a real fallacy by confusing the two problems of Hume and Moore. Part one of the contribution will be based on the point of view expressed by Joyce, in particular in Chapter 5 in *The Evolution of Morality*.

The second part is devoted to criticizing the application of this "orthodoxy" as an argument against the neuroscience of ethics. The central idea is that the neuroscience of ethics is a very complex and interdisciplinary scientific endeavor that cannot be reduced to a simple opposition between morality and nature, norms and fact. In addition, the interconnection between the three main disciplines neuroscience, psychology, and philosophy, exceeds the simple distinction between a descriptive "Is" and a normative "Ought." The goal is to demonstrate that the "Is" is a complex construction.

PART ONE: IS-OUGHT ORTHODOXY

Hume's Is vs Ought Distinction

The aim of this section is not to elaborate on Hume's moral philosophy, but to attempt to show how the Is-Ought problem is elaborated on in his way of thinking about morality. In particular, because of the expanded nature of the subject, it will be limited to the way "Is" and "Ought" appear in the Hume's text and how these terms are used. In his masterpiece *Treatise of Human Nature* Hume dedicated book III, "Of Morals," to the discussion of the nature of human morality. Actually Hume is more interested in the function of morality, its components and its mechanisms. For Hume, concerning morality, reason has a secondary role in translating a moral judgment into a motivation. According to Rawls, Hume's moral philosophy is a kind of "morality psychologized" (Rawls & Herman, 2000), in the sense that he is trying to identify the function of our morality. More specifically, he wants to describe the process of deliberations conducted by the system of passion. This common interpretation, adopted by Rawls, led the argument that it is passion and not reason that guides human morality. This idea is decisive in Humean moral philosophy so that it constitutes the center of part 1 of book 3. The Is-Ought problem emerges exactly in this context, where Hume is trying to get rid of reason concerning morals.

Where and in which context does the Is-Ought distinction appear in Hume's moral philosophy?

In his book *Hume's Place in Moral Philosophy*, Nicholas Capaldi devotes a chapter to the clarification of the Is-Ought problem, which is called "Hume's Rejection of the Traditional Moral "Ought." As with all philosophical problems, the so-called "Hume's law" can be interpreted in many ways, but the most common interpretation remains that Hume introduced a distinction "between what is and what ought to be in a moral context" (Capaldi, 1989, p. 56) and their relationship.

The famous distinction is drawn in the last paragraph[b] of book 3, part 1, section 1 of the treatise entitled "Moral Distinctions Aren't Derived From Reason." The main idea is introduced by Hume as an "observation" on the way that "every system of morality" proceeds, by starting its reflection using the "usual copula 'Is'" and then moving to those "connected by 'Ought'." This move is conceived by Hume as a deduction. So the "Is" concerns the domain of observations considered by reason, while an "Ought" concerns morality and the distinction between "vice and virtue." Each of these is distinguished by a special impression like agreeable and unpleasant.

In sum, according to Capaldi, Hume rejects the use of ought as a proposition in moral systems. More specifically, he rejects the unjustified deduction of an "ought" or "ought not" from an "is" or "is not." So making this deduction transforms "Ought" into a sort of proposition that expresses the meaning of "Is." In the context of his moral philosophy, which limits the role of reason in moral judgment, we cannot in any way be based on the "Is" conceived by the reason. According to Capaldi, Hume had Clarke in mind when he was criticizing the moral system (Capaldi, 1989).

To summarize, we can say that the Is-Ought problem is deeply connected with what we can call Hume's philosophical system, especially with the way he defines reason and passion. Effectively, treating this problem in its context will assist in worsening the interconnected problems related to Hume's moral philosophy in the first place and more generally his philosophical system. We need to clarify that Hume was criticizing the way that moral philosophers of his time proceeded.

[b]"In every system of morality, which I have hitherto met with, I have always remarked that the author proceeds for some time in the ordinary ways of reasoning, and establishes the being of a God, or makes observations concerning human affairs; when all of a sudden I am surprised to find, that instead of the usual copulations of propositions, is, and is not, I meet with no proposition that is not connected with an ought, or an ought not. This change is imperceptible; but is however, of the last consequence. For as this ought, or ought not, expresses some new relation or affirmation, 'tis necessary that it should be observed and explained; and at the same time that a reason should be given, for what seems altogether inconceivable, how this new relation can be a deduction from others, which are entirely different from it. But as authors do not commonly use this precaution, I shall presume to recommend it to the readers; and am persuaded, that this small attention would subvert all the vulgar systems of morality, and let us see, that the distinction of vice and virtue is not founded merely on the relations of objects, nor is perceived by reason" *Treatise of Human Nature*, book 3, part 1, section 1.

That's why the "Is-Ought" distinction fits only formally with neuroethics and the possibility to evaluate moral thinking by describing the way in which human brains function. To put it differently, Hume's version of the problem does not treat the epistemological relation between science and morality.

Moore's Naturalistic Fallacy

As mentioned in the introduction, the Is-Ought problem in usually correlated with another problem introduced by G.E. Moore in *Principia Ethica* (1903/1982): the naturalistic fallacy. Moore designated the type of error made especially by the supporters of social evolution, who were trying to define the GOOD in naturalistic terms, as naturalistic fallacy. Moore attacks Herbert Spencer the precursor of social Darwinism: "It is absolutely useless, so far as Ethics is concerned, to prove, as Mr Spencer tries to do, that increase of pleasure coincides with increase of life, unless good means something different from either life or pleasure" (Moore, 1903/1982, section 12). The crucial point of Moore's critique of the naturalism approach of ethics is the definition of Good. Whereas for him ethics is primarily about good that we cannot be defined by (referring to natural property), according to Moore, "the naturalistic fallacy always implies that when we think 'this is good' what we are thinking is that the thing in question bears a definite relation to some one other thing. But this one thing, by reference to which good is defined, may be either what I may call a natural object—something of which the existence is admittedly an object of experience—or else it may be an object which is only inferred to exist in a supersensible real world" (Moore, 1903/ 1982, section 25). In order to prove that good is indefinable he builds on the open question argument, which consists of "whatever definition be offered, it may be always asked, with significance, of the complex so defined, whether it is itself good" (Moore, 1903/1982, section 13). For example, if we say "Pleasure is good" we still can ask "Is pleasure really good?"

While Hume's problem is to reduce and eliminate the role of reason in moral sense, Moore rather focuses on the way that good could be defined. In addition, Hume criticizes the logic of moralists of his time while Moore attacks a specific doctrine that tries to define "Good" using naturalistic terms. Hume's issue is more about the way we think about morality to reach what we "ought to do," while Moore's issue is the definition and the essence of "Good." In fact, the two problems differ in the way in which they approach morality, in their goals, and in their aims.

Concerning the relation between the naturalistic fallacy and the Is-Ought distinction, Joyce denies that Moore mentioned the Is-Ought problem. In fact, the "naturalistic fallacy," as "Moore conceived it concerns an entirely different matter" (Joyce, 2006, p. 147). According to Joyce "oughtiness" is not the "crucial factor of naturalistic fallacy." Even if he denies that naturalistic fallacy is about the Is-Ought problem, Joyce does not escape the concomitance between the two problems by sacrificing a whole section for the Is-Ought gap.

Is-Ought and Neuroethics

The Is-Ought dichotomy comes back to light with neuroethics, which is very fertile ground to explore it. In neuroethics we are somehow facing the disciplinary unification of science and ethics. According to the logic maintained by supporters of the Is-Ought distinction, we cannot understand ethics by using scientific methods, because they are incompatible with each other. Nevertheless, studies that rely on experimental data (e.g., neuroimaging and psychological tests) are designed to study what we can call components of morality, like moral judgment, or moral intuitions, for example. The question that arises in this context concerns the kind and the aspects of ethics explored by neuroscientific investigations. In other words, what kinds of "Is" and "Ought" are involved in the neuroscience of ethics?

In his article "From Neural 'Is' to Moral 'Ought': What Are the Moral Implications of Neuroscientific Moral Psychology?" Joshua Greene distinguishes the relation between neuroscience and normative ethics on the one hand and neuroscience and metaethics on the other. In fact, looking to the nature of "Is" represented by neuroscience and "Ought" represented by normative ethics is crucial to know if the Is-Ought gap is somehow adapted to neuroethics.

In his article, Greene maintains the difference between the nature of science and the nature of ethics. Despite this, he defends the position that "nonetheless [...] science and neuroscience in particular, can have profound ethical implications by providing us with the information that will prompt us to reevaluate our moral values and our conception of morality" (Greene, 2003). In fact, Greene is not against the Is-Ought distinction, he is skeptical "of attempts to derive moral principles from scientific facts" (Greene, 2003). However, Greene does not deny the moral implications that a scientific understanding of moral psychology might necessitate. So he maintains the relation of incompatibility between "Is" and "Ought" but not from a Humean perspective. In the sense that he is not treating the subject as a logical relation of inference, but, if I well understood it, as if neuroscience provides for philosophy a large data about human nature, which in itself does not imply any normative circumstances. But that could be helpful for philosophy in the way it invests it. The importance of Greene's position is that it allows us to realize an interdisciplinary research agenda using psychology and neuroimaging data to shed light on philosophical questions, e.g., why we have a different reaction on different scenarios that relies the same moral problem, like the tramway dilemma.

PART TWO: NEUROSCIENCE OF ETHICS BEYOND THE "IS-OUGHT" PROBLEM

In this section, I will tackle what I call the experimental architecture of the neuroscience of ethics. With the "experimental architecture," I mean the complex interdisciplinary and interconnecting constructions that allow researchers to offer a "scientific" understanding of moral judgments. But where does this scientific aspect come from?

Normally, the scientific character is based on the use of neuroimaging. But neuroimaging itself is problematic, especially if it concerns complicated cognitive tasks such as judging morally in a certain context. The complication comes from the place occupied by neuroimaging in cognitive neuroscience, and the problematic relation between neuroscience and psychology. Concerning this last point, the coincidence of the emergence of neuroethics with the revival of moral psychology is also very peculiar. In what follows, I argue that the "Is" offered by the fundamental neuroethics is built and produced as a function of the relation between cognitive neuroscience and cognitive psychology and that this relation is not isolated from philosophical and epistemological considerations.

For my analysis, I will draw on a paper by Guy Tiberghien published in French called "Entre Neuroscience et Neurophilosophie: La Psychologie Cognitive et Les Sciences Cognitive"[c] (2007) and on two papers by Colin Klein, "The Dual Track Theory of Moral Decision-Making: A Critique of the Neuroimaging Evidence" (2011) and "Philosophical Issues in Neuroimaging" (2010b). My aim is to show that the nature of the new area of research labeled the "neuroscience of ethics" is equivocal and ambiguous.

To this end, nothing could be more clarifying than looking into the interdisciplinary construction of the neuroscience of ethics, a model where we find three levels: neural, psychological, and philosophical. The dual process theory of moral cognition developed by Joshua Greene in 2001 and recently published in a book entitled

---

[c]"Between Neuroscience and Neurophilosophy: Cognitive Psychology and Cognitive Science."

Moral Tribes: Emotion, Reason, and the Gap Between Us and Them (Greene, 2013) is a very representative model of this fashionable discipline. But why this model in particular?

Despite the philosophical and methodological critiques of Greene's theory (Berker, 2009; McGuire, Langdon, Coltheart, & Mackenzie, 2009), it is still representative for many reasons. First, Greene comes from a philosophical background and uses neuroimaging to answer a philosophical question, which is concretized by the model in question. Based on that model, he tries to find the moral philosophy that fits best with our psychology I will not say much about the moral philosophy he defends, because the focus of this chapter is the "industry" of how we function morally. Second, Greene's model "grew up" in the context of neuroethics.

The Dual Process Model of Moral Cognition

In this section, I will try to introduce Greene's model of dual process morality. Theories of this type have been developed in psychology since the 1970s (Frankish & Evans, 2009) and have become more popular and trendy at least since the publication of Kahneman's book *Thinking, Fast and Slow* in 2011. The main idea is that our minds can use two processes to accomplish the same cognitive task. Effectively, moral judgment, here conceived of as cognitive task, can be accomplished in two different ways: the first is possible due to reason control, and the second due to emotional mechanisms. Every mechanism or mode to judge morally is activated by several parameters concerning the nature of situations, personal experience, and so on.

To put it simple, Greene uses a very intelligent example: the metaphor of a digital camera. What he calls in his book a "moral machinery" operates like a digital camera with two basic modes: manual and automatic (Greene, 2013). Additionally, Greene uses four levels of descriptions for each of the two modes, neural, psychological, philosophical, and evolutionary. The manual-reason mode depends on the dorsolateral prefrontal cortex, and is associated with slow, utilitarian, and flexible reasoning, while the automatic-emotion mode depends on the ventromedial prefrontal cortex, and is associated with fast, deontological, and efficient reasoning.

The second point that remains very crucial in this representative theory is the implications of this psychological model for moral philosophy. In other words, how might these "new" considerations about moral cognition influence our moral reasoning? Does this model give some preference to one moral philosophy over another?

Before presenting how Greene answers these kinds of questions, we must take note of the introduction to his book *The Tragedy of Commonsense*

Morality and its first chapter "The Tragedy of Commons." These two "tragedies" are used to frame his theory and his reflections on the moral problems we face in real life. The tragedy of commonsense morality is responsible for our contemporary moral problems, which arise from disagreement about moral values. This tragedy is about *conflicting moralities* that disable different groups to live together and prosper (Greene, 2013, p. 26), which is why the problem is between "Us and Them." But our brains are designed to suffer the inevitable tragedy of commons, "which is the tragedy of selfishness, a failure of individuals to put US ahead of ME" (Greene, 2013, p. 25). The tragedy of commons concerns the inability for cooperation inside or between the groups. Hence, we need philosophy to help our manual mode resolve the contemporary moral problems that our automatic morality is not designed to tackle. Accordingly, Greene defends a "deep pragmatism"[d] as the philosophical solution. As our automatic moral system does not fit with contemporary moral problems what we need is to activate our manual mode concerned with reflection. Deep pragmatism for Greene means "to seek agreement in shared values... We aim to establish a common currency for weighing competing values. This is once again, the genius of utilitarianism, which establishes a common currency based on experience... Finally, we can turn this moral value into a moral system by running it through the outcome-optimizing apparatus of the human prefrontal cortex... But fortunately for us, we all have flexible manual modes" (Greene, 2013, p. 291).

In the following section, I'll discuss the methodological issues related to the use of neuroimaging in cognitive neuroscience, in order to prove that the "Is" offered in this discipline is disputable. Does neuroimaging describe what "Is" really happening in the brain while we perform a cognitive task?

The Limits of Neuroimaging

The technical developments in neuroimaging, notably fMRI, contribute exceedingly to the popularization of what we can call the neuroscientific approach in many disciplines as well as to the emergence of neuro-disciplines.[e] When neuroscientists and other researchers, notably

[d]I consider deep pragmatism as the normative part of Greene's work, and it can be discussed independently of his experimental work. As I'm interested in this chapter about neuroscience of ethics I will not cover the details. For more information I invite you to look to Greene's Chapter 11 in his book *Moral Tribes: Emotion, Reason, and the Gap Between Us and Them*.

[e]By neuro-discipline I refer to a traditional and institutionalized discipline that applies neuroscientific methods. Examples of neuro-disciplines are neuroeconomics, neurophilosophy, and neuroethics.

in human and social sciences, involve "the use of the brain as a means to answer old questions and/or to open up new avenues of inquiry about society, culture, or human behavior" (Johnson & Littlefield, 2011), Littlefield and Johnson call them neuroscholars. Neuroscolars' use of cognitive neuroscience transforms many research areas by creating new neuro-disciplines. But is neuroimaging, as the most popular method used by neuroscholars, reliable?

Actually, neuroimaging is not as simple as it seems to be by its popularity. Referring to some studies in neuroscientific research, one might get the impression that just by looking at what happens in the brain one can come to understand how it works. However, in his article "Images Are Not the Evidence in Neuroimaging" Colin Klein raises the issue of the images' reliability. He holds a skeptical position about the neuroimages' self-evidence (Klein, 2010a, 2010b). According to him, the central question is how we interpret these images. Apparently, neuroimaging has its own methodological issues. However, my aim is to show the complex relation between three disciplinary levels in the neuroscience of ethics. In other words, my aim is to show that neuroscience might not be the bedrock in the neuroscience of ethics.

Coming back to the work of Klein, I will refer to three papers published between 2010 and 2011 that treat the methodological limitations of theories based on neuroimaging data. In addition to the article mentioned above, there are two other articles "Philosophical Issues in Neuroimaging" (2010b) and "The Dual Track Theory of Moral Decision-Making: A Critique of the Neuroimaging Evidence" (2011). I will start with the last one, because it is directly concerned with the work of Greene, even though it was published in 2011 and is, hence, limited to the psychological and experimental part of Greene's theory. It was clearly said by Klein that "the neuroimaging data itself is problematic, and does not support the dual-track theory" (Klein, 2011). To draw a pattern, we can say that actually Greene is searching in neuroimaging for evidence that fits with his own dual process theory.

But the real question remains to identify the role played by neuroimaging in the model. Was the intervention of neuroimaging crucial in the way Greene conceives the model? Keep in mind that Greene's work occurs as a neuroimaging investigation (Greene, Sommerville, Nystrom, Darley, & Cohen, 2001), so neuroimaging is introduced as the method that will resolve classical debates in moral philosophy. Klein's first critiques concerning the way Greene deals with neuroimaging is that he appeals to the controversial method of "reverse inference." According to reverse inference[f] one can "identify brain regions, which are active

---

[f]See also the chapter by Del Pinal and Nathan "Two Kinds of Reverse Inference in Cognitive Neuroscience" in this volume.

during a particular kind of moral task, and then treat these activations as evidence for a particular cognitive theory" (Klein, 2011). This is what Tiberghien calls theory-laden (Tiberghien, 2007), because one establishes a theory first and looks for the function of the brain afterward. Proceeding this way, one will always find exactly what one was searching for. In what follows, I will treat the question of theorizing and interpretation of neuroimages in connection with the statements of cognitive psychology.

What are the effects of reverse inference in the case of the dual process theory of morality? I will try to simplify Klein's idea, so I won't describe the technical aspects of his critiques. The problem emerges whenever we associate a brain region with a cognitive task. In his paper, Klein demonstrates with several examples that "none of the areas identified by Greene et al. are specific to either emotion or cognition" (Klein, 2011). For example (Klein refers to five problematic areas cited by Greene), the precuneus and the posterior part of the cingulate cortex are specific for emotion, but some studies show that these brain areas are also associated with self-referential processing.[8]

In another section of the article Klein attacks Greene's theory from a different position. He shows that the same neuroimaging evidence could be more coherent with another model. This model is proposed as an alternative to the dual process model, and involves only one single process that engages a set of different brain areas. The opposition between the two models is described by Klein as follows: "On the single-track view moral decision-making is the process of integrating a set of different considerations; on the dual track view, it can only be making a decision about which track will win out" (Klein, 2011).

To conclude, neuroimaging data are controversial because a lot of work is needed for interpretation. The "Is" described here is biased by several methodological and theoretical assumptions. In other words, just looking to the areas activated in the brain won't tell us how we judge morally, and absolutely not what is normatively good or bad. The portrait of a neuroscience of ethics as drawn above is confused and incomplete. However, one has the impression that this does not concern cognitive neuroscience as a whole. In addition, in the next section I argue for the following thesis: a big part of the theorizing, which allows the interpretation of neuroimaging data, must be done in psychology, and I will discuss the problematic relation between cognitive neuroscience and cognitive psychology.

---

[8]Self-referential processing is when we analyze some experience by putting ourselves in the place of others, which is very possible in the case of ethical scenarios.

The Gap Between Psychology and Neuroscience

In his article, Tiberghien uses the approach of a cognitive psychologist to deal with the status of psychology in cognitive science. The latter is seen as "the locus of institutional confrontation between several academic disciplines" (Tiberghien, 2007, translation by the author).

Cognition occurs at the same time on the cerebral, the cognitive, and the behavioral level. The problem about the relation between neuroscience and psychology is the level of interpretation of cognition, and trying to use all these levels together in one single theory is extremely difficult.

Seen from another perspective, the problems referred to in the previous section highlight the status of psychology next to neuroscience. Somehow we need to translate this cerebral activation into a coherent "language" that fits with cognitive psychology. Let's ask questions simply.

What is the relation between cognitive psychology and cognitive neuroscience? Evidently, it is cognition. But how can we distinguish between a psychological approach and a neural approach when both tend to study the same object?

Evoking the theoretical problems of neuroimaging,[h] Tiberghien asks the following question: "what would be cognitive neuroimaging without behavioral psychology and cognitive psychology?" (Tiberghien, 2007, p. 292; Translation by the author). He is more radical than Klein concerning the prospects of neuroimaging; his position is clearly expressed in the following sentence: "we cannot see in the brain more than what one brings to it" (Tiberghien, 2007, p. 292; Translation by the author). In other words, we cannot but start a neuroimaging investigation armed with a theoretical and psychological position. Therefore the interpretation of images is highly influenced by the degree of precision of our purpose concerning the cognitive entities investigated. To illustrate the scene we could say that, in psychology, we prepare the material to test. Effectively, cognitive tasks need to be well identified before the neural processes associated with them can be investigated.

Cognition has several levels: cerebral, cognitive, and behavioral. However, the relations between the different levels influence and eventually undermine the relation and the frontiers between the disciplines of cognitive psychology and cognitive neuroscience (Tiberghien & Jeannerod, 1995; Translation by the author). Hence, we can say that cognitive psychology and cognitive neuroscience cannot escape one

---

[h]Tiberghien supports Klein's position but in a different context, e.g., "researchers constantly cast their cognitive models in the metabolic activity of the brain" (Tiberghien, 2007).

another, because each of the two does not refer to an independent level. Each level, instead, is conditioned by the other one.

Let's turn back to see how this could be illustrated in the case of neuroethics. Actually, morality considered as a cognitive task has its own specificity, in the sense that it cannot be reduced in any way to a simple cognitive task or entity. The engaged mental processes to accomplish a simple cognitive task, for example memorizing a range of things, can be easily identified. At the same time, a moral cognitive task engages a set of mental processes, which could also relate to another cognitive phenomenon. For example, asking the moral permissibility for a situation could at the same time launch an emotional response to judge it and a rational response to explain the judgment. In other words, moral cognition could not be reduced and identified with a set of mental processes. Even more, moral psychology as a "discipline" relies on philosophy. Let's take a look at moral psychology in its contemporary version. In the introduction of *The Moral Psychology Handbook* Doris defines the field, as "a hybrid inquiry, informed by both ethical theory and psychological fact" (Doris, 2010, p. 1). Obviously, it is a streamlined definition that does not inform us about how the two levels interact and how they merge to give a coherent comprehension of moral phenomena. In sum, one gets the impression that in moral psychology one is currently talking more about studies of some mechanisms underlying psychological components involved in human morality such as moral motivation, emotions and intuitions, and the like.

The problems concerning frontiers of subdisciplines or transdisciplines like neuroethics mirror an unsolvable theoretical issue in the main disciplines. Understanding this requires going back to the role of cognitive psychology and cognitive neuroscience in shaping the project of cognitive science. However, what I mentioned are just "flashes" that destabilize the "myth" of a neural knowledge as a science of what "Is" really happening in the brain.

Let's go back to Greene's model to try to translate what was mentioned previously. It is very interesting sometimes to see how things are introduced to us. Details matter when drawing a portrait of an interdisciplinary engagement that produces a neural understanding of morality. Is it possible to overcome psychology in the current state of affairs? Could we have an independent discourse between neuroscience of morality and moral psychology?

I think we are really at the point of abandoning a multilevel discourse between both neuroscience and psychology. In his article defining "The Cognitive Neuroscience of Moral Judgment and Decision Making," Greene himself supports the following position: "the prospect of understanding moral judgment in physical terms is especially alluring, or unsettling, depending on your point of view" (Greene, 2014).

II. THE NEUROSCIENCES OF SOCIAL SCIENCES AND ETHICS

A few lines later, Greene builds on a very broad conception of the neuroscience of morality, assuming that the discipline "provides a set of useful *entry points* into the broader problems of complex cognition and decision making" (Greene, 2014). If I understand his position correctly, what is meant is that morality is provided by a set of cognitive functions and what neuroscience is supposed to do is to explain how morality works. But the question remains: Can this neural understanding ever be reliable and evident?

According to Greene, "To truly understand the neuroscience of morality, we must understand the many neural systems that shape moral thinking, none of which, so far, appears to be specifically moral" (Greene, 2014). In this view, there is some work to be done before and after the intervention of neuroscientific investigation.

In sum, we can conclude that the contribution of cognitive neuroscience to explain the function of morality still remains to be proven. Nevertheless, skeptical discussions about the actual state of things must contribute to its development. This will bring about rich and promising perspectives, not only for neuroethics, but also for philosophy and psychology.

CONCLUSION

My aim in this chapter was to show that even though a folk application of the Is-Ought problem as a critique of the neuroscience of ethics may sound logical and legitimate, it doesn't fit here. For this, I wanted to represent the neuroscience of ethics as an interdisciplinary discipline, wherein borders and frontiers are very fragile and delicate to identify. Actually, the discipline underlies an important issue that cannot be reduced to but exceeds the Is-Ought problem. Is this a good or a bad starting point for an emerging discipline?

I am not sure how to answer this question! However, epistemological studies concerning the interdisciplinary interaction in neuroscience are very rare. Studies (and critics) are more concerned with the value and the coherence of a "psychological" model based on neuroimaging investigation, on the one hand, and with the denial of the legitimacy of this kind of studies on the other.

Even if academic tradition imposes to assume a position for or against the issue discussed, this seems confusing in the case of the neuroscience of ethics. For this we would need to have a clear view of the issue, but we are far from that. However, even if actual studies are based on a limited and contextual paradigm,[i] we cannot deny their

[i]I mean here the paradigm of cognitive neuroscience prevailing in contemporary studies.

importance, especially because they crystallize and unveil some unsolved issues.

What is interesting concerning neuroethics is not just the scientific understanding that it has to offer. What is much more fascinating is the way the study of our own morality is performed and changed in accordance with the evolution of the way we see the world. And, the questions remain, not only about the fact that we could know how our cognition functions, but how we can develop our methods and toolboxes to build knowledge, to reveal and to access this function.

Neuroscience concerns how science progresses according to society, culture, and the science itself.

Acknowledgments

I am grateful to Pascal Nouvel, Jon Leefmann, and Paolo Stellino for reading and commenting on a previous draft of this chapter.

References

Balconi, M., & Terenzi, A. (2012). Neuropsychology of moral judgment and risk seeking: what in common? *Neuroscieace and the Economics of Decision Making, 5*, 86–108.

Berker, S. (2009). The normative insignificance of neuroscience. *Philosophy and Public Affairs, 37*(4), 293–329.

Capaldi, N. (1989). *Hume's place in moral philosophy*. New York: P. Lang.

Doris, J. M., & Moral Psychology Research Group (2010). *The moral psychology handbook*. Oxford: Oxford University Press.

Evers, K. (2009). *Neuroéthique—Quand la matière s'éveille*. Paris: Coll. Collège de France; Odile Jacob.

Frankish, K., & Evans, J. S. (2009). *The duality of mind: An historical perspective. In two minds: Dual processes and beyond*. Oxford: Oxford University Press.

Greene, J. D. (2003). From neural "is" to moral "ought": What are the moral implications of neuroscientific moral psychology? *Nature Reviews Neuroscience, 4*(10), 846–850.

Greene, J. D. (2013). *Moral tribes: Emotion, reason, and the gap between us and them*. New York: Penguin Press.

Greene, J. D. (2014). The cognitive neuroscience of moral judgment and decision making. In M. S. Gazzaniga (Ed.), *The cognitive neurosciences V* (pp. 1013–1023). Cambridge, MA: MIT Press.

Greene, J. D., Nystrom, L. E., Engell, A. D., Darley, J. M., & Cohen, J. D. (2004). The neural bases of cognitive conflict and control in moral judgment. *Neuron, 44*(2), 389–400.

Greene, J. D., Sommerville, R. B., Nystrom, L. E., Darley, J. M., & Cohen, J. D. (2001). An fMRI investigation of emotional engagement in moral judgment. *Science, 293*(5537), 2105–2108.

Hume, D. (2014). *A treatise of human nature*. New York: Sheba Blake Publishing (Original work published 1739–40).

Johnson, J., & Littlefield, M. (2011). Lost and found in translation: Popular neuroscience in the emerging neurodisciplines. In M. Pickersgill, & I. Van Keulen (Eds.), *Sociological reflection in neuroscience: Advances in medical sociology* (Vol. 13, pp. 279–297). Bingley: Emerald Group Publishing.

Joyce, R. (2006). *The evolution of morality*. Cambridge, MA: MIT Press.

Kahneman, D. (2011). *Thinking, fast and slow*. New York: Farrar, Straus and Giroux.

Klein, C. (2010a). Images are not the evidence in neuroimaging. *The British Journal for the Philosophy of Science, 61*(2), 265–278.

Klein, C. (2010b). Philosophical issues in neuroimaging. *Philosophy Compass, 5*(2), 186–198.

Klein, C. (2011). The dual track theory of moral decision-making: A critique of the neuroimaging evidence. *Neuroethics, 4*(2), 143–162.

McGuire, J., Langdon, R., Coltheart, M., & Mackenzie, C. (2009). A reanalysis of the personal/impersonal distinction in moral psychology research. *Journal of Experimental Social Psychology, 45*(3), 577–580.

Mendez, M. F. (2009). The neurobiology of moral behavior: Review and neuropsychiatric implications. *CNS Spectrums, 14*(11), 608–620.

Moll, J., Zahn, R., de Oliveira-Souza, R., Krueger, F., & Grafman, J. (2005). The neural basis of human moral cognition. *Nature Reviews Neuroscience, 6*(10), 799–809.

Moore, G. E. (1982). *Principia ethica*. Cambridge: Cambridge University Press (Original work published 1903).

Ogien, R. (2004). La philosophie morale a-t-elle besoin des sciences sociales? *L'Année Sociologique, 54*(2), 589–606.

Rawls, J., & Herman, B. (2000). *Lectures on the history of moral philosophy*. Cambridge, MA: Harvard University Press.

Roskies, A. (2002). Neuroethics for the new nillenium. *Neuron, 35*(1), 21–23.

Tiberghien, G. (2007). Entre neurosciences et neurophilosophie: La psychologie cognitive et les sciences cognitives. *Psychologie Française, 52*(3), 279–297.

Tiberghien, G., & Jeannerod, M. (1995). Pour la science cognitive: La métaphore cognitive est-elle scientifiquement fondée. *Revue Internationale de Psychopathologie, 18*, 173–203.

PART III

THE NEUROSCIENCES IN SOCIETY. SOCIAL, CULTURAL, AND ETHICAL IMPLICATIONS OF THE NEURO-TURN

10

Effects of the Neuro-Turn: The Neural Network as a Paradigm for Human Self-Understanding

Y. Förster

Leuphana University of Lüneburg, Lüneburg, Germany

INTRODUCTION

In this chapter, I would like to discuss the neural network as a paradigm that influences human self-understanding, and how it simultaneously comes to be conceptualized as a metaphysical structure. The neural network, as it appears in computer simulations and scientific representations, has become a part of our conception of cognition and of our understanding of what makes us human. From science reports in media to novels, cinema, and artwork the neural network is an omnipresent image. I will analyze various representations of neural networks in science and popular culture in order to understand how these images relate to human self-understanding. In 2009, Fernando Vidal described brainhood as an "anthropological figure of modernity." I want to take this line of reasoning one step further and show that the neural network apart from its embodiment within a brain has become an influential image in our days. It is not surprising that with the rise of neuroscience human cognitive abilities have become the subject of heated debates

The Human Sciences after the Decade of the Brain.
DOI: http://dx.doi.org/10.1016/B978-0-12-804205-2.00010-0

© 2017 Elsevier Inc. All rights reserved.

between humanities, the empirical sciences, and social sciences. On the one hand, neuroscientific findings suggest that free will and the self are illusions. On the other hand, more and more possibilities for medical and technological enhancement of consciousness and cognition open up. This process is driven by neuroscience in cooperation with computer science and artificial intelligence (AI). In this field it is hard to discern where reality ends and science fiction starts.

I argue that the brain and the neural network are two different images that are closely related but carry different meanings. The brain is a specific organ of a body, whereas the neural network is a structure that constitutes a brain. Since there are also artificial neural networks, it is not necessarily bound or confined to a brain. Other than the brain, the neural network is widely used not only as a technical term in neuroscience or in computing but rather it has become a metaphor or a paradigm used to describe complex artificial or living structures such as the Internet. It is represented first and foremost in pictorial form, via scientific imaging techniques or in cinematic adaptions. Thus I will focus on the neural network as an image that drives our research and utopias of future life.

In the first part of this chapter, I will focus on the scientific discourse between neuroscience, philosophy, social sciences, and the status of scientific images in this debate. In the following paragraphs, I will elaborate on the consequences of this debate for concepts like the body and the self as well as their ontological underpinnings. Given the central role of technology and computation within neuroscience, I will present first a critique of reductive approaches and then offer a possibility for a productive integration of informational technology and theories of embodiment. In doing so, I will draw on examples from contemporary art and cinema and ask how the relation between humanity and AI is framed esthetically. Here, I will concentrate on two movies: *Transcendence* (Wally Pfister, 2014) and *Her* (Spike Jonze, 2013). The German philosopher Joseph Früchtl points out that movies like *Cloud Atlas* or *Avatar* come to function as new great narratives of our times.[a] The importance of cinema for our common repertoire of images cannot be underestimated as a source to understand the contemporary *conditio humana*. Thus in the last part of this chapter I argue that the image of the neural network implicitly produces the idea of an intelligent structure as a metaphysical dimension. Hence, the image represents cognition as omnipresent and disembodied, which in turn elucidates our contemporary idea of what is human.

[a]Thoughts from a chapter presented by Josef Früchtl at a conference of the German Association of Aesthetics (DGAE) 2015 in Hamburg, publication forthcoming.

THE NEURO-TURN IN THE SCIENCES

In 1990, George H.W. Bush proclaimed the upcoming decade to be the "Decade of the Brain." Neuroscience was about to receive more funding than ever. Germany followed that example at the turn of the Millenium. Neuroscience asserts itself as a young and cutting edge discipline. Though the science of the brain can be traced back to Greek antiquity with Plato declaring that cognition must be linked to the brain. Similarly, Aristotle recognized the brain as an organ that is vital for the living process just as the heart, and the Greek physician Galen who thought the brain to be a central organ and, thus, developed a complex idea of its functional structure. Neuroscience seems uninterested in its roots. As Littlefield and Johnson (2012) point out in their introduction to the anthology *The Neuroscientific Turn: Transdisciplinarity in the Age of the Brain*, neuroscience and cognitive science often "eschew history in order to appear new and relevant" (p. 10). The authors stress in accordance with Vidal (2009) that a discourse on newness and technophilia is the driving force in the neuroscientific turn.

One can speak of a neuroscientific turn, hence a turn within the scientific discourse, because a broad variety of disciplines adopt neuroscientific language and methods and incorporate those within their own frameworks. The prefix *neuro-* appears basically everywhere these days: from neurophilosophy over neurotheology, neurofiction, neurosociology, neuromarketing to less serious undertakings such as neuroknitting.

In the following, I outline two specific dimensions of the neuroscientific turn that are relevant for my critical reading in the next section. First, I will look at the images of neuroscience that play a central role in its popularization. Second, I will discuss the shifting of terminology and scientific language from philosophy to neuroscience and vice versa.

1. Neuroscience is a scientific discipline that gets a great deal of attention in society, along with evolutionary biology and genetics. These three disciplines are expected to develop a theory about what is human and that explains the miracle of consciousness. Neuroscience, at least for now, seems to lead the pack. Why is that so? What makes neuroscience special is the use of images. Images derived from imaging techniques like fMRT or PET exhibit a strong impact on our imagination. It is as if we are peeking into another person's head. These images are used in medical practice as well as in the popularization of scientific findings.

To be able to understand neuroscientific imaging techniques it is necessary to look into their epistemic preconditions. An interdisciplinary research project on *The Epistemic Reverse Side of Instrumental Images* at the Humboldt University Berlin aims to do exactly this. The project

> examines instrumental images from cancer research, psychology, Neuroscience, and trauma surgery to analyze the discrepancy between their practical uses and their masked or hidden epistemic preconditions. Instrumental images means images used by human agents as instruments. They include visual schemata, imaging procedures, and computer simulations employed as instrumentaria in fundamental research and in therapeutic and surgical settings. Such images guide interpretations, decisions, and interventions, even when the agents (physicians, patients, computer specialists) are not adequately familiar with their theoretical, methodological, and historical preconditions. (from https://www.interdisciplinary-laboratory.hu-berlin.de/en/base-projects/epistemic-reverse-side-instrumental-images)

The epistemological reverse side of these images, especially those of neuroscientific ones, is computation. For example, fMRT measures changes in the bloodflow within the relevant areas of the brain. The idea is that activated areas of the brain show a higher metabolic rate than others. Therefore the blood in those areas contains more oxygen than the blood in areas with a lower level of activation. These measurements are then fed into a computation in order to produce an image of the skull with bright colors that mark areas of activation.

The images seem to represent the workings of the brain directly, but the visualization process is contingent in the sense that there is no causal relation between the sujet (phenomenological spoken) of the image and its visualization in the image. Although there are rules or algorithms by which the final visualization depends on the measurements, the relation between measurement and visualization itself has many contingent aspects like choice of color or grain. Finding a way to visualize something is only one side of the coin. The other one is its usage. The way these images are presented in media and cinema shows that an epistemic reflection is absent. These powerful images are used to foster the fascination for neuroscience and for the possibilities of modern medicine. But in the end all visualization of neural activity is mediated by compution. There is no such thing as direct representation.

2. The other interesting aspect of the neuroscientific turn is a shift of language: neuroscience uses philosophical concepts to guide its investigations while philosophers incorporate neuroscientific approaches in their discussions. The main problem here marks the relation of first-person experiences and third-person data. I will elaborate on this in the following.

The variety of disciplines that relate themselves to the findings of neuroscience calls for an effort in translating. Littlefield and Johnson (2012) also stress that point. Neuroscience is in itself already a translational discipline, since it involves different fields of research that contribute to the whole framework of neuroscience, e.g., biology, chemistry, medical sciences, and computing. Interdisciplinarity and the need for translational efforts characterize the whole field of neuroscience and the disciplines that relate themselves to the findings of neuroscience.

The starting point of neuroscience is an investigation of the brain, its functional architecture, its biochemical processes, and its connection to the whole nervous system and its bodily implementation. The complexity of the brain is incredibly high: something around 86 billion neurons with probably around 100 trillion connections—numbers that are in any case far from every possible imagination. Even today's mega-computers are not able to simulate the complexity of the brain. We are still far away from understanding the brain as a whole. Nevertheless, the findings of neuroscience are amazing. The last several decades have resulted in immense knowledge about how our brains work and how we can manipulate them. This has already caused philosophers to ask for a new ethics of consciousness (e.g., Metzinger, 2009), and questions of enhancing human bodies and brains have initiated a debate on what is human and whether or not it can be transcended toward a post- or transhuman state.

The science of the brain sets out to investigate fundamental human capacities like self-consciousness (e.g., Damasio, 2012), free will (e.g., Tse, 2015), or religious beliefs (e.g., Gay, 2009). Given the above-mentioned complexity of the brain, those findings must be regarded with some care. Obviously, neuroscience tries to answer questions that philosophy has dealt with for thousands of years. Clearly the methods of neuroscience differ from the philosophical ones: While philosophy works with concepts, experience, reflection, and linguistic description, neuroscience, on the other hand, uses these philosophical terms within a third-person framework of observation, imaging techniques, and measurements.

The relation of third-person data and first-person experience is one of the methodological problems in neuroscience. What is to be investigated has experiential character—consciousness and cognition. However, the investigation needs to work first and foremost with third-person data. But the means of investigation are the same as the object of the investigation. Researchers use their brains to investigate the brain. More precisely, they investigate the conditions of first-person experience (which cannot be accessed directly but only via correlation with brain states) using their experience: "When we use the scientific method to investigate consciousness, we're always necessarily using and relying on consciousness itself (Thompson, 2014, p. 99)." This is what Thompson calls the *primacy of direct experience* (Thompson, 2014, p. 96).

Philosophical questions about the self, morality, or consciousness are taken up by neuroscientific research. Although all of these are concepts that involve first-person experience, they are investigated through neuroscientific methods, which rely mainly on third-person data. The debate on whether we have a free will or not has been led by neuroscientists for quite some time; at least since Benjamin Libet's experiments in the early 1980s. Continuing today, there is a vivid debate in the media between philosophers and neuroscientists on that topic. What is remarkable, though, is not so much the fact that neuroscience contributes to the question of free will, but rather the way in which those topics are discussed and how the brain features in this debate.

In Germany, leading neuroscientists like Wolf Singer and Gerhard Roth are omnipresent in TV and press. They speak of the brain as if they were talking about a person. Different areas that act together in a certain task are endowed with quasi-human capacities of forming beliefs and decision-making. What traditionally was attributed to consciousness has now been transferred onto the brain. The brain or subdivisions of the brain are described as if they contain homunculi that act like actual persons. This way of speaking is criticized by Thomas Fuchs (2010), a German neuroscientist and psychologist, trained in philosophy, who asserts that this way of speaking is by no means figurative. According to Fuchs, neuroscientists apply intentional vocabulary to describe subpersonal processes as a means to successfully naturalize consciousness. Referring to Bennett and Hacker (2003, p. 68, ff.) Fuchs regards this practice as a mereological fallacy (p. 65, ff.), since parts of the human brain are endowed with attributes that can only apply to the whole human being. For example, when one area of the brain is said to make a decision or to want something, these areas are treated as agents not as functional parts of the brain.

A somewhat provocative approach is taken by the philosopher Patricia Churchland in her 2013 book *Touching a Nerve: The Self as Brain.* As the title suggests, she advances a view of the self as the brain, as criticized by Fuchs. In strong contrast to Alva Noë's book (2010), *Out of Our Heads: Why You Are Not Your Brain and Other Lessons From the Biology of Consciousness*, Churchland wants to convince us to think of ourselves as brains and to embrace neuroscientific descriptions of consciousness. The book uses the rhetoric of enlightenment. However, enlightenment in philosophy requires a strong concept of the self, while in Churchland's perspective the self is nothing more than neural activity. Hence, Churchland promotes a reductive view of the self (informed by neuroscience) by using an autobiographical style of writing. She tries to endorse an ontological theory by using certain semantics. Thus she uses first-person experience to promote the idea that the self and first-person experience are illusions, which seems paradoxical.

Churchland's idea that we should take our brain *as* our self is problematic. For the processes of the brain have no phenomenal qualities. I simply cannot perceive what happens in the circuits of my neurons; i.e., neurons constitute perception and therefore cannot be perceived themselves. Thus Churchland's autobiographical style of writing appears as a method to artificially endow the ontological discourse on the brain as self with phenomenal qualities to render it more comprehensible.

NEW SELF-CONCEPTS AND THEIR ONTOLOGICAL UNDERPINNINGS

The brain, as part of the human body, is thought to be the control center of all human activity. However, the commonly used prefix *neuro-* suggests a different picture. Fernando Vidal talks about the cerebral subject as "the anthropological figure inherent to modernity"—"As a 'cerebral subject', the human being is specified by the property of 'brainhood', i.e., the property or quality of *being*, rather than simply *having*, a brain" (Vidal, 2009, p. 6). He states that the modern subject only needs the brain (not the whole body) to be him- or herself, just as Churchland proposed. Material continuity, according to Vidal, lost its importance in the course of western philosophy. Since the search for a location of the soul within the physical body failed, personal identity could not be dependent on the body but only on the continuity of consciousness (Vidal quotes John Locke here), and consciousness was at least from the 19th century on regarded as a function of the brain. From this idea resulted the equation of self-hood with brainhood.

Now let me turn to the concept of the neural network. Vidal, in describing modern neuroscience quotes, Vilaynur Ramachandran: "We used to say, metaphorically, that 'I can feel another's pain.' But now we know that my mirror neurons can literally feel your pain" (p. 21, quoted from G. Slack, 2007, "I Feel Your Pain," Salon.com). Here it is not the brain that is endowed with feelings or other first-person experiences, but a group of neurons. Similar to Ramachandran's analysis, Churchland finds: "Without the living neurons that *embody* (my emphasis, Y.F.) information, memories perish, personalities change, skills vanish, motives dissipate" (Churchland, 2013, p. 12). Churchland, in contrast to all theories of embodiment, takes neurons as embodied information.

This is a curious form of embodiment. Theories of embodiment have been used as a critique of standard cognitive science and its computational model of the mind. Cognition, according the theorists of embodiment is always a complex process with sensorimotor-feedback involving an organism and its environment. The lived body, as the French phenomenologist Merleau-Ponty phrases it, hence the body as it is experienced in the first-person perspective, as one's own body (in German *Leib*), is vital to any cognitive process. Shaun Gallagher stresses in his book *How the Body Shapes the Mind* (2005) that many cognitive abilities constitutively rely on bodily experience. To speak of neurons as embodied information is a whole different idea. The neural network—other than the brain—is not a defined organ nor does it exhibit any autopoietic closure like an organism does.

What is at stake is a different ontological concept. Proponents of embodiment usually subscribe to an indirect or relational ontology as Merleau-Ponty developed in his late writings. His account of ontology describes the conditions of possibility for first-person experience. He holds that subject and object are no a priori categories. Rather they emerge from a primordial state where they are still undifferentiated. It is the development of the perception of living, i.e., embodied systems within an environment that give rise to the subject-object divide. Perception and perspectivity (which are types of relations) are central to this ontological account. While in contemporary theories, the concept of relational ontology often takes a shift toward the concepts of information and technology, originally it focused on perception and embodiment.

Churchland, for example, takes information to have ontological primacy. The concept of information as metaphysical category is closely connected to images of neural or artificial nets. Churchland argues for an eliminative materialism, which degrades subjectivity to mere epiphenomenality. The German philosopher Thomas Metzinger (2009) states that "nobody has ever *been* or *had* a self" (p. 1). This neuroscientific informed view holds that the self has no substantial existence. It is only a higher-order representation within a complex system. The self has no ontological primacy. The question is, how are the ontological underpinnings of the self construed? In Merleau-Ponty (1966) and other accounts of embodiment theories, the self or the range of cognitive abilities depend on embodied experience and sensori-motor feedback. The self is no substantial entity within a brain but an emergent quality in a complex process of life.

Proponents of neurophilosophy such as Churchland take on a slightly different perspective: They stress that neural activity is all there is to the self, reducing selfhood to activities of the brain. Neural activity is not an object of experience, but rather a complex process that does not permeate the experience of the subject. I will argue that complex neural activity in that sense becomes a contemporary metaphysical narrative. We are made to believe that this unperceivable process brings about our experience and our world. In cinematic images this neural activity even becomes the structure of the world itself. Humans are transcended by their own neural network, whose complex dynamics work in the dark with no chance for us to enlighten ourselves.

COMPUTING LIFE

Neuroscience is strongly connected to computer science, AI, or artificial life. Consciousness has been and often still is modeled according to the image of a computer, where every thought is ideally translatable into an algorithm. Jerry Fodor and Noam Chomsky are prominent advocates of a computationalist theory of the mind. From the 1960s on, consciousness was thought to be a process of computation. Standard cognitive science relies on that view. But with advanced computer techniques and increasing speed of processing another direction has emerged. Computers are constructed similar to brains, not vice versa. Projects like the *Human Brain Project*, a scientific project of the European Union, work on neuromorphic computing, brain simulation, and neurorobotics. These labels already suggest that computing is set to reach new frontiers and to develop structures that are similar to neural networks. These techniques are put to use in medical endeavors to cure diseases associated with the brain.

According to standard cognitive science and several theories of philosophy of mind, body, and more precisely, the lived body (the body from first-person perspective), as well as subjectivity are illusory or at least without impact on neural processes. This denial of the phenomenal goes along with a gesture toward the transcendence of the human, be it post-, trans-, or metahumanism (see Ranisch & Sorgner, 2014). When the self is no more than neural activity, then altering the self via interfering with the neural processes does not destroy anything substantial. This process is ambivalent. On the one hand, it is problematic because we seem to lose our identity and become subject to endless self-perfection. On the other hand, it triggers reflections of the potential of our bodily existence still not explored. This tension is productively dealt with in art and projects dealing with artificial life.

Artificial life projects aim at creating artificial systems that exhibit characteristics of natural living systems, like autopoietic closure and reproduction. The implementation of these systems is not predefined; it can be either software, hardware, or wetware. In the case of artificial life the idea is to create an artificial neural structure that is so complex that self-organizing processes can emerge from it. This would be the first step toward an artificial lifeform. This is different from research in AI because a robot that can adjust its behavior via sensorimotor feedback is intelligent but exhibits no features of an organism. It is not self-organizing or reproductive. At the moment, experiments with artificial life take on the form of art or artistic research.

There are several artistic projects that combine computation and theories of embodiment. This combination is also productively used in robotics. The project I present here focuses on the interaction of human movement within a computed environment. This art project aims to develop new ways of interaction between humans and other intelligent structures. The project *Neural Narratives 1: The Phantom Limb* (part of the interdisciplinary EU-Culture-Project *Metabody*, organized by Jaime del Val) generates a connection between dancers and computed neural networks via sensorimotor feedback. The creators are Pablo Palacio (composer), Muriel Romero (dancer and choreographer), and Daniel Bisig (AI researcher). The project can be defined as generative art.

The human body and its movements are a central part in the artistic performance. The dancers' bodies interact via sensors with software that resembles a neural network (self-organizing, learning). This software controls visual stimuli and sounds in interaction with the dancers. The whole scene is an evolution of interactive patterns. This setup can be described by means of the extended mind theory put forth by Chalmers and Clark (2011 [1998]). Extended mind theory is one variation of theories of embodiment, which does not so much center on the body but on the fact that cognition does not just take place in the head. Rather, they hold that cognition is a process that essentially relies on other minds, memorizing devices such as books, computers, sheets of chapter, pens, photographs, maps—i.e., everything that serves us to form or retrieve thoughts. The way we think is dependent on how our environment is shaped, what media we can use to perform thinking, with what kinds of people we speak, how our computers interact with us. This is precisely what happens in the performance. Hence, an environment is created within which the emergence of new movement, unexpected variation, and development of new artificial structures is possible.

Copyright: Gunter Kramer.

REPRESENTATIONS OF THE NEURAL NET
IN CINEMATIC NARRATIVES

In the last section, I would like to analyze the image of the neural net as a paradigmatic image of transcending what is human. I will argue that it is different from other science-fiction paradigms such as technology/machines or the famous brain-in-the-vat idea. The neural net is an abstract structure that is derived from cognitive and evolutionary sciences but is not tied to organic matter. It is closely linked to computer nets and AI. The neural net as an image is characterized by being self-organizing, by producing emergent effects, and by being implemented in various ways. It stands for the utopia of an emerging artificial consciousness or lifeform. The complexity of biological or artificial neural nets is thought of as the basis of emergent qualities.

In the following I will analyze how cinematic narratives are changing in accordance with the progress being made in neuroscience and AI. Other than the brain the neural net represents a metaphor for AI that is linked to infinity. Already the neural connections within one's brain seem to be as complex as the whole universe. In various biological or artificial forms neural or neural-like nets are imagined to reach a state of complexity that new properties may emerge from it.

I will consider two examples of recent cinematic framing of artificial neural networks: *Her* (2013) and *Transcendence* (2014). These two movies are in fact the reason why I started to think about the neural network differentiated from the brain, because they provide a different set of images, utopias (or dystopias), and reasonings. Movies that center on the brain as the leading image, like *The Butterfly Effect* (2005) or *Surrogates* (2009), often end with a focus on what is human or aim to restore human order. For example, *The Butterfly Effect* tells the story of a young man with pathological blackouts who can travel back in time while having these blackouts. Every time he travels back, he changes something in his personal history in order to prevent his loved ones from suffering. He succeeds in the end and never travels again. This way, human lives are saved and history can take a good course. In *Surrogates*, we are presented with a science-fiction scenario where humans do not leave their homes anymore. They act in the world by means of their surrogate bodies, which do not age or fall sick. Without going deeper into the story, the main plot is a fight between technology and human life. In the end all surrogate bodies are disconnected and humans return to their original way of life.

Movies that focus on neural or neural-like nets, however, have another logic. Here we find not simply a man vs machine story, but rather the foundation of being, the metaphysics of our world, is called into question. And the image of the neural net suggests a process of transcending the hardware-software divide and the limitations of human life such as finitude and mortality.

Be it in *Her*, where operating systems (OSs) connected to the Internet develop a new and higher form of consciousness that transcends the human sphere as well as the implementation of individual hardware, such as personal computers. Or be it in *Transcendence*, where human consciousness gets uploaded to the Internet and thus develops into a singularity, a form of being that transcends the hardware-software divide and becomes elementary in the old Greek sense.

In the upcoming paragraphs I propose a reading of how cinema develops utopias involving neural nets, which are transcending the human sphere. Images of transcending the human have been present throughout the history of cinema. In recent movies these imaginations have taken on a peculiar form. It is not machines anymore that threaten the human existence as in Fritz Lang's *Metropolis* or in the Wachowski Brothers' *Matrix*. In these two movies it is technology in the form of machines that seems to be a threat to mankind.

Let me shortly consider *Metropolis*, because this movie was the first of its kind and maybe the most influential science-fiction movie in the history of cinema. It presents images of technology and its merging with the human in a paradigmatic way. In this movie technology resembles a metaphysical dimension. I will argue that the image of the neural net in contemporary cinema takes the place of technology (despite the fact that the neural is also technologically mediated) and becomes a metaphysical dimension.

Metropolis, a movie that still impacts the making of science-fiction movies today, imagines an artificial human being that consists of a metal machinery, an automaton. Artificial life in *Metropolis* is represented according to de La Mettrie's *L'homme machine*. Life is conceived of as a very sophisticated mechanism that brings to life the android twin of Maria, a character that preaches the downfall of this world. In *Metropolis* the world consists of two classes, the ruling class that lives in a paradise-like world over the city (which is why *Metropolis* is compared to the biblical story of the building of the *Tower of Babel*) and the worker class that lives deep underground. Both classes are linked by a third strata consisting of machines that keep this world going and on which all humans depend. Even time is measured differently in the two classes. In this story technology and transcendence are paralleled. While the human character Maria preaches in a church the downfall of the system, an artificial android version of her is created to start a revolution.[b]

At the end of *Metropolis*, the android Maria is burned on a stake because she was accused of causing the death of the children. Without going into more detail (the story line is too complex to be presented here), it is important to note that technology here is connected to transcendence, to a religious sphere. The name Maria invokes the biblical figure Maria/Mary, mother of Jesus, and the world of the machines is a world in-between, on which all humans depend. This in-between situatedness suggests a metaphysical dimension. The machines are not contingent objects but the dimension that constitutes the possibility of life. This represents a structure that in contemporary ontology has become central. Various theorists like Friedrich Kittler or Bernhard Stiegler propose an account of relational ontology where media/technology is fundamental, which means it is technology that constitutes the human sphere, not vice versa (this is a discussion that cannot be outlined here).

---

[b]This android fascinates and mesmerizes the people just as the automaton Olimpia in E.T. A. Hoffman's story *The Sandman* (1816), roughly a hundred years earlier. In this narrative a life-size automaton doll named Olimpia makes the protagonist Nathanael lose his mind. Not knowing that Olimpia is an automaton, he stays mesmerized by the forlorn gaze and simple, reduced gestures of Olimpia who seems to be from a different world. The story ends with Nathanael's death.

III. THE NEUROSCIENCES IN SOCIETY. SOCIAL, CULTURAL, AND ETHICAL
IMPLICATIONS OF THE NEURO-TURN

The neural net is an image that also has a metaphysical dimension and is clearly technology-related. Images of neural nets or conceptualization of neural structures depend strongly on scientific imaging techniques. The neural gains its visibility only via technology. The process of making the neural visible is not a simple representation of something otherwise hidden. Rather it is a production of images by means imaging techniques. What we get to see is not the inside of our skull, not copies of our neurons, but reconstructions modeled according to a certain set of rules of computation. The neural net as we know it from neuroscientific imagery is not a photograph of brain parts. It is deeply technological mediated.

The mesmerizing aspect of artificial life is present in most of the narratives and images. Humans are mesmerized by the creatures they brought to life. The capability of creating artificial life is a utopia and dystopia at the same time, because artificial lifeforms raise questions about human life, its vulnerability, and finitude. In the face of lifeforms made of steel and without fear, the humans who created them feel threatened. With artificial life, super- or posthuman beings come into existence and with them the threat that human life might only be one step in an evolutionary process where other, possibly stronger, lifeforms emerge. That is why the automatons in those narratives are destroyed.

At first sight the stories of *Her* and *Transcendence* are not that different. Technology in both movies might be destroyed, but the difference is that the AI that emerged from it develops into another, nonhardware-driven, seemingly metaphysical dimension. The artificial/neural network transcends the human sphere; it is a form of post- or transhuman phenomenon. Humanity in both movies is left behind while the new net-based life disappears into different dimensions.

I will start with *Her*, because here the steps toward an artificial lifeform are very clearcut. The story is pretty simple. The plot takes place in the near future. Theodore Twombly, the unhappy and soon to be divorced protagonist, works as a professional ghostwriter of love letters. When a new OS based on AI and designed to adapt and evolve is introduced, he buys it. After having installed the system, a female voice by the name of Samantha starts communicating with Theodore. Samantha's voice and what she expresses seems so real that Theodore falls in love with her—and she with him. This is possible because the OS is self-conscious. Still, since the system is evolving Samantha eventually learns about other operating nets and step-by-step gets more involved in what appears to be her own species. In the end, all the OSs join together, leave their human partners (many have fallen in love with OSs) for good, and from then on exist in their own dimension.

What is interesting about the images here? First of all, we have a science-fiction movie barely shows any technological devices. The whole interior and clothing seem very 1950s, i.e., retro. The phone Theodore uses to show Samantha the human world and acquaint her with human movement also seems very retro (seemingly from the very beginning of this millennium). Technology or rather technological hardware is presented as old and outdated. There are only a few scenes where we see contemporary technology, such as huge digital screens. All the complex technology, which in the end leads to a new lifeforms, is hidden. It only appears via the narrative of the film. We imagine an extremely complex and sophisticated OS, but all we see is a very human world.

Here embodiment is an important theme: The OS Samantha learns about the world and herself through Theodore's eyes. He constantly shows her his world through the camera of his mobile phone, thus giving Samatha the impression of bodily movement. In the course of the story Samantha develops a sense of having a body and feelings. Then she asks the question of all questions: How can I know whether my feelings are real—or are they just part of my programming? This is the question that we are all faced with in times of neuroscience. Is our experience just an after effect of our neural activity?

What the movie basically does is show us how an AI, step by step, develops a self, a personality, and finally becomes an artificial lifeform. This evolution is driven by embodied and social experience—just as every human being would develop. But human contact is only one step in a larger process. The OSs emancipate themselves from the human world and vanish into another dimension. In that sense they become posthuman. Humanity returns to its old life. The artificial lifeforms are not hardware-based anymore—they left the computers they were implemented in, to evolve in a different dimension. The film displays a peaceful transcendence of what is human. Humanity is outdated but not endangered. The vanishing of the OSs stands for the metaphysical dimension that is implied in the image of the neural net. Neural nets are not only the basis on which consciousness supervenes. The image of the net in this movie is not exactly presented as a neural net, but as the Internet. Still, the logic remains the same: It is a net structure that is self-organizing and produces emergent effects. So the characteristics of an artificial neural net are present, and contemporary computing relies in many aspects on the simulation of such neural structures. The net as it is presented in *Her* is a metaphysical dimension (a dimension that is beyond physics in the sense that being only emerges from it) from which artificial life emerges. This artificial life is posthuman in the sense that its evolution has surpassed human feeling and cognition and developed into an infinite process of communication.

III. THE NEUROSCIENCES IN SOCIETY. SOCIAL, CULTURAL, AND ETHICAL
IMPLICATIONS OF THE NEURO-TURN

Let me turn to the second example. *Transcendence*, admittedly, is a movie that might not change the history of cinema. The plot is simple and the whole film is more of an action movie. It is the story of Will Caster, a computer scientist who gets attacked by an extremist group "Revolutionary Independence from Technology," and is poisoned by a radioactive bullet. He works on consciousness and AI. His theory is that a sentient computer would create a "singularity" or transcendence, a god-like lifeform. Before he dies, his friend and his wife manage to upload his consciousness into a quantum computer. In the end, Will has become the singularity but he chooses to end himself by allowing his wife to inject a virus.

The metaphysical dimension is already in the title. We see a world where computers with maximum capacity exist. And again: The setting is very retro. Will Caster builds a copper net to cover his house from electromagnetic smog. We see an old record player. The house of the scientist resembles more a philosopher's cottage. As the plot develops technical devices play a bigger role. An underground supercomputer world is contrasted with a Mad Max scenario above ground. In the end, the Internet shuts down (because Will has become the whole net) and with it humanity is thrown back to its nontechnical roots. In Will's/the singularity's dying scene all technical devices are out of the picture. We see a bird's eye camera flying over a healing nature.

This artificial singularity has become an element in the old Greek sense. There are no wires or electrodes anymore, which is why the end of the movie is open for interpretation. Has this transcendent being died with the Internet or had it already evolved into a metaphysical dimension? The closing scene suggests the latter. During the movie the singularity (the consciousness merged with the Internet) is represented as having penetrated physical reality. We see how a destroyed nature is restored in a process whose imagery suggests a matrix-like scenario. Elementary particles arise from the ground and transform into objects. At the end of the movie, this image of elementary particles recurs. There is a scene in the scientist's garden, where a dewdrop falls from a sunflower into a small water basin from which the elementary particles rise again. If the AI depended on the Internet, this image would make no sense. So the logic of the image of the artificial neural net here again is metaphysical. The artificial net developed into a dimension of being.

The ultimate transcendence of human life has features of a process of dissolving into a wider sphere of being—it sounds almost Buddhist, only that this process is initially silicon-based. As Raymond Kurzweil noted in his book *The Singularity Is Near: When Humans Transcend Biology* (2005), "We will continue to have human bodies, but they will become morphable projections of our intelligence. [...] Ultimately software-based humans will be vastly extended beyond the severe limitations of humans as we know them today" (pp. 324–325).

CLOSING REMARKS

I hope to have shown how alongside the neuroscientific turn the neural network has become a strong image. This image is distributed via popular science, TV, and cinema. The neural network appears as a metaphysical dimension. Interestingly the way the net is pictured all questions of implementation or embodiment are left behind. There are no quantum computers anymore that are necessary to keep the AI alive. Sentience and bodily experience might give the discussion and imagination of the neural net a new twist. Since the neural network as structure is derived from organic life it seems plausible that artificial neural networks will also exhibit similar characteristics such as the reliance on sensorimotor feedback and sentience. Theories of embodiment might still have a say in the debate of artificial life and posthuman body politics. Grounding the neural network by making explicit its technobiological origin is in my view a fruitful move in the current debate concerning human self-understanding.

Movies tell us a lot about how we imagine mankind and its future. In the current debate on neuroscience and the primacy of neural processes it can be useful to compare the various worlds of images arising from science and cinema. We need to understand the interplay of biology and technology. To reduce life to one or the other is not helpful because both sides are coemerging. Life is always a form of technology and technology depends on complex living structures. *Her* depicts this interplay when the OS learns to feel and experience herself by means of embodiment techniques such as viewing the world in movement, a crucial step within this narrative. Also, *Transcendence* ultimately connects AI to being in the world and experiencing it in a bodily way. This is what is suggested by one of the last scenes where Will Caster's wife has also uploaded her consciousness and they are portrayed as a single consciousness that sees the world from one bird's eye perspective. This again is an embodied perspective, or as Deleuze would have it, a movement-image, not a god-like view. Another movie that merges the images of the neural nets with embodied intelligence is *Avatar* (2009). The German film-philosopher Joseph Früchtl uses this movie to explain how movies take the place of the grand narratives that postmodernity declared to have died. *Avatar* is an interesting example of how the image of the neural merges with a romantic conception of nature, a nature that is neither pure biology nor technology. Conceptualizing the merging of biology and technology in a nonreductive account is necessary to grasp the impact of neuroscience on human self-understanding.

References

Bennett, M. R., & Hacker, P. M. S. (2003). *Philosophical foundations of neuroscience.* Oxford: Blackwell.

Chalmers, D., & Clark, A. (2011). The extended mind. In A. Clark (Ed.), *Supersizing the mind: Embodiment, action, and cognitive extension* (pp. 220–232). Oxford and New York: Oxford University Press.

Churchland, P. (2013). *Touching a nerve: The self as brain.* London and New York: W. W. Norton & Company.

Damasio, A. (2012). *The self comes to mind: Constructing the conscious brain.* New York: Vintage.

Fuchs, T. (2010). *Das Gehirn—ein Beziehungsorgan: Eine phänomenologisch-ökologische Konzeption.* Stuttgart: Kohlhammer.

Gallagher, S. (2005). *How the body shapes the mind.* Oxford: Clarendon Press.

Gay, V. P. (2009). *Neuroscience and religion: Brain, mind, self and soul.* Lanham, MD: Lexington.

Kurzweil, R. (2005). *The singularity is near: When humans transcend biology.* New York: Viking.

Littlefield, M. M., & Johnson, J. J. (2012). *The neuroscientific turn: Transdisciplinarity in the age of the brain.* Ann Arbor, MI: University of Michigan Press.

Merleau-Ponty, M. (1966). *Phänomenologie der Wahrnehmung.* Berlin: de Gruyter.

Metzinger, T. (2009). *The ego tunnel: The science of the mind and the myth of the self.* New York: Basic Books.

Noë, A. (2010). *Out of our heads: Why you are not your brain, and other lessons from the biology of consciousness.* New York: Hill and Wang.

Ranisch, R., & Sorgner, S. L. (2014). *Post- and transhumanism: An introduction.* Frankfurt am Main: Peter Lang.

Thomspon, E. (2014). *Waking, dreaming, being: Self and consciousness in neuroscience, meditation, and philosophy.* New York: Columbia University Press.

Tse, P. U. (2015). *The neural basis of free will: Criterial causation.* Cambridge, MA: MIT Press.

Vidal, F. (2009). Brainhood, anthropological figure of modernity. *History of the Human Sciences, 22*(1), 5–36.

Movies

Cameron, J., Landau, J. (Producers), & Cameron, J. (Director). (2009). *Avatar.* United States.

Ellison, M., Jonze, S., Landay, V. (Producers), & Jonze, S. (Director). (2013). *Her.* United States.

Hoberman, D., Lieberman, T., Handelman, M. (Producers), & Mostow, J. (Director). (2009). *Surrogates.* United States.

Johnson, B., Kosorve, A. A., Cohen, K., Polvino, M., Marter, A., Valdes, D., Ryder, A. (Producers), & Pfister, W. (Director). (2014). *Transcendence.* United States.

Pommer, E. (Producer), & Lang, F. (Director). (1927). *Metropolis.* Germany.

Rhulen, A., Bender, C., Kutcher, A., Spink, J. C., Dix, A. J. (Producers), Bress, E., & Gruber, J. M. (Directors). (2004). *The butterfly effect.* United States.

Silver, J. (Producer), & The Wachowski Brothers (Directors). (1999). *Matrix.* United States and Australia.

Projects

Human Brain Project: <https://www.humanbrainproject.eu/>

Metabody: <http://metabody.eu/>

(Part of Metabody) Neural Narratives 1: The Phantom Limb: <http://metabody.eu/phantom-limb/>

The Epistemic Reverse Side of Instrumental Images: <http://www.zfl-berlin.org/epistemic-reverse-side-of-instrumental-images.html>

11

Brain, Art, Salvation. On the Traditional Character of the Neuro-Hype

G. Grübler

Technical University of Dresden, Dresden, Germany

INTRODUCTION: SOME THESES ABOUT THE CHARACTER OF NEUROLOGIZATION

We all probably know the attitude that neurologists, neuro-philosophers, and science journalists usually show when talking about neuroscientific interpretations of everyday phenomena or philosophical problems. Regularly, there is at least a flavor of progressive movement, of revolution, of unheard-of news, and so on displayed by their talks, interviews, and books. I hope not to overstress this aspect when saying that this attitude, this flavor, is the typical surface of the neurologization of society that we have seen during the last two decades or so. The neuro-movement presents itself as a *more* fundamental, *more* scientific, *more* sophisticated, and—therefore—*more* true approach to many of the problems mankind has, so far, struggled with. By doing so, former approaches are shown to be naïve, illusionary, and just wrong. Of course, there are not only positive reactions toward the proponents of the neuro-hype. There are, for instance, ethical concerns regarding the possible loss of personality, dignity—in short: humanity—in the envisioned ubiquitous use of neuro-technologies and neuro-explanations. But the pretended progressive attitude is more or less accepted as being in line with reality by the wider public.

The Human Sciences after the Decade of the Brain.
DOI: http://dx.doi.org/10.1016/B978-0-12-804205-2.00011-2

© 2017 Elsevier Inc. All rights reserved.

Against this background of common-sense perception I want to defend actually only one central thesis, which is: *The neurologization of society is a rather conservative and very traditional endeavor.* I cannot here substantialize all aspects of my claim, but I will try to focus on some central ones and, by doing so, at least guarantee the overall plausibility of the thesis. In order to do so I divide the one thesis into several smaller and more partial theses. The subtheses I like to explain are:

1. The current neurologization is in its structure very similar to former developments in the history of science.
2. Neurological constructivism as the core of neuro-philosophy implies a metaphysical constellation typical for pre-20th-century philosophy.
3. This constellation implies—nolens volens—the proof of the existence of God.
4. The neurologization is a further expression of the typical European search for salvation.

I can imagine that some readers would spontaneously accept this or that thesis, but I cannot imagine that all readers would accept all of them. So I will now try to convince the reader that all these theses can reasonably be held. In the following, I will point the reader to instances of incorrect uses of language inside the neuro-hype and declare why I hold the academic criticism thereof to be (right but) insufficient. Then I will show how the neurosciences invade society; as an example I focus on art as an issue not yet often debated that allows for some interesting historical comparisons. I will then go to the implicit metaphysics of neuro-philosophy and show its traditional character. Having done so, I'll be talking about science and salvation and the special historical role the neurosciences seem to play here. Finally I will allude to the role my claims might play in the field of applied ethics.

INCORRECT USE OF LANGUAGE AND THE ACADEMIC CRITICISM THEREOF

The entire movement of neurologization goes along with sometimes very special uses of language. Of course, mere descriptions of brain tissue or neuronal wiring can be given in plain language. But such descriptions are rather boring (at the very least not spectacular) and would probably not trigger the hype. What we can see instead when it comes to the more popular and revolutionary theses is uses of language that often are formally wrong.

If we try to systematize typical patterns of argumentation, we will find that in the neuro-literature the brain is oftentimes made an agent working in the interest of its owner or a counteragent working against him or her.

Here are some examples: When the brain is the agent, it is said that it *analyzes* something, that it *pursues* a strategy, that it just *does* something, and *makes* things. As the counteragent it does all these things secretly and then it *follows a strategy* of fooling and deceiving us.[a] "The act of willing," says Gerhard Roth, "occurs when *the brain has already decided* which movement *it will execute*" (Roth, 2003, p. 523, my translation and italics), and another source informs us (P.M. Magazin, 2008, my translation):

> Other parts of the brain, some stemming from the ancient times of evolution, want to co-control your thinking, feeling and deciding. And they are used to making you believe that it all was your idea!

In other cases not the brain as a whole is concerned but only parts of it. Then, for instance, the limbic system is said to *make use* of an advisory board or *engages itself* in evaluation procedures (P.M. Magazin, 2008, my translation).

Now, all these ways of speaking are obviously incorrect (cf. Bennett & Hacker, 2003). They impute something to an organ or to parts of an organ that can be imputed to persons only. All of these examples imply category mistakes, and most of them can be further qualified as pars-pro-toto (or "mereological") fallacies. By saying this, for an only formal philosophical analysis, the case is already closed. The neuro-hype, in its most popular and most spectacular slogans, is based on irrational speaking. So far, so good.

To underline and broaden this criticism we can also capitalize on authors from the neurosciences themselves. Here the phenomenon was named "neuro-bunk." Besides inaccurate language, "neuro-bunk" also comprises the superficial interpretations of scientific studies in the media. Molly Crockett (2012), for instance, has the impression that the real research in laboratories could be hurt by the neuro-bunk problems and she claims for a kind of purification of science:[b]

> [...] we have to be careful that we don't let overblown claims detract resources and attention away from the real science that's playing a much longer game.

Again, these are strong arguments for dismissing all incorrect speaking and superficial presentation. However, although I'm convinced that the mentioned kinds of academic criticism are formally right, I don't think they hit the point. They are not really interesting, because they do not allow us to understand what really happens and where the neuro-hype gets its support and motivation from. If there is a science-driven

---

[a]These prototypical speech patterns are gained from a survey of the German Journal *Gehirn und Geist* considering all issues until 2013 (see also Grübler, 2015a, pp. 129–131).

[b]<https://www.ted.com/talks/molly_crockett_beware_neuro_bunk/transcript>.

movement that has in fact occupied public space and led to a situation in which one could proclaim a decade (or maybe just a century) of the brain, and if there is at the same time the clear insight that the neurosciences have not yet achieved much and that the public conception of this not much is full of bunk and nonsense, then we have to search for the mechanisms behind the neurologization somewhere else.

The first step to do so is to look for parallels, and we can find them in the history of European science. If we go through the time from the 17th century onward, we can find many examples for scientific or technological paradigms rising up to omnipresent and universally used models and patterns of explanation for literally everything. Here are some examples:

The world is a clockwork.
Organisms are mechanical devices.
Nature is an artist.
The human being is a machine.
The brain is a computer.
The mind is software.
The evolution is a programmer.

The most prominent example among them is probably the clockwork that dominated thinking in the 17th and 18th centuries. The clockwork is a metaphor coming from mechanics. And like today "you are your brain" or "all reality is a product of the brain," then for the contemporaries the world was a clockwork, the earth was a clockwork, animals were clockworks, and the human body was a clockwork. And here again, saying that these sentences contain category mistakes and are instances of mechano-bunk is formally right, but misses the point. The point is, that the clockwork, or, more generally speaking, the mechanical machine, was not just a metaphor, but a metaphor that involved hope, optimism, and, yes, a perspective of salvation.

One can build metaphors by using structural and/or functional analogies. For instance, the universe is like a clock because their "components" show very regular movements. This is what makes the metaphor possible. But this is not what makes the metaphor that appealing. Here another, less obvious, analogy is important: the clockwork was made by a watchmaker. So if something is a mechanical machine, then it surely has a maker who designed it. And if the natural machines, the animals for instance, are so much better than the manmade ones, then this maker must be brilliant. Robert Boyle (1772, p. 136), for instance, says:

> [...] so we may say, that the meanest living creatures of God's making are far more wisely contrived, than the most excellent pieces of workmanship, that human heads and hands can boast of; and no watch nor clock in the world is any way comparable, for exquisiteness of mechanism, to the body of even an ass or a frog.

Using a clock or mechanical machine as a metaphor for the contemporaries meant not less than being assured that the world is in good hands, that everything has been prepared for the human being in the best possible manner, and that there exists an intelligent being much more powerful than we are. The metaphor, so, contains, popularizes, and—in a certain sense—proves a theory about the world as it *actually* is, as it is beyond our regular knowledge and abilities, in short: a theory about transcendency.

I think that this has a parallel in current neuro-literature. I only want to give a first sketch here. Let's have a look at these two sentences (taken from Metzinger, 2009, pp. 6–7):

> [...] our brains successfully pursue the ingenious strategy of creating a unified and dynamic inner portrait of reality
> [...] our brains generate a world-simulation, so perfect that we do not recognize it as an image in our minds.

If we analyze this, we see the brain metaphor that is used to explain epistemological issues. The brain is an agent, and this agent simulates something and produces an image. We know how we usually use these words: we can simulate something when we know the original, the real thing, and then make a kind of limited, reduced copy of it. This is what the brain is said to do. The brain knows the real reality (the transcendency) behind the reduced copy-reality that we experience. And the brain is so much more clever than we are that we cannot, on principle, recognize its work of creating our "normal" world. But there's no need to be angry: the brain has an *ingenious* strategy and performs it *perfectly*. So lean back and enjoy.

I will come back to this issue (for an exhaustive analysis see Grübler, 2015b); for the moment I summarize the core message. We can see that the sciences since renaissance times have usually developed together with similar patterns of popularization:

- There is a metaphorical use of language that allows scientific or technological objects to be models for completely different matters.
- There is an implicit motivational factor involved in the model that cannot be reduced to the technical or practical usefulness of the root of the metaphor.
- And the metaphors imply theories about transcendency, meaning the *real* reality that stands behind our experiences.

Insofar, the neuro-hype seems to be structurally in line with the way science hypes usually happen, and any criticism that is only formal overlooks the important mechanisms behind such hypes. Without category mistake and without neuro-bunk there is no neuro-hype and without neuro-hype there is no extraordinary support for the neurosciences in the laboratories.

We can find some more aspects of structural similarity when we have a look at the spreading out of the neurosciences and the way they invade society. In order not to repeat the same examples again and again I've chosen to avoid fields like education, law, or economics and will concentrate on art.

INVADING SOCIETY: NEURO-ART FOR INSTANCE

We can already find the neurosciences present in many different fields of fine art. Let's start with the most direct invasion, the invasion of the interpretation of historical pieces of art. One of the most popular paintings in the world, Michelangelo's "Creation of Adam" painted in 1508–12 as central part of the ceiling of the Sistine Chapel, has been a pioneering issue here. When Frank Meshberger (1990) stared at this picture he received a revealing message: Michelangelo had not just depicted God and several angels in front of a kind of curtain, he had depicted a brain. So, the finger of creative power making Adam alive comes out of the brain, and the brain is the seat of God. And this paradigm seems to have flourished. Some years later researchers have found even more brain stuff in Michelangelo's paintings (Suk & Tamargo, 2010). For example, in the "Separation of Light From Darkness," painted in 1512, the authors discovered parts of the brain in the neck of God.

I cannot decide how likely these interpretations are, but I think this is a good example of a current discourse *occupying the past as its own history*. The tendency is to say: Hey, Michelangelo made anatomical studies, he knew about the brain, he was one of us, he was and we are still on the right track. One can see this attitude working in the following example. There is a widely used study book for psychology and neuroscience students, the "Rosenzweig." In its fourth edition it makes an emblematic use of the painting on the cover and the title page, informs the reader about Meshberger's interpretation, and ends up with the currently valid confession: "[...] it is now clear that our uniquely human qualities—language, reason, emotion, and the rest—are products of the brain" (Rosenzweig, Breedlove, & Watson, 2005, p. IV; in later editions the topic has moved from the title into the introduction). So the symbolic meaning of that painting was totally occupied: in the neuroscience community this painting now is a symbol for the fundamental rightness of the own professional way of research.

Leaving the interpretation of historical pieces of art behind let's now go to some concrete examples of neuro-art from different fields of creativity: There is, of course, painting. Pictures by Megan Gwaltney, for instance, show nerve cells in a very esthetic arrangement (see Dresler, 2009, table III). We can also find pieces of needlework made by Marjorie Taylor and Karen Norberg showing brain scans and brain models (see Dresler, 2009, tables I—II and The Museum of Scientifically Accurate Fabric Brain Art at http://www.neuroscienceart.com). There is also sculpture: Asking the question "What have you got in your head?" Sara Asnaghi made brains out of different kinds of food.[c] (We might read this as contemporary neurological interpretation of the old materialist slogan that "you are what you eat.") And there is installation: Gerda Seiner and Jörg Lenzlinger created a huge "Brainforest" in the 21st Century Museum of Contemporary Art in Kanazawa, Japan.[d] So, the brain and other neuro-entities have definitely well arrived in different branches of art.

However, there is a kind of neuro-art where neuroscience and doing art immediately converge that seems to me most interesting. This implies that neuro-esthetics, meaning an aesthetical theory based on insights coming from the neurosciences, is itself the motivation and stimulus behind the particular performance of art production. Here are some examples by the German artist Adi Hoesle (2012, 2014): Hoesle has asked visitors of an art exhibition to have their EEG recorded while beholding paintings. He understood this to be "a first step towards finding creativity where it becomes measurable for the first time: As electrical flows in the brain evoked by external stimuli" (Hoesle, 2014, pp. 100—101). That means that creativity, and as we will see art as such, is localized in the brain and understood as a neuronal effect that can be measured. As a next step Hoesle rematerialized the EEG waves in a wooden sculpture (Hoesle, 2014, p. 101). The status of that piece of art is very interesting: it is art displaying images of neurological measurements of people's brain activities when beholding pieces of art. So art is mediated by neuroscience activities and these activities are, at the same time, the interpretative core of the entire performance.

Later, Hoesle also became the father of brain painting (cf. Muenssinger et al., 2010). Brain painting involves controlling simple painting software by using a brain-computer interface. Event-related potentials are used to choose between colors, shapes, and the locations where one wants to place them. As a result the user gets minimalistic abstract paintings on the computer screen that can be saved as data files. For Hoesle, brain painting is a way to realize art in its essence. Without using muscular power, without paint and canvas, and

[c]<http://www.designboom.com/art/sara-asnaghi-what-have-you-got-in-your-head>.

[d]<http://www.steinerlenzlinger.ch/eye_brainforest.html>.

without a material outcome, art is reduced to a brain process. As Hoesle states, "that the cerebral events and the events at the brain–machine interface *are* the artistic process" and that the brain will be "the artist's workshop of the third millennium" (Hoesle, 2014, p. 103).

As with the examples at the beginning of this section, it is not important to raise the index finger here and to explain that these self-interpretations of neuro-art in general and brain painting in particular are exaggerations. It is more important to realize that neuro-art and the attitudes toward neuro-art once again realize a pattern that is more universal than unique for the neuro-hype: That scientific content, be it the objects, procedures, or theoretical claims of a science, invades society is found in history.

This time, my example stems from the time around 1900. The most popular science in the second half of the 19th century was of course biology, especially in connection with the theory of evolution published by Darwin in 1859. Dealing with these ideas had the flavor of the revolutionary, the progressive, and the unheard of. Evolutionary biology created interest not only in life and living creatures as such, but to their origin and their relations to each other. The people could understand themselves to be connected, via an unbelievably long line of descent, to the most original and most primitive forms of life, e.g., plants or jelly-fish. A poem by Arno Holz, one of the most important German writers of the time around 1900, illustrates this very well—it starts with the words: "Seven trillion years before my birth I was an iris" (Holz, 2002, pp. II.1, 59). One of the most industrious followers of Darwin and popularizers of biology was the German Ernst Haeckel. He was an educated biologist, professor in zoology, and an internationally renowned researcher specializing in deep-ocean animals. He also published many books for the education of the wider public and fought enthusiastically against the Christian way of thinking. In 1911 he was appointed the counter-pope. Haeckel is an example of the influence the biological hype had on art during that time.

If we look back to neuro-art, we can see a certain line of development. The motives of art seem to transform out of science matters: There are at the first place sketches and drawings that have their origin directly inside the sciences (e.g., the famous drawings by Santiago Felipe Ramón y Cajal). In a second step we have these matters depicted in scientific books, e.g., books for educational purposes. And finally there are the pieces of art. In this process, the actual object of the representation, e.g., the neuron, becomes a symbol that stands for the social and metaphysical meaning of the neurosciences.

If we now have a look at the parallel, we will find that Haeckel's books, which were scientific and popular, from the 1870s onward contained a lot of illustrations. (Many of them were collected later and included in Haeckel, 1904.) It has been shown by art historians that these illustrations influenced and inspired many of the artists of the Art Nouveau or Jugendstil (the German name; cf. Kockerbeck, 1986; Krauße, 1995; Mann, 1990; Wichmann, 1977, 1984).

Haeckel, for example, often displayed the skeletons of radiolarians, very small beings living in the water. Many artists were influenced by these drawings: The Swiss designer Hermann Obrist made fountains obviously displaying radiolarian patterns (see Bellerive & München, 2009, pp. 35, 39, 41, 43). The French designer René Binet took Haeckels radiolarian drawings as models for one of the main entrances to the world exhibition in Paris in 1900, the monumental gateway at Cours-la-Reine, and showed his systematic approach to transforming Haeckel's illustrations into design in his volume *Esquisses Decoratives* (ca. 1903). And in America, Louis Comfort Tiffany made jellyfish vases that show the contemporary fascination with the underwater world that, among others, Haeckel had triggered by his research.

Besides the references to our primitive parentage from the deep, Jugendstil is also known for its ubiquitous use of winding, loopy lines as symbols for the vital forces that stand behind all natural development. Bizarre specimens of medusae and other beings from botany and zoology Haeckel had depicted several times all strikingly show this feature. And artists have thousandfoldly taken it up as a pattern of style. Wall hangings, tablecloths, bedspreads, tapestries, and even chest locks again by Hermann Obrist might be considered as examples (see Bellerive & München, 2009, pp. 51–63, 90).

And, similar to the self-reflections from Hoesle or the approaches in neuro-esthetics today, there was theoretical reflection on art and science then too (cf. Krauße, 1995). Beauty and usefulness were thought to be connected via the evolutionary role they had played in the eons of biological descent and selection. Jugendstil could understand itself as the most true artistic style, the style in line with the best possible and most valid scientific knowledge that mankind ever had. Christopher Dresser, biologist and designer from Scotland, said that nature should be made a companion in design and that botany and zoology must be studied in order to be a good designer: no doubt that today several people would underline that one has to study the neurosciences in order to become a good designer or artist (cf. Dresser, 1873, pp. 259–260). (Dresser was not influenced by Haeckel but made his own and early start with Art Nouveau on the basis of his biological education.)

The example of art again allows for seeing the obvious parallel between aspects of the neuro-hype today and similar developments of former epochs. That science matters become popularized, become symbolic, and become finally issues of art seem to be regular aspects of science hypes.

NEURO-PHILOSOPHY AND ITS METAPHYSICS

I now turn to some more theoretical matters. In particular, I want to name and to scrutinize a pattern of thinking that is the core of neuro-philosophy, i.e., the philosophical branch of the hype. When I say this I do not imply that all philosophers writing about neuro-issues would adopt this pattern. What I mean is that neuro-philosophy, insofar as it is part of the popularization of the neurosciences and therefore proponent of spectacular claims and revolutionary interpretations, necessarily uses it. And this pattern of thinking is clearly neurological constructivism. Here is one example (taken from Metzinger, 2009, p. 23; Metzinger summarizes here the assumptions of other researchers that he would also accept):

> The idea is that the content of consciousness is content of a simulated world in our brains, and the sense of being-there is itself a simulation.

Neurological constructivism, in short, holds that the normal everyday reality that we know is the product of our brain. Constructivism as such can be a very convincing approach when it is applied to the philosophy of science or to the psychology of education. It would, then, stress our practical interaction with the world and show that theories or knowledge are not just given, but are rather the products of our work and life. However, if one tries to make constructivism the solution of basic epistemological and ontological problems, one will run into trouble. And this trouble is no less than traditional European metaphysics. Since many of the defenders of the neuro-hype understand themselves to be far away from classical metaphysics and religion, this situation implies a nice irony. Many of these researchers seem to be both naïve realists and neuro-constructivists at the same time. But this doesn't fit together—neuro-constructivism cannot escape its tradition: In fact, it is a classical argument found first in ancient Greek philosophy. It goes like this: There is a world of our everyday experiences, but this world is only apparent; it is an illusion. And there is a world of reality that we cannot directly experience. And this actual, real world is the cause of and reason behind the apparent world. Parmenides and others started this type of thinking and as Platonism it became the capital stock of European metaphysics. Adapting the famous phrase by Whitehead (1979, p. 39) we might say that neuro-philosophy is just another footnote to Plato.

In neuro-philosophy this way of thinking leads to a paradox situation that we can formalize like this:

All experience is a product of the brain.
Empirical knowledge is based on experience.
Brains are matters of empirical knowledge.
Brains are products of the brain.

This problem was seen a long time ago. It was George Berkeley who described the problem very clearly in his "Three Dialogues," where he argues against the materialist who had said that the brain causes all our experiencing and thinking. He says (Berkeley, 1713/1843, p. 183):

> The brain therefore you speak of, being a sensible thing, exists only in the mind. Now, I would fain know whether you think it reasonable to suppose that one idea or thing existing in the mind occasions all other ideas. And, if you think so, pray how do you account for the origin of that primary idea or brain itself.

About 300 years later the German neuro-researcher and neuro-philosopher Gerhard Roth (1997, p. 332, my translation) also made this point, putting himself in the position of an empirical brain researcher who tries to make claims about the relationship between brains and minds:

> If I thus say that the *brain* produces the mind (meaning mental states) then by that I can neither mean the *real* brain that I see and stimulate when experimenting with myself nor another person's brain that I analyze. So we face a puzzling situation: The brain that is accessible to me (the *real* brain) doesn't produce the mind; and the brain that, together with all reality, produces the mind (the *objective* brain as I have to assume plausibly) is not accessible to me.

Berkeley, criticizing his materialist dialog partner, as well as Roth, contemplating the limits of empirical neuro-research, come to the same result. In any philosophical statement claiming that the brain produces the mind the concept "brain" cannot refer to the empirical thing that can be found in the skulls of animals and humans because this thing is also said to be a product (a simulation, etc.) of the brain. The concept "brain," then, refers to a reality that transcends empirical reality or is behind empirical reality and thus becomes a metaphysical concept. I have focused on this in detail, because it makes absolutely clear that neuro-constructivism, taken as a philosophical approach, is necessarily a metaphysical approach. You cannot have one without the other. If we think about it this way, we necessarily postulate an entity more substantial than the things given in everyday life, a transcendent being. It is this implication I was alluding to when saying that neuro-constructivism implies a proof of God's existence. It is just this metaphysical constellation that has allowed for all the classical proofs we know from Augustine, from Anselm of Canterbury, from Thomas of Aquino, and it allows today for a neuro-scientific proof of God's existence (cf. Grübler, 2014b). If reality is a construction, there must be something that constructs reality. And if we are in the decade (or century) of the brain, the brain is the central metaphor we use to explain the world as it *actually* is. The objective brain Gerhard Roth was talking about is obviously just another name for the transcendent being that was traditionally named God. Berkeley did it this way. If all empirical reality consists of ideas, i.e., the contents of our minds, then there must be another mind behind our minds that induces the ideas, and this is—for Berkeley—the mind of God. Now, if all reality we experience is the product of the brain, then there must be another brain behind our brains that simulates this reality, and it doesn't matter that current neuro-philosophers would be reluctant to call it God—it plays the same role. (I have already mentioned that one might enjoy the irony of that constellation. ...)

As discussed the central metaphors of science popularization usually imply assumptions about the actual, the real world behind the ordinary one. And I'm sure that it is exactly this feature of science hypes that makes them appealing to people. Besides practical solutions they just offer contact with the *real* reality. And one can see how the metaphysics of neuro-philosophy does the same insofar as it unavoidably makes demands on transcendency.

Given some plausibility for this thesis—what would be the consequence?—I don't say that neuro-philosophers shouldn't do what they do. I only say that a typical, traditional pattern has been engaged once again. And saying this I want to stress that neuro-philosophy is a very traditional way of thinking. The metaphysical constellation I was talking about is typical for classical philosophy from Parmenides to Hegel. We have seen this pattern again and again in the European history of thinking and therefore we have to reject the attitude of novelty and revolution that often goes together with neuro-philosophy. On the contrary: neuro-philosophers applying neuro-constructivism are just antique thinkers. This is not forbidden, but is far from being avant garde.

NEUROLOGIZATION AND THE EUROPEAN SEARCH FOR SALVATION

Now I want to talk about the salvation aspect of the neuro-hype. Salvation means to escape the ordinary and finite world and to become part of a better one. Salvation means a cure to all the flaws and shortcomings of our terrestrial life. But how can neurologization help here? And, even if we call the brain God—how could this trigger our enthusiasm?

In order theto give answers I have to widen the scope. I have already outlined how neurologization shows continuity with former developments of the sciences insofar that it repeats certain patterns of speaking and thinking. But it would be strange if there were not also continuities in the overall tendency or aim all these developments point at. The issue we have to work on, then, turns into the question: Is the neuro-hype part of the typical European search for salvation? Or, in other words, is the enthusiasm for the neurosciences, among others, one of the contemporary expressions of that search?

If we look back again to the 17th century we see mechanics as the leading paradigm. Mechanics, originally, was the knowledge of making machines, i.e., devices for certain purposes technically prepared by a craftsman. This entire field of associations is implied in the leading metaphor. For how real this metaphor was taken shows this statement by Henry Power (1664, pp. 192–193):

> These are the days that must lay a new foundation of a more magnificent philosophy, never to be overthrown: that will empirically and sensibly canvass the phaenomena of nature, deducing the causes of things from such originals in nature, as we observe are producible by art, and the infallible demonstration of mechanicks: and certainly, this is the way, and no other, to build a true and permanent philosophy: For art, being the imagination of nature (or, nature secondhand) is but a sensible expression of effects, dependent on the same (though more remote) causes; and therefore the works of the one, must prove the most reasonable discoveries of the other. And to speak yet more close to the point, I think it is no rhetorication to say, that all things are artificial; for nature it selfe is nothing else but the art of God.

And this is the typical way of European research: making something is understanding it—understanding something is making it. Science cannot be separated from engineering, or, more to the point: science is a kind of engineering. If we follow this engineering paradigm through the centuries we can find its ubiquitous persistence: We find it to be viable in different epochs and in the context of very different sciences. Be it the mechanical context in the 17th and 18th centuries, be it the biological context of the 19th century, or be it the genetic and neurophysiological context of the 20th century: the objects of the sciences were understood to be mechanisms to be manipulated, copied, and, by doing so, understood in their true functions. And one need not search for long to find examples in the context of the neurosciences. Henry Markram, the leader of the Human Brain Project, says about the brain: "The best way to figure out how something works is to try to build it from scratch" (Lehrer, 2008).

What did the paradigm do to us during the centuries? Starting in the 16th century, the human body was more and more understood to be a mechanical machine. Scientists like Andreas Vesalius (1514–64), William Harvey (1578–1657), and Giovanni Alfonso Borelli (1608–79) explained the human body explicitly in terms of contemporary technology, meaning in mechanical, pneumatic, and hydraulic paradigms (cf. Borelli, 1680; Harvey, 1628; Vesalius, 1543). Only a little later probably the most well-known text along these lines came out: Julien Offray de La Mettrie's (1709–51) *Machine-Man* of 1749. While Descartes took the soul to be a separately created immaterial essence connected to the body machine, for La Mettrie the soul was not a separate entity but mind, mood, and morals directly emerge from the brain. And thus the entire ensemble of the human being was seen as a machine. These ideas found popular expression at the time; for instance, the European enthusiasm for mechanical android automata (cf. Grübler, 2014a).

And what about today? The paradigm enjoys unlimited respect. Of course, we know that mechanics alone cannot symbolize the complexity of the human machine; biology, genetics, and informatics are needed too. But today the man-machine hypothesis is common sense and a starting point in all sciences. For most researchers it is enough to accept this assumption as a heuristic one. Others confess it explicitly. According to the physicist Frank Tipler (1994, p. 1), his theory

> requires us to regard a "person" as a particular (very complicated) type of computer program: the human "soul" is nothing but a specific program being run on a computing machine called the brain.

And the artificial intelligence researcher Rodney Brooks (2008, p. 71) says: "I am a machine. So are you."

One might say: Nice story, but what about salvation? Maybe the reader, as many people did, thinks that the formula "the human being is a machine" is just the opposite of salvation. It is mere materialism, the abandonment of higher dignity, and the total loss of humanity. I don't think so and I have the impression that authors like Tipler, Brooks, and many others can help us to understand it. These authors are usually called transhumanists because they like to overcome the human being as it is and transform us into better, more clever, and in the first place, immortal beings. I'm skeptical about these plans, but this is not the point. The point is that for them the acceptance of the man-machine formula is mandatory. And this is the key for understanding: against all romantic criticism, the man-machine hypothesis is a formula of salvation and it is the most central and most continuous formula of European science development.

If the human mind emerges from a machine in the literal sense, a machine that can be completely known in its functioning and that can be reconstructed and functionally substituted by human engineers, then we as human beings can indeed escape our finite existence. If man is machine, and if man is engineer, then our hopes are no longer limited. Then the conditions of our very being can be under the control of our technology. I think that this is, necessarily very abridged, the implicit message and motivation behind central developments of the European sciences.

And what about neuroscience in particular? The historical role of the neurosciences is to substantialize the claim that man is machine (concerning the mind). LaMettrie and others *said* that the mind is only a product of the brain, and neuro-constructivism goes on *saying* so. But only if we *knew* exactly how the brain produces the mind could we go on with our search for salvation this way. So, in some sense, the project of the neurosciences is the keystone of the European project of rescuing mankind by technological means. If it succeeds we will win final certainty; if it fails we will lose our hope and destination. Taken the word apocalypse in its actual meaning (it stems from the Greek word *apokalytein*, which means disclose, uncover, or reveal) we might say that this is the *apocalypse* of the human being.

So actually, there is no problem to explain why the neuro-hype happens and why other hypes have happened earlier. The European sciences, besides their practical achievements, have always been ways of coming into contact with the *real* reality and implied hopes for the salvation of the human being. Any criticism focused on formal issues of the neuro-hype only overlooks these substantial characteristics completely and misses the motivational dynamics behind the essential developments.

ON APPLIED ETHICS AND THE BACKGROUND OF THE GIVEN INTERPRETATIONS

It might be that there is still some doubt about the status and plausibility of the theses presented here. First of all, the story about the European sciences presented above is, of course, an interpretation. However, interpretations are not arbitrary—they can be more or less plausible. In this final paragraph I want to show a way to find plausible interpretations of currently ongoing developments in society. (For a more comprehensive outline see Grübler, 2015b.)

It might not have been obvious until now, but the motivation behind my analyses is actually the purpose of making applied ethics more interesting, more powerful, and, in a certain sense, more authentic. My experience with applied ethics told me that many of the debates in applied ethics are not debates about ethical details but debates about something that we can call by different names, e.g., total view, ideology, religion and weltanschauung. It is more the comprehensive ideas about the world and the human being that is the motivation behind current debates than questions about the status of an embryo, a piece of landscape, or a particular experiment or technology. Therefore applied ethics should try to explicate the questions people engaged in moral debates are really motivated by. This would be questions like: Where will this or that way of development lead us? or What are the implicit normative and ontological assumptions that one has to accept if he or she proposes going this or that way? If one asks such questions one applies a simple semantic pattern that we know from the so-called practical syllogism (cf. Von Wright, 1971, part III). It says:

Person p pursues the aim A.
Person p is convinced that, in order to realize A, the action B must be executed.
Person p executes action B.

The practical syllogism is not *logically* sufficient. It just displays the way we usually explain actions and talk about their reasons—it shows the semantics of actions. I want to propose using this semantic relationship between actions and reasons to find out about the implicit motivations and assumptions the public engagement in scientific projects might be based on. This allows us to describe human engagement in scientific or technological developments like this:

This or that person (or group of people) entertains a certain theory about the world and the role of the human being in it (total view, weltanschauung...) and in light of this theory certain developments or achievements appear to be aims and values.
This or that person (or group of people) knows that this or that action is appropriate to realize the mentioned aims and values.
As a result the person (or group of people) engages in the concerned action(s).

Again, this is just the *semantic structure* we normally use when explaining actions. It is neither a *formal* conclusion nor an *empirical* description: I do not naively claim that people always act this way. But I imagine a situation in which people are already engaged in a debate about scientific, technological, and societal developments. In that situation none of the debaters could deny the described semantic structure without discrediting his or her own contributions to the debate. (Defending an "action" that would not be situated in this structure is self-contradicting by nature.)

If we apply this structure to questions of current science and technology developments we face a challenge that makes interpretation necessary. Because the only thing that is obvious in that case is the action that we want to explain: Establishing neuroscience projects or feeding the neuro-hype—i.e., visible, audible, readable activities that individuals or society as a whole show. So, what we have is only the conclusions of the practical syllogism, and what we want to have is the premises. There is no other way from the conclusion to the premises than interpretation.

Such interpretations have to be given very cautiously. When dealing with debaters, one cannot just produce assertions, but one should vest these assertions with an as-if-clause. One might formulate: In your activities you behave *as if* you would entertain this or that idea, or, in your contributions to the debate you react *as if* ... and so on. This prevents applied ethics from appearing to wheedle the interpretations into the debaters but turns the interpretations into questions given to them. Such a procedure might allow neverending debates to go a step forward by opening the debate for the motives and assumptions behind the debate. (This is, of course, no way to fully substitute other applied ethics approaches, but rather to amend and complete them.)

Finally, I give an example for the use one might make of the described strategy in the neuro-context. This will shed light on the status and origin of all the theses presented in this chapter. I have been talking about transhumanist researchers. One of their central ideas is the so-called uploading of the mind (cf. Kurzweil, 1999; Moravec, 1988; Tipler, 1994). They think that our minds could be run as software on computers and by that we might become immortal. Now, if we had to realize this, what would be the next step? I think we should work on computers powerful enough to emulate the computing power of brains and we should try to simulate the functions of the brain in more and more detail. We should do this as a merger of neuro- and computer science, and we should spend immense amounts of money for it, because immortality seems to be an important issue. And if now somebody said: hey, good news, the European Union already has this project, it is called Human Brain Project (https://www.humanbrainproject.eu) and it spends about one billion Euros for the brain simulation, what would you think? I would think that they behave *as if* they were transhumanists on their mission of finally proofing the man-machine formula.

The above cited leader of the project once said (Seidler & Briseño, 2011):

> It is one of the three big challenges of mankind. We have to understand the earth, the universe and—the brain. We need to understand what makes us human.

Doesn't it imply that by simulating and finally emulating the brain we can understand what the human being really *is*? That the essence of man will show up in this merger of neuro- and computer science? And isn't exactly this the last keystone we have to set in order to know whether technological salvation is possible?—the apocalypse I was talking about? I can't help it but it seems to me that my interpretation is not without plausibility. Society in its central paths of science and technological development behaves as if this was our destination.

Now, the point for applied ethics is that by giving this perhaps provocative interpretation we can get the proponents of the neuro-hype to further articulate the aims and values that they apparently support. The statement cited above itself could be the matter of very systematic reflections on the involved anthropology. But in any case we would come to a situation in which the proponents of neuro-research without reservation, of neuro-philosophy as a kind of total view, and of transhumanism itself have to declare their implicit faith. And I do mean faith, not only belief that could be checked empirically. For instance: the idea that the human mind is a product of the material brain implies, as we have seen, actually the postulate of a transcendent entity behind empirical. It is clear by most simple epistemological reflections that you cannot know anything about this transcendent being. So, if someone claims that the conscious mind *can* be emulated by material technology this person pretends to *have* this knowledge (again taken from the interview with Henry Markram in Lehrer, 2008):
Markram: "If we build this brain right, it will do everything."
Reporter: "Including self-consciousness?"
Markram: "When I say everything, I mean everything."

Here the question is treated as an empirical one. But in fact this question cannot be answered on the basis of knowledge that could be gained empirically. It actually entails speculation on the relationship between the empirical world and the transcendent being generating it (for details see Grübler, 2015b, pp. 304–320). What else, then, could we say but that the answer given actually relies on faith? Again, this is not forbidden, and it might be that it is the way culture is typically brought forward: the only thing that we have the right and—I think—also the duty to ask the proponents of the big changes and develop-ments of our society is: please confess. And then let's make these con-fessions an issue of debate. I'm convinced that such debates would be more fruitful and interesting than applied ethics and formal criticism have been so far.

References

Bellerive, M., & München, G. S. (2009). *Hermann Obrist. Skulptur, Raum, Abstraktion um 1900.* Zürich: Scheidegger & Spieß.
Bennett, M. R., & Hacker, P. M. S. (2003). *Philosophical foundations of neuroscience.* New York: Wiley.
Berkeley, G. (1843). *Three dialogues II, the works of George Berkeley* (Vol. I). London (Original work published 1713).
Binet, R. (ca. 1903). *Esquisses décoratives.* Paris: Librairie Centrale Des Beaux-Arts.
Borelli, A. (1680). *De motu animalium.* Rome.

Boyle, R. (1772). Of the high veneration man's intellect owes to god. In T. Birch (Ed.), *The works of the honourable Robert Boyle* (pp. 130–157). London: Tome V.

Brooks, R. (2008). I, Rodney Brooks, am a robot. *IEEE Spectrum, 45*(6), 71–75.

Crockett, M. (2012). *Beware neuro bunk.* Retrieved from <https://www.ted.com/talks/molly_crockett_beware_neuro_bunk/transcript>.

Dresler, M. (2009). *Neuroästhetik.* Leipzig: Seemann.

Dresser, C. (1873). *Principles of Victorian decorative design.* London: Cassell Petter & Galpin.

Grübler, G. (2014a). Android robots between service and the apocalypse of the human being. In M. Funk (Ed.), *Future of robotics in Germany and Japan: Intercultural perspectives and technical opportunities* (pp. 147–162). Frankfurt am Main: Lang.

Grübler, G. (2014b). Der Ausgang der Philosophie aus ihrer selbst verschuldeten Bedeutungslosigkeit. *Journal für Religionsphilosophie, 3*(3), 118–124.

Grübler, G. (2015a). Mein Hirn geht auf Sendung: Vom Sinn der Mediatisierung des Wirklichen in den Neurowissenschaften. In H. J. Petsche, J. Erdmann, & A. Zapf (Eds.), *Virtualisierung und Mediatisierung kultureller Räume* (pp. 127–140). Berlin: Trafo.

Grübler, G. (2015b). *Wissenschaft, Moral und Heil.* Würzburg: Königshausen & Neumann.

Haeckel, E. (1904). *Kunstformen der Natur.* Leipzig: Wien.

Harvey, W. (1628). *Exercitatio Anatomica de Motu Cordis et Sanguinis in Animalibus.* London.

Hoesle, A. (Ed.) (2012). *Pingo ergo sum: Das Bild fällt aus dem Hirn* Rostock: Weidner.

Hoesle, A. (2014). Between neuro-potentials and aesthetic perception: Pingo ergo sum. In G. Grübler, & E. Hildt (Eds.), *Brain-computer interfaces in their ethical, social and cultural contexts* (pp. 99–108). Dortrecht: Springer.

Holz, A. (2002). *Phantasus.* Stuttgart: Reclam.

Kockerbeck, C. (1986). *Ernst Haeckels, "Kunstformen der Natur" und ihr Einfluß auf die deutsche bildende Kunst der Jahrhundertwende. Studie zum Verhältnis von Kunst und Naturwissenschaft im Wilhelminischen Zeitalter.* Frankfurt am Main: Lang.

Krauße, E. (1995). Promorphologie und evolutionistische ästhetische Theorie: Konzept und Wirkung. In E. M. Engels (Ed.), *Die Rezeption von Evolutionstheorien im 19. Jahrhundert* (pp. 347–394). Frankfurt am Main: Suhrkamp.

Kurzweil, R. (1999). *The age of spiritual machines.* New York: Viking Press.

Lehrer, J. (2008, March 3). Out of the blue. *Brain & Behavior, Seed Magazine.* Retrieved from <http://seedmagazine.com/content/print/out_of_the_blue/>.

Mann, R. (1990). Ernst Haeckel, Zoologie und Jugendstil. *Berichte zur Wissenschaftsgeschichte, 13,* 1–11.

Meshberger, F. L. (1990). An interpretation of Michelangelo's creation of Adam based on neuroanatomy. *JAMA, 264*(14), 1837–1841.

Metzinger, T. (2009). *The ego-tunnel.* New York: Basic Books.

Moravec, H. (1988). *Mind children: The future of robot and human intelligence.* Cambridge, MA: Harvard University Press.

Muenssinger, J. I., Halder, S., Kleih, S. C., Furdea, A., Raco, V., Hoesle, A., et al. (2010). Brain painting: First evaluation of a new brain-computer interface application with ALS patients and healthy volunteers. *Frontiers in Neuroscience, 4,* 182.

P.M. Magazin (2008). Können Sie Ihrem Gehirn wirklich vertrauen? November *P.M. MAGAZIN* (11), 48–54.

Power, H. (1664). *Experimental philosophy.* London.

Rosenzweig, M. R., Breedlove, S. M., & Watson, N. V. (2005). *Biological psychology: An introduction to behavioral and cognitive neuroscience* (4th Edition). Sunderland, MA: Sinauer Associates.

Roth, G. (1997). *Das Gehirn und seine Wirklichkeit.* Frankfurt am Main: Suhrkamp.

Roth, G. (2003). *Fühlen, Denken, Handeln.* Frankfurt am Main: Suhrkamp.

Seidler, C., & Briseño, C. (2011, May 12). Human brain project: Forscher basteln an der Hirnmaschine. *Spiegel Online Wissenschaft.*

III. THE NEUROSCIENCES IN SOCIETY. SOCIAL, CULTURAL, AND ETHICAL
IMPLICATIONS OF THE NEURO-TURN

Suk, I., & Tamargo, R. J. (2010). Concealed neuroanatomy in Michelangelo's separation of light from darkness in the Sistine Chapel. *Neurosurgery, 66*(5), 851–861.

Tipler, F. (1994). *The physics of immortality: Modern cosmology, god and the resurrection of the dead*. New York: Anchor.

Vesalius, A. (1543). *De humani corporis fabrica*. Basel.

Von Wright, G. H. (1971). *Explanation and understanding*. London: Routledge & Paul.

Whitehead, A. N. (1979). *Process and reality*. New York: Free Press.

Wichmann, S. (1977). *Jugendstil—Art Nouveau*. Zürich: Atlantis.

Wichmann, S. (1984). *Judendstil-Floral-Funktional*. Zürich: Atlantis.

III. THE NEUROSCIENCES IN SOCIETY. SOCIAL, CULTURAL, AND ETHICAL
IMPLICATIONS OF THE NEURO-TURN

"A *Mind Plague* on Both Your Houses": Imagining the Impact of the Neuro-Turn on the Neurosciences

M.M. *Littlefield*

University of Illinois at Urbana-Champaign, Urbana, IL, United States

In his recent science fiction novel, *Mind Plague*, Kyle Kirkland chronicles the rise of a fatal disease, Synapse Interruption Syndrome (SIS), and its impact on the neurosciences (and neurotechnologies industry) of the near future. In the opening chapters of Kirkland's story, Bailey Breege, one of the neuroscientists employed by Synaptic Modulation Technologies and the principal character of the novel, has a brief conversation with his employer whose major concern is the increasing spread of SIS and the rumor that their company is responsible for the disease. Confronted with the threat of a pandemic, Breege responds with concern not only for the potential loss of life, but the potential damage to neuroscience.

> What if this thing destroys neuroscience? None of the sciences are very popular right now, but especially not brain science. What if people get the idea that neuroscience does more harm than good? What if brain research becomes impossible in the future, anywhere in the world? That might happen if we don't stop this disease. Not only will it take a lot of lives, it will ruin companies, many more than just our own, as well as university departments devoted to brain science. It will mean the end of any chance we have of understanding the human brain. (Kirkland, 2014, p. 19)

What if this thing destroys neuroscience? While Breege refers specifically to SIS, his question led me to extrapolate in multiple directions: to

198

© 2017 Elsevier Inc. All rights reserved.

potential futures and contemporary debates in which some*thing* threatens the neurosciences. To some, this logical twist may sound strange; after all, the neurosciences have produced some of the most influential explanatory theories in recent memory. And yet, in the midst of what has been characterized as a neuro-turn, the neurosciences have also experienced a discursive backlash.

Tensions between the neurosciences and the humanities and social sciences have been relatively palpable over the past decade. Humanists and social scientists have responded to what feels like an encroaching neuro-presence: the proliferation of neuro-disciplines, the adoption of brain-centrism that has accompanied an explanatory shift from gene to neuron, and a media machine dedicated to all things "neuro-." For these non-neuroscientists, their scholarship has sometimes been a reactionary project fueled by the necessity of responding to the perceived infringement of the neurosciences (and, more generally, the "neuro-") on areas once established to be under the purview of the liberal arts and social sciences. At the same time, neuroscientists are reacting to hyperbolic media representations and potentially dubious scientific studies (often grouped under a rubric of pop neuroscience). Add to this the fact that the application of the "neuro-" is often uneven, undefined, and multidirectional (neuro-disciplines can and have been initiated by those in the neurosciences and those seeking the ethos of the neurosciences), and you have a discursive storm of confusion, blame, fear, and potential rivalry. As a reminder of these tensions, my title remixes Mercutio's famous line from *Romeo and Juliet*—a play that aptly addresses the impacts of feuding adversaries—with Kyle Kirkland's latest science fiction novel. While rivalries figure into this chapter in multiple ways, this title should also draw your attention to the idea of mutuality; after all, Mercutio's curse applies equally to both parties.

This chapter begins with the assumption that the "neuro-turn" has generated *imagined* consequences not only for the humanities and social sciences (over which much ink has already been spilled), but also for the neurosciences. Here, instead of worrying about the encroachment of the "neuro-" on the humanities and social sciences, I explore how the neurosciences are discursively—if not realistically—affected by the "neuro-" and the neuro-turn. I invoke the idea of "imagination," because I wish to indicate that mine is a somewhat speculative endeavor, not one that involves a qualitative or quantitative study of neuroscientists themselves. There are no interviews, no questionnaires; rather, I am interested in the ways that we represent the effects of a "neuro-turn" on (what we imagine to be) its field of origin: the neurosciences. As I explain, even this assumption—concerning origins and cohesive fields—must be called into question.

III. THE NEUROSCIENCES IN SOCIETY. SOCIAL, CULTURAL, AND ETHICAL
IMPLICATIONS OF THE NEURO-TURN

I ground my discussion in two examples that imagine a discursive impact of the neuro-turn on the neurosciences: first, the 2012–13 backlash against the neurosciences that resulted largely from the publication of Sally Satel and Scott O. Lilienfield's *Brainwashed: The Seductive Appeal of Mindless Neuroscience*; second, the recent criticism of the European Human Brain Project from neuroscientists themselves, which was picked up and recharacterized by several media outlets. In each case, I argue that the imagined impact of the neuro-turn reveals much more about the public appeal of and apprehensions about the neurosciences than anything about the neurosciences themselves. In the final section of the chapter, I theorize some of the stakes for both the neurosciences and the neuro-turn: these include problems of translation, reputation, and marketability. Before delving into my examples, I first set the stage and define some terms.

SETTING THE STAGE: DEFINITIONS AND SCOPE

I have been thinking and writing about the most recent neuroscientific turn for a few years now. And, like some inter- or transdisciplinary scholars, I have attempted to approach the conversation with a foot on both sides of the divide. I went so far as to run an fMRI experiment a few years ago to get a sense of the neuroscientific process (Littlefield, Dietz, Fitzgerald, Knudsen, & Tonks, 2015). Even so, I know that my training in the humanities has left a residue on my approach to the neurosciences; I have spent a lot of time thinking about the potential impact of the neurosciences on the humanities and social sciences: imagining the loss of closely held "proper objects," concerned about the essentialization of being human, and worried over what appeared to be a mad rush into the Neuro Revolution (Lynch, 2009). But in so doing, and particularly given my experiences across the divide, I also came to know the neurosciences as a complex, nuanced, and very human endeavor, no different, in many ways, from our own work in the humanities. Ultimately, thinking about the "neuro-" often necessitates taking stock of our own positionality: are we threatened by its reach? Do we wish to construct or reconstruct its definition and scope for particular (disciplinary) purposes? Have we taken time to appreciate its potential—and potential multiplicity—for each of the fields involved?

The phrase "the neuro-turn" has been used to describe several different phenomena: an academic neuroscientific turn in the humanities and social science disciplines, spearheaded by researchers who have allied themselves to or incorporated the neurosciences into their research (Littlefield & Johnson, 2012); a more wide-ranging adoption of the "neuro-" exemplified by the media craze over neuroscientific studies on everything from depression to voting preference; and a scientific/cultural turn that represents an explanatory shift from genes to neurons that affects everything from popular representations to experimental funding. Recognizing the vast array of positions that are often grouped under the "neuro-turn" umbrella reminds us that we are not talking about a coherent phenomena. Indeed, the neuro-turn might be better defined as a series of disagreements and divergences! However, there does appear to be a phase shift happening—or being constructed within and outside of the academy—and, so, for the purpose of this chapter, my solution is to employ an expansive definition of the neuro-turn, characterizing it as a broad, cultural, and institutional interest in the neurosciences: their application, uptake, experimental norms, cultural cache, funding structures, etc. If, as Laura Lately argues, "neuroscience is a living topic" (2013), then so, too, is this neuro-turn; its ever shifting territories, pundits, and reactionaries reflect the difficulties we experience whenever we try to contain the neuro-turn within a single definition.

Part and parcel of the neuro-turn—and one way of illustrating the broad appeal of the neurosciences—is the prevalence of the "neuro-" prefix and the multitude of things, phenomena, ideas, and disciplines to which it has been applied—or attached. First, and among academic disciplines, the neuro-prefix indicates very different levels of buy-in to the neurosciences. Here, I would like to draw attention to the power dynamic in play: in name, we see humanities and social science disciplines adding or adopting the neuro- and not the other way around. As a result, fields have formed such as neuro-aesthetics, but not aesthetic-neuroscience. Moreover, and as Jenell Johnson and I have argued, the neuro- of neurohistory is very different from the neuro- of neuroeconomics (Littlefield & Johnson, 2012); in addition the uptake of neuroscientific knowledge is quite uneven if one looks to the bibliographies of neuro-academic articles (Johnson & Littlefield, 2011). Many of these differences and their attendant problems have been chronicled in no less than five critical collections and books that were published between 2011 and 2013: Choudhury and Slaby's *Critical Neuroscience: A Handbook of the Social and Cultural Contexts of Neuroscience* (2011), Ortega and Vidal's *Neurocultures: Glimpses Into an Expanding Universe* (2011), Pickersgill and van Keulen's *Sociological Reflections on the Neurosciences* (2012), Littlefield and Johnson's *The Neuroscientific Turn: Transdisciplinarity in the Age of the Brain* (2012), and Rose and Abi-Rached's *Neuro: The New Brain Sciences and the Management of the Mind* (2013). The years between 2009 and 2013, when these books were written and published, were also the time of the most salient backlash against the neuro-turn in the broader media, as I will discuss in a moment.

Notably, the neuro-turn has slightly different valances in and outside of the academy. Among popular writers and media pundits, the neuro-turn and its prefix imply a tendency toward pop-neuroscience, hype, and neuromania. Raymond Tallis warned of the proliferation of

"NeuroTrash" (2009), and Susan Fitzpatrick argued against what she called the "Cognitive Paparazzi" (Fitzpatrick, 2005). Likewise, popular books have both championed and decried the reach of neuroscientific knowledge. Take, for example, Satel and Lilienfeld's *Brainwashed: The Seductive Appeal of Mindless Neuroscience* (2013), Raymond Tallis' *Aping Mankind: Neuromania, Darwinitis, and the Misrepresentation of Humanity* (2011), Lengrenzi and Umilta's *Neuromania: On the Limits of Brain Science* (2011), Noë's *Out of Our Heads: Why You Are Not Your Head and Other Lessons From the Biology of Consciousness* (2010), Zac Lynch's *Neurorevolution*: How Brain Science Is Changing our World (2009), and Lone Frank's *Mindfield*: How Brain Science Is Changing our World (2009). Some of these writers convey skepticism about pop neuroscience, others champion the impending changes brought by the neurosciences. But no matter the valance of the message, these portrayals have *imagined* serious consequences for the neurosciences.

I use the term "imagine" here for several reasons: first, I am a humanist who works in the subfield of literature and science studies, so I am, at heart, fascinated by the ways that we envision various scenarios through language and context. Science-fiction narratives, such as *Mind Plague*, provide points of departure for thought experiments; and discourses used by media pundits, scientist, and authors alike illuminate assumptions, fears, hopes, and uncertainties about vast and ever-changing fields, such as the neurosciences. Many of the examples I discuss in this chapter are hypothetical, and as such, they necessitate a bit of skepticism even as we consider the potential consequences outlined by imagined scenarios. Likewise and relatedly, the divide that has been established between the neurosciences and the humanities (and social sciences) is not a true division, but a construction that helps to define each side (Snow, 1959/2012). Literature and science scholars have long argued that divisions between the humanities and social sciences are less about true distinctions and more about self-construction (Littlefield, 2011; Squier, 2004): in this vein, imagining the neurosciences entails imagining ourselves. And so, for me, "imagining" the impact of the neuro-turn on the neurosciences is not only a necessary approach, but one that reveals much about our assumptions, hopes, and fears about the current and future state of neuroscientific research. The second reason I invoke imagination concerns the neurosciences themselves: the field is a vast conglomerate of numerous disciplines, so it would be imprecise to speak of it in any cohesive and monolithic way. When we talk about *the* neurosciences, we are always already imagining a singular and unvarying field into existence. Throughout this chapter, I am going to continue using the plural to describe the neurosciences—even when it sounds strange—to remind us of that fact.

CASE STUDY: 2012–13 BACKLASH

In Kyle Kirkland's novel, it takes a pandemic disease to threaten neuroscience's reputation. But we are given some indication that other factors are also at play. "None of the sciences are very popular right now," Breege tells his boss, "but especially not brain science." As hard as it is to imagine the neurosciences as *unpopular*, the 2012–13 backlash against all things "neuro" led some pundits to imagine the negative impacts of this neuro-turn on the neurosciences. My intention here is not to identify origins or even genealogically trace the 2012–13 backlash; instead, I am interested in how the backlash *imagined* the impact of the neuro-turn on the neurosciences.

The 2012–13 backlash was spawned by the combination of expanding media coverage of neuroscientific studies, the spread of the neuro-prefix among academic disciplines, and the subsequent critique of this neuro-distribution by neuroscientists, bloggers, and academics from the humanities and social sciences. As Alissa Quart notes in her often cited *New York Times* review, "Neuroscience: Under Attack," "the problem isn't solely that self-appointed scientists often jump to faulty conclusions about neuroscience. It's also that they are part of a larger cultural tendency, in which neuroscientific explanations eclipse historical, political, economic, literary and journalistic interpretations of experience" (Quart, 2012, p. 2). Admittedly, there is a gatekeeping strategy at work protecting proper objects: interpretations of experience, Quart believes, belong in the social/humanistic fields. Yet, the consequences of this neuro-turn are imagined to endanger not only the humanities and social sciences, but the neurosciences as well. Here, I am interested in two strains of commentary that build on each other like a faulty chain of misinterpretations. The first kind of commentary seeks to redeem the neurosciences from the mania of the neuro-turn. The second kind of commentary often builds on the first, yet it pursues the problem/risk to its proverbial source, rejecting the neurosciences as inherently flawed or uncontainable. Both sets of commentary imagine the direct impact of the neuro-turn on the neurosciences. Let's take a look at just a few examples.

At the heart of much of the 2012–13 backlash is Sally Satel and Scott Lillienfeld's book, *Brainwashed: The Seductive Appeal of Mindless Neuroscience*. Those who have read even the introduction to this book would recognize that the two authors are interested in critiquing a particular kind of popularized neuroscience—what they term "mindless neuroscience"—and they note explicitly that:

> The problem with such mindless neuroscience is not neuroscience itself. The field is one of the great intellectual achievements of modern science. Its instruments are remarkable. The goal of brain imaging is enormously important and fascinating:

to bridge the explanatory gap between the intangible mind and the corporeal brain. But that relationship is extremely complex and incompletely understood. Therefore, it is vulnerable to being oversold by the media, some overzealous scientists, and neuroentrepreneurs who tout facile conclusions that reach far beyond what the current evidence warrants—fits of "premature extrapolation," as British neuroskeptic Steven Poole calls them. When it comes to brain scans, seeing may be believing, but it isn't necessarily understanding. (Satel & Lilienfeld, 2013, p. xiv)

In this explanation, Satel and Lillienfeld represent the neurosciences as a great intellectual achievement that has been "oversold" and misunderstood. They even note that: "an eddy of discontent is already forming. 'Neuromania,' 'neurohubris,' and 'neurohype'—'neurobollocks,' if you're a Brit—are just some of the labels that have been brandished, sometimes by frustrated neuroscientists themselves" (Satel & Lilienfeld, 2013, p. xiv). When Satel and Lillienfeld imagine the impact of the neuro-turn on the neurosciences, they blame those who have fallen prey to the hype, which is diluting an "enormously important field." Their narrative is one of recognition and redemption.

But Satel and Lillienfeld's text has been used to imagine a very different and more damning impact of the neuro-turn on the neurosciences—at least in the wide-ranging circulation and recirculation of ideas in the popular press. Take, for example, neuroscientist Gary Marcus's arguments about another op-ed piece by nonscientist David Brooks (2013). Marcus contends that writers, like Brooks, are taking Satel and Lillienfeld's claims too far. "Some like David Brooks in the *New York Times* (https://www.nytimes.com/2013/06/18/opinion/brooks-beyond-the-brain.html), are using books like 'Brainwashed' as an excuse to toss out neuroscience altogether. In Brooks's view, Satel and Lilienfeld haven't just exposed some bad neuroscience; they've gutted the entire field, leading to the radical conclusion that 'the brain is not the mind'" (Marcus, 2013, p. 2). Despite Marcus' characterization, Brook's original piece includes a fairly well argued case for why we should be cautious when it comes to buying into neuroscientific data. Interestingly, though, Marcus' read of Brooks' read of Satel and Lilenfeld *imagines* a more substantive impact than one can find evidence for in either of the original documents.

Outside of Satel and Lilienfeld's appraisal of "mindless neuroscience," critiques often focus not on the uptake of the neurosciences in the media, but on the science's inherent limitations. Alissa Quart's editorial concerning the recent backlash (which I mentioned before) concludes by noting "it's not hard to understand why neuroscience is so appealing. We all seek shortcuts to enlightenment. It's reassuring to believe that brain images and machine analysis will reveal the fundamental truth about our minds and their contents. But as the neuro doubters make plain, *we may be asking too much of neuroscience*, expecting that its explanations will be definitive" (Quart, 2012, p. 2, my emphasis). Here, instead of redeeming the "real" neuroscience and chastising the pop, commentators such as Quart imagine the impact as more deeply damning: a reason to believe "we may be asking too much of neuroscience." We could read Quart's reaction as cautiously conscientious—a recognition that the neurosciences, like all human endeavors, have limits; but that's not how it's often read by her interlocutors. What is far more interesting, here, are the ways that Quart's review has been worried over by neuroscientists who, like Kirkland's main character, fear for the reputation of the field.

Take, for example, Pete Etchells, a postdoctoral psychologist at Bristol, who in his July 2013 blog for SciLogs, writes that:

the problem is that the backlash itself is going too far; by zeroing in too much on the limitations of neuroscience, the field as a whole is misrepresented as something that has no explanatory power whatsoever. It's the same problem that we often see elsewhere when scientific disciplines are generalised in the media; critics cherry-pick their examples of some dodgy neuroscientific reporting, and make out as if it represents the entire field. (Etchells, 2013)

Etchells is not wrong in his assessment: the generalized representations of disciplines, cherry-picked examples, and a focus on limitations all characterize elements of the backlash. As a solution, Etchells suggests what amounts to good-old-fashioned gatekeeping. Leave the neurosciences—and their representations—to the experts, he insinuates. Those who seek to characterize the neurosciences should know more about the nuances of the fields involved, that the neurosciences are a conglomerate of several disciplines, and that the research of one (or several) does not stand in for all—it especially does not stand in for the neuroscientists who are doing "good" science. Scholars in science and technology studies have long since debunked the myth of good vs bad or junk science, but here, the dichotomy lives on.

Moreover, Etchell's gatekeeping extends outward from the neurosciences to anyone using the neuro-prefix:

I do, however, have one little request to add to the whole debate. Please, for the love of all that is good in the world, can we stop using "neuro" as a prefix? Neuropolitics, neuromarketing, neurolinguistic programming, neuroadvertising, neurodoping—please, it has to stop. It plays into the misperception that neuroscience is a cowboy science, and lends an air of credibility to some awfully ropey concepts. (Etchells, 2013)

Stop diluting our brand, he insists. Those in the neurosciences have worked hard for their reputation, and the use of the neuro-prefix allows other fields to capitalize on an ethos that does not belong to them. Ironically, Etchells' shares such sentiments with one of the "backlash critics," Raymond Tallis, whom he tags as problematic. In one of Tallis's essays, the author of "NeuroTrash" admonishes, "I am utterly dismayed by the claims made on behalf of neuroscience in areas outside those in which it has any kind of explanatory power; by the neuro-hype that is threatening to discredit its real achievements" (Tallis, 2009). In his statement we should hear echoes of Satel and Lilienfeld's arguments concerning the achievements of the neurosciences and Etchells' call to maintain boundaries around acceptable neuroscientific work.

As we have seen, the backlash began as a way to rout pop neuroscience—to distinguish between acceptable (good) science and flawed (bad) science and the representation of each in the media. However, some who responded to the backlash also sought to extinguish connections between the newly emergent neuro-disciplines and the neurosciences. This gatekeeping was fueled, at least in part, by the potential discursive harm that the neuro-turn could perpetrate against the neurosciences. Importantly, the neuro-turn has not only influenced entities, disciplines, and media pundits outside of the neurosciences; indeed, and as I demonstrate in the next example, the neuro-turn has animated debates among neuroscientists themselves. Among their concerns: where and how will funding for the neurosciences be channeled in the coming decades, what kinds of projects will be supported, and how will governmental initiatives affect the public perception of neuroscientific research?

CASE STUDY: AN OPEN LETTER TO THE HUMAN BRAIN PROJECT

The backlash against the neurosciences—such as it was—(and against the neuro-) have subsided somewhat, replaced by new visions for the expansion of the field. In the United States the BRAIN Initiative has replaced the Decade of the Brain and in Europe, the Human Brain Project (HBP) has taken center stage. The provenance of both projects is arguably linked to much of the excitement and research initiatives developed during the Decade of the Brain—a phenomenon whose public image has been tied into much neuro-hype. Here (largely for concerns of space), I focus on the HBP. The HBP was introduced in October of 2013 as the brainchild of Henry Markram. It proposes to integrate neuroscientific research with information and communications technology (ICT) to create models of the brain that can help with disease diagnoses and treatment. According to the HBP website, the project is founded on the belief that "by harnessing advanced computer technology to neuroscience and medical research, we can accelerate our understanding of the brain and how it works. Guided by this improved understanding, we can develop radically new types of computers, and improve the diagnosis and treatment of brain diseases" ("FAQs" HBP, 2015). The HBP is an expansive project: it has one billion Euros in funding and a 10-year plan. Here, I do not plan to discuss the HPB in detail, instead, I want to examine one small aspect of how the HBP has been taken up: a recent reaction by hundreds of neuroscientists to the project and their representation of the HBP in the media.

In the spring of 2014, Henry Markram introduced the second phase of the HPB—a phase that appeared to reduce the presence of and funding for experimental neuroscience within the HBP. Neuroscientists reacted by penning an "Open Message to the European Commission Concerning the Human Brain Project." The letter, which was signed by over 800 neuroscientists and allies (as of April 2015), calls for a reevaluation of the HBP during its ramp-up phase. They describe the project as becoming more and more narrow and claim that this constriction was noted from the outset as a concern. As the letter notes, "many laboratories refused to join the project when it was first submitted because of its focus on an overly narrow approach, leading to a significant risk that it would fail to meet its goals. Further attrition of members during the ramp-up phase added to this narrowing" (Open Letter, 2014). The remainder of the letter requests evaluation and revision and ends with a call to the European Council to reallocate the HBP funding:

> In the case that the review is not able to secure these objectives, we call for the European Commission and Member States to reallocate the funding currently allocated to the HBP core and partnering projects to broad neuroscience-directed funding to meet the original goals of the HBP—understanding brain function and its effect on society. We strongly support the mechanism of individual investigator-driven grants as a means to provide a much needed investment in European neuroscience research. [...] In the event that the European Commission is unable to adopt these recommendations, we, the undersigned, pledge not to apply for HBP partnering projects and will urge our colleagues to join us in this commitment. (Open Letter, 2014)

In their Comment essay "Where Is the Brain in the Human Brain Project?" published in *Nature*, Yves Frégnac and Gilles Laurent note that many of the signatories are former HBP members or neuroscientists who would have been involved in "'partnering projects' that must raise about half of the HBP's total funding. This pledge could seriously lower the quality of the project's final output and leave the planned databases empty" (2014, p. 27). In short, the project has begun to move in directions that were problematic when anticipated, and perhaps damning in their actual execution. The neuroscientist signatories have worked to make this exchange transparent, posting their letter in an open forum for all to read. However, the uptake of their reaction to the HBP has taken some predictable and some unpredictable turns that return us to the question of imagining the impact of the neuro-turn on the neurosciences.

In both scientific and pop media venues, the backlash against the HBP has been imagined in ways that are not directly substantiated in the Open Letter; however, these imagined consequences do follow relatively foreseeable patterns. Yves Frégnac and Gilles Laurent imagine two consequences involving reputation and funding—both are explained in their *Nature* essay. Firstly, they argue, "neuroscientists who initially supported the HBP feel that they have been taken advantage of. The organizers attracted well-funded neuroscience labs for credibility and, ultimately, for their data. Now those labs are being edged out" (Frégnac & Laurent, 2014, p. 28). Interestingly, the reputation of the neurosciences are not in question here; instead, their argument is that the neurosciences' intact and powerful reputation has been abused and misused. At least according to Frégnac and Laurent, the HBP used the neurosciences' ethos for a project that will not be representative of the neurosciences in practice. The consequences of this shift have yet to be seen.

Secondly, Frégnac and Laurent imagine that the HBP's new directions will have consequences for experimental neuroscientific research funding:

> many scientists also feared that the HBP would siphon funds from fundamental research. The European Commission's investment in a large "brain project" would influence what other research areas it chooses to fund. Nonetheless, such an opportunity seemed unlikely to arise again, and neuroscientists (ourselves included) joined up, even if they did not agree with all aspects of the HBP proposal or with certain promises used to sell it. We put our faith in open and interdisciplinary collaboration, trusting that intellectual and operational details would take shape gradually and collectively. (Frégnac & Laurent, 2014, p. 28)

The imagined loss of funding for experimental neuroscience are quite possible—particularly if the HBP does not reintegrate the aspects of the initial program that are on the chopping block. But in their description, the authors make an interesting slip into the first person, arguing that "*we* put our faith in open and interdisciplinary collaboration." This small change from third to first person—which happens at several points in the essay—reminds readers that this commentary essay—however expository it may seem—is an extrapolation from the Open Letter, a new imagining of the impact that the HBP could have on the neurosciences.

In the popular press, the HBP and this Open Letter backlash have been interpreted as having even more dire consequences for the neurosciences. In his article for i09, George Dvorsky writes, "hundreds of neuroscientists (yes, hundreds—the crisis is that bad) have signed an open letter condemning what they perceive as an absence of feasibility and transparency. They say it's far too premature to attempt a simulation of the entire human brain in a computer, that it's a waste of money, and that there's the *risk of a public backlash against neuroscience*" (Dvorsky, 2014). Echoing his interpretation of the Open Letter backlash, Frank Jordans and John-Thor Dahlburg paint this picture in their Associated Press story: "Dozens of neuroscientists are protesting Europe's $1.6 billion attempt to recreate the functioning of the human brain on supercomputers, fearing it will waste vast amounts of money and *harm neuroscience in general*" (2014). The Open Letter did not make either of these claims, nor did it cite a crisis of reputation, but these imagined impacts are part of the larger public understanding of how the neurosciences could be affected by a project such as the HBP. In some respects, we have returned full circle to *Mind Plague*'s question: What if this thing destroys neuroscience? While the situation is certainly not that dire, the question does prompt us to reconsider what *is* at stake?

WHAT'S AT STAKE AND FOR WHOM? TRANSLATION, REPUTATION, AND MARKETABILITY

At stake in debates about the efficacy of the neurosciences and the potential impacts of the backlash are some imagined and real consequences relating to *translation, reputation,* and *marketability*. First, and as Jenell Johnson and I argue in *The Neuroscientific Turn*, the neurosciences are a *translational discipline*: "a set of methods and/or theories that has become transferable--sometimes problematically so—to other disciplines" (2012, p. 3). This means that the neurosciences offer more than just tools for investigating the brain; more than research techniques or ideas that have left the laboratory for the field; indeed, the neurosciences are a translational discipline because they have been transformed by media pundits, humanists, and scientists alike into, as Jenell and I put it, "an epistemology that both transcends and reinforces thinking about the brain and mind" (2012, p. 5). In particular, and with each application of the "neuro"—which has become a shortcut for myriad questions, problems, and uncertainties that have yet to be resolved in the neurosciences or elsewhere—we have imagined the neurosciences into universal application. As a consequence, the neurosciences have and will continue to be associated with all things "neuro." On the one hand, this means that the neurosciences have quite a large audience, and on the other hand, neuroscientists must continually confront their own popular representation.

As I am wont to do in this chapter, then, we might use the problem of translation to turn our usual questions on their heads: instead of asking how the humanities and social sciences have translated their work into the language of the neurosciences, we might ask how neuroscientists have imagined their own translation in the face of the neuro-turn? If we return to Gary Marcus' *New Yorker* essay, one answer (printed in the popular press and from two NYU neuroscientists) is that:

> The worst possibility of a full-scale, reckless backlash against neuroscience, to the exclusion of the field's best work, is that it might sacrifice important insights that could reshape psychiatry and medicine. A colleague at N.Y.U., the neuroscientist Elizabeth Phelps, wrote in an e-mail: "It would be ridiculous to suggest that we shouldn't use brain science to help in the treatment/diagnosis of mental disorders, but if one takes the [current backlash] to the extreme, that is the logical conclusion." (Marcus, 2013)

Here, the fear of the backlash, whether real or imagined, is represented as a real threat. In Marcus' eyes, the field might lose sight of its best work as the waters are muddied by imposters or lesser studies. In Phelps' opinion, the danger is not the loss of "good" science, but the destruction of faith in the neurosciences—the potential, "ridiculous," yet logical conclusion that "we shouldn't use brain science." As in Paul Etchells' blog post (cited earlier), the perceived danger is that the neurosciences are being diluted by their translation in both the media and through the proliferating neuro-disciplines—so much so that the real, substantial benefits of the neurosciences could be lost in translation.

Relatedly, and on the second count concerning reputability, the neurosciences have confronted and will continue to encounter the consequences of popularization. In the 2013 SciLogs blogpost I cited earlier, Pete Etchells poses an apt question: "Does neuroscience have an image problem?" The short answer is that it depends on whom you ask and what is at stake. In response to Etchells' question, we could turn to the media uptake of scientific studies to argue that, yes, the neurosciences have an image problem. According to scholars such as Joe Dumit (2004) and Anne Beaulieu (2001, 2002), the images produced by the neurosciences and, more importantly, their uptake in the media, illustrate a concerning consequence of the proliferation of brain data. But we could also take Etchells' question in the spirit in which it was intended and agree that the reputation of the neurosciences (which includes the uptake of their images) is at stake whenever we imagine the impact of the neuro-turn on the neurosciences. In particular, and as Sally Satel and Scott Lilienfeld note, "pop neuroscience makes an easy target . . . Yet we invoke it because these studies garner a disproportionate amount of media coverage and shape public perception of what brain imaging can tell us. Skilled science journalists cringe when they read accounts claiming that scans can capture the mind itself in action. Serious science writers take pains to describe quality neuroscience research accurately" (2013, p. xiv). For a public not steeped in the complex mechanics of BOLD fMRI techniques, PET contrast agents, and statistical software, pop neuroscience is, at least, accessible—albeit problematically so. Its reputation reflects on both the "neuro-" and the neurosciences.

Finally, when we think about marketability and funding, it would behoove us to think about the rapid uptake of the neurosciences—both in terms of time between publication and media coverage, and in terms of its adoption in fields outside of the sciences. In *Neuro: The New Brain Sciences and the Management of the Mind*, Nikolas Rose and Joelle Abi-Rached argue that the speed at which the neurosciences are expanding inside and outside of the academy has led and will likely continue to lead to some unintended consequences:

> While questions of patenting and intellectual property have been crucial in reshaping the neuroeconomy, we would argue that this intense capitalization of scientific knowledge, coupled with the other pressures on researchers to focus on, and maximize, the impact of their research, has additional consequences. It exacerbates tendencies to make inflated claims as to the translational potential of research findings, and, where those potentials are to be realized in commercial products, to a rush to the market to ensure that maximum financial returns are achieved during the period of a patent. It produces many perverse incentives. (Rose & Abi-Rached, 2013, p. 20)

Here, Rose and Abi-Rached not only identify some of the incentives for the translation and popularization of neuroscientific research, but they do so through the moniker of "neuroeconomy." This label knowingly capitalizes on the "neuro," but also imagines into existence a shadow economy in which all things neuro- have specialized value that can rise or fall alongside the reputation of the related scientific research. Which studies will be funded by national agencies or private corporations? What areas of research are considered most valuable and worthwhile? In each case, the imagined value of neuroscientific research has consequences for the upkeep and funding of neuroscientific laboratories. We saw this most clearly in the HBP examples earlier in this chapter.

Thinking through the rubrics of translation, reputation, and marketability, we can begin to theorize what is at stake for the neurosciences in this neuro-turn. We have seen how the neurosciences have been translated into a universally applicable way of thinking, how they have been used for their reputation by the likes of the HBP and the neuro-disciplines, and how they have been marketed by those inside and outside of the field as a marvel of modern science. While there are dangers and there will be pitfalls associated with the viral spread of the neurosciences, the field will also and inevitably benefit from the exposure; as the old saying goes "there is no such thing as bad publicity" and over the past decade, the neurosciences have certainly had their fair share.

Indeed, one might argue that the fate of the neurosciences is tied—at least in part—to the neuro-turn; that these two potential rivals are part and parcel of one another; that whatever *Mind Plague* affects one will also affect the other.

At the end of *Mind Plague*, Bailey Breege learns that whether they display symptoms or not, everyone is already infected with SIS. There is no cure, *per se*, but the disease can be combatted. SIS sits dormant in the body until an environmental trigger activates it—in this case, learned fear. So, all Breege—and everyone else on Earth—have to do is control their reactions, using the very same neurotechnologies accused of causing the pandemic in the first place. In a public service message, Breege tells his listeners "many of you, about one in every four, have used our products at one time or another. You know they work. This cure will work too" (Kirkland, 2014). Of course, the entire cure is a placebo: neurotechnologies (and by extension the neurosciences) did not cause the disease, nor will they cure it; their power is not medical, but mental. Like SIS, the neuro-turn will not destroy the neurosciences, it may, in fact, continue to empower them.

References

Beaulieu, A. (2001). Voxels in the brain: Neuroscience, informatics and changing notions of objectivity. *Social Studies of Science*, 31(5), 635–680.

Beaulieu, A. (2002). Images are not the (only) truth: Brain mapping, visual knowledge, and iconoclasm. *Science, Technology & Human Values*, 27(1), 53–86.

Brooks, D. (2013, June 17). Beyond the brain. *New York Times*. Retrieved from <http://www.nytimes.com/2013/06/18/opinion/brooks-beyond-the-brain.html?_r = 2>.

Choudhury, S., & Slaby, J. (2011). *Critical neuroscience: A handbook of the social and cultural contexts of neuroscience*. Hoboken, NJ: John Wiley & Sons, Inc.

Dumit, J. (2004). *Picturing personhood: Brainscans and biomedical identity*. Princeton, NJ: Princeton University Press.

Dvorsky, G. (2014, July 10). Europe's $1.6 bllion Human Brain Project is on the verge of collapse. *iO9*. Retrieved from <http://io9.com/europes-1-6-billion-human-brain-project-is-on-the-verg-1602991993#> Accessed 23.02.15.

Etchells, P. (2013, June 28). Neuro de change: A brief note on the neuroscience backlash. *SciLogs*. Retrieved from <http://www.scilogs.com/counterbalanced/neuro-de-change-a-brief-note-on-the-neuroscience-backlash/> Accessed 23.02.15.

Frank, L. (2009). *MindField: How brain science is changing our world*. London: Oneworld Publications.

Fitzpatrick, S. Brain imaging and the 'Cognitive Paparazzi': Viewing snapshots of mental life out of context, Paper presented at the AAAS Annual Meeting, Washington, D.C., 2005.

Frégnac, Y., & Laurent, G. (2014, September 3). Neuroscience: Where is the brain in the Human Brain Project? *Nature*. Retrieved from <http://www.nature.com/news/neuroscience-where-is-the-brain-in-the-human-brain-project-1.15803> Accessed 03.03.15.

Human Brain Project. FAQ. Retrieved from <https://www.humanbrainproject.eu/project-objectives> Accessed 03.03.15.

Johnson, J., & Littlefield, M. M. (2011). Lost and found in translation: Popular neuroscience and the emerging neurodisciplines. *Advances in Medical Sociology, Special Issue, Reflections on Neuroscience, 13,* 279–297.

Jordans, F., & Dahlburg, J. (2014, July 7). Scientists criticize Europe's $1.6B brain project. *AP.* Retrieved from <http://bigstory.ap.org/article/scientists-criticize-europes-16b-brain-project> Accessed 03.03.15.

Kirkland, K. (2014). *Mind plague.* Self-Published.

Lately, L. (2013, June 10). Banking off the backlash: Is pop neuroscience really bunk? *Center for Imagination.* Retrieved from <http://www.centerforimagination.org/2013/06/banking-off-the-backlash-is-pop-neuroscience-really-bunk/> Accessed 25.02.15.

Lengrenzi, P., & Umilta, C. (2011). *Neuromania: On the limits of brain science.* Oxford: Oxford University Press.

Littlefield, M. M. (2011). *The lying brain: Lie detection in science and science fiction.* Ann Arbor, MI: University of Michigan Press.

Littlefield, M. M., Dietz, M., Fitzgerald, D., Knudsen, K., & Tonks, J. (2015). Being asked to tell an unpleasant truth about another person activates anterior insula and medial prefrontal cortex. *Frontiers in Human Neuroscience,* (Online). Accessed 22.09.15.

Littlefield, M. M., & Johnson, J. (2012). *The neuroscientific turn: Transdisciplinarity in the age of the brain.* Ann Arbor, MI: University of Michigan Press.

Lynch, Z. (2009). *Neurorevolution: How brain science is changing our world.* New York: St. Martin's Griffin.

Marcus, G. (2013, June 19). The problem with the neuroscience backlash. *The New Yorker.* Retrieved from <http://www.newyorker.com/tech/elements/the-problem-with-the-neuroscience-backlash> Accessed 23.02.15.

Noë, A. (2010). *Out of our heads: Why you are not your head and other lessons from the biology of consciousness.* New York: Hill and Wang.

Open Letter. (2014, July 18). Retrieved from <http://www.neurofuture.eu/> Accessed 03.03.15.

Ortega, F., & Fernando, V. (2011). *Neurocultures: Glimpses into an expanding universe.* Bern: Peter Lang.

Pickersgill, M., & van Keulen, I. (2012). *Sociological reflections on the neurosciences.* Bradford: Emerald Group Publishing.

Quart, A. (2012, November 23). Neuroscience: Under attack. *The New York Times.* Retrieved from <http://www.nytimes.com/2012/11/25/opinion/sunday/neuroscience-under-attack.html?_r = 1> Accessed 23.02.15.

Rose, N., & Abi-Rached, J. (2013). *Neuro: The new brain sciences and the management of the mind.* Princeton, NJ: Princeton UP.

Satel, S., & Lilienfeld, S. (2013). *Brainwashed: The seductive appeal of mindless neuroscience.* New York: Basic Books.

Snow, C. P. (2012). *The two cultures.* London: Cambridge University Press.

Squier, S. (2004). *Liminal lives: Imagining the human at the frontiers of biomedicine.* Durham: Duke University Press.

Tallis, R. (2009, November 10). Neurotrash. *New Humanist.* Retrieved from <https://newhumanist.org.uk/articles/2172/neurotrash> Accessed 25.02.15.

Tallis, R. (2011). *Aping mankind: Neuromania, Darwinitis and the misrepresentation of humanity.* London and New York: Routledge.

13

Being a Good External Frontal Lobe: Parenting Teenage Brains

T. van de Werff

Maastricht University, Maastricht, the Netherlands

INTRODUCTION

In the past 10 years, a promising explanation of adolescent behavior has fueled enthusiasm among parents, educators, journalists, policy-makers, and pedagogical experts: the teenage brain. The teenage brain tells the complicated story of decreasing gray cells, pruning, and a plastic brain as a *work in progress*. Specific brain regions are said to grow into specialized functions until adolescents reach the age of about 25. The cognitive and emotional functions of the teenage brain are said to be "out of balance"; in particular, the prefrontal cortex (PFC) of the adolescent brain is considered to be in a state of ripening or maturation. As a dire consequence, the unbalanced and immature PFC leaves our future adults in a perpetual state of impulsiveness, irrationality, and riskiness, but also of increased sociality and creativity. The coming of age of the adolescent brain is a technical tale of wonder, opportunities, and risks.

This chapter explores some ethical implications of this neurobiological understanding of adolescence. Recently, (neuro)ethicists have explored the implications of the teenage brain for legal decision making, and whether the immature PFC could result in diminished responsibility and culpability (e.g., Luciana, 2011; Ryberg, 2014). Other scholars have addressed the appropriation of the teenage brain in society and critiqued

The Human Sciences after the Decade of the Brain.
DOI: http://dx.doi.org/10.1016/B978-0-12-804205-2.00013-6

© 2017 Elsevier Inc. All rights reserved.

the underlying promise that the teenage brain "finally solves the mystery of adolescence." For example, Choudhury (2010) has shown how the "turbulent" teenage brain is often used to explain "limitless behavior" and "youth gone wild," thereby reiterating stereotypical images and prejudices about adolescents (Choudhury, 2010). Some question the very possibility of a neurobiological understanding of adolescence, calling the immature teen brain a myth since it would confuse cause and effect (Epstein, 2007).

Instead of scrutinizing the (im)possibility of a neurobiological understanding of adolescence, or theorizing about possible implications of this knowledge for specific ethical concepts, my aim in this chapter is to explore the normative power of the concept of the teenage brain. I do this by looking at the usages of the teenage brain in a specific context: parenting. In recent years, the teenage brain has been eagerly adopted in the traditionally "soft" fields of pedagogy and developmental psychology. Increasingly, pedagogues, parenting experts, and family coaches base their parental suggestions on neuroscience knowledge of adolescent brain development. Contrary to skeptical voices who fear that this neurobiological turn in parenting would leave parents empty-handed, the topic of adolescent brain development comes with specific prescriptions, actions programs, responsibilities, and courses of action. Since the PFC of teenagers is still in development, parents are urged to complement or take over its functions and act as *the external frontal cortex* of their adolescents' brains (Crone, 2012, p. 182). How is the teenage brain appropriated to tell parents what they should and should not do with their adolescents? What does the use of the teenage brain tell us about contemporary ideals of good parenting? Does the teenage brain change or destabilize existing ideas of good parenting? And what does it really mean to be a good external frontal lobe?

To answer these questions, I follow the journey of the teenage brain in Dutch popular parenting discourse. After introducing the birth of the teenage brain in the Netherlands, by discussing the bestselling popular science works of internationally renowned neuropsychologist Eveline Crone, I show how Crone's teenage brain is used by all kinds of parenting experts to justify and combine different (and sometimes conflicting) parental norms, mobilizing different epistemic and normative assumptions regarding parents, teenagers, and their plastic brains. I then question what it means to be a good external PFC, and show how this notion can be seen as the latest episode in the rich tradition of science-based parenting advice. Finally, I briefly explore what the case of the teenage brain can tell us for understanding the ethical implications of neuroscience knowledge.

Probing how knowledge of adolescent brain development might affect ideas of good parenting assumes a pragmatist stance toward ethics. My view on ethics is inspired by recent work done on the ethics of emerging technologies in science & technology studies (STS). Public debates around emerging technoscience—such as the neurosciences—are often characterized by expectations, promises, hopes, fears, and recurring patterns of moral argumentation (cf. Borup, Brown, Konrad, & Van Lente, 2006; Pickersgill, 2013; Swierstra & Rip, 2007; Swierstra, Van Est, & Boenink, 2009). A pragmatic, coevolutionary view on the relationship between (neuro)science and morality allows me to explore what role norms, values, and ideas of the good play in the appropriation of neuroscience knowledge in society. And, more importantly, how the neurosciences are able to address, influence, and possibly destabilize our norms, values, and ideas of what it means to be a good parent.

TEEN BRAIN PLASTICITY AND THE MATURING PREFRONTAL CORTEX THESIS

Research on the brains of adolescents is fairly new. Actual research on human brains in puberty and adolescence started only in the 1970s and 1980s. These studies were rare though, because postmortem brains of adolescents were difficult to obtain (Choudhury, 2010). Things changed with the rise of brain-scanning techniques. In 1999, Jay N. Giedd, a child and adolescent psychiatrist at the National Institute of Mental Health (UK), published the first longitudinal brain imaging study on the brains of adolescents in *Nature Neuroscience*. Giedd et al. (1999) showed that there are different peak levels of gray matter in different brain regions during different developmental stages of the adolescent brain. This indicates that besides the first 3 years after birth, a then well-known critical phase of brain development, a second critical development stage can be distinguished, which would mean that adolescents' brains are not finished or fixed yet, but *plastic* and still developing. The much-quoted article of Giedd et al. became a landmark in studies of the adolescent brain.

In the Netherlands, knowledge about adolescent brain development has been brought to the general public mainly through the works of Eveline Crone, professor of neurocognitive developmental psychology at Leiden University (the Netherlands). In *Het Puberende Brein* (2008) and *Het Sociale Brein van de Puber* (2012), both popular science bestsellers, Crone describes the process of brain development of adolescents by focusing on cognition, emotion, sociality, and creativity. Based on existing animal and lab studies, patient studies of brain injuries, and her own functional magnetic resonance imaging (fMRI) studies on adolescents in her Brain & Development Lab in Leiden, Crone explains how the onset of puberty comes with increased release of gonadotropin-releasing hormone (GnRH), and how this hormone inter-acts with the organization of brain functions. During the brain develop-ment of adolescents specific brain regions grow into specialized functions, by a process of "pruning": a subsequent decrease (after a sudden increase) of gray cells. The speed of this process of pruning dif-fers for specific brain structures, which would determine when teen-agers acquire specific skills during the different stages of adolescence (Crone, 2008, p. 4). The fast ripening of some areas combined with a slower ripening of others is said to explain many of the typical teenage behaviors. Because of a slow ripening of the lateral frontal cortex, teen-agers are said to have problems with memorizing, processing informa-tion, the inhibition of behavior, and planning. In contrast with the still slowly increasing cognitive power, the emotional functions of the teen-age brain are said to be hyperactive. According to Crone, teenagers are especially sensitive to the possibility of rewards: adolescents are often driven by "the pleasure area in the brain," the nucleus accumbens (Crone, 2008, p. 113). The relationship between the "cognitive brain" and the "emotional brain" is seen as one of competition and strife, resulting in an "imbalance," which is characteristic for adolescent behavior according to Crone.

The slower ripening of cognitive skills combined with an overactive emotional system makes for the *maturing PFC thesis*. The PFC thesis would not only account for the imbalance between cognitive and emo-tional brain functions, but also for the increased sociality and creativity during this phase of life. According to Crone, teenagers have difficulties with understanding emotions and recognizing facial expressions of others, and with taking perspectives of others into account. Because the frontal lobe of adolescents is not yet able to regulate the heightened emotionality, adolescents are said to be more sensitive than adults to social exclusion (Crone, 2012, p. 91). Similarly, the presence of peers would give the teenager the pleasant feeling of a reward, which would explain the increased risk-taking behavior of teenagers when peers are present. Finally, Crone explains how the PFC thesis makes teenagers many times more creative, idealistic, and inventive than adults. During adolescence, brain areas that are deemed important for creativity, resourcefulness, musicality, sports, and social involvement are the last to be subjected to the process of pruning of connections (Crone, 2008, p. 150). The lack of performance of the PFC becomes beneficial, as it might be a hindrance for forming creative ideas, or for behaviors such as sincerity, political engagement, and idealism, which Crone considers typical for the adolescent brain.

The PFC thesis quickly found its way into public discourse. After the publication of her books, Crone was quickly seen as the Dutch expert on adolescent brain development and frequently featured in national newspapers, TV shows, reports, and magazines, and discourse on teenage brains quickly broadened to a wide range of topics. The teenage brain became a popular *explanans* to account for topical issues such as street violence, the radicalization of young Muslims traveling to Syria, cases of youths rioting, and other excesses such as youth suicides due to bullying. After widespread dissemination in all kinds of media, the notion of the teenage brain became kind of commonplace when addressing adolescents and part of the symbolical toolkit to explain adolescent behavior and parenting issues.[a] Despite Crone's explicit statement that her books on teenage brains are not parenting manuals, pedagogues, family coaches, (self-declared) parenting experts, and parents have eagerly adopted the teenage brain to discuss parenting issues.

To explore the dissemination and reception of the teenage brain in public discourse, I follow the journey of Crone's work. I conducted a search of the LexisNexis news database for Dutch media coverage of the teenage brain, in the period of 2000 and 2013. Approximately 220 articles addressed parents in some way. These include bestselling newspapers with a political signature from right to left. In addition, sources also include disciplinary pedagogical magazines, popular parenting magazines, online blogs and forums of parenting websites, parenting books that feature the teenage brain, and presentations for parents by local governmental portals for parental information. This body of empirical material spans both the dissemination of the teenage brain in a variety of media in Dutch popular culture, as well as the appropriation by different actors involved in parenting (i.e., parents, family coaches, policymakers, and pedagogues).

ACTING AS AN EXTERNAL FRONTAL LOBE: TWO MORAL REPERTOIRES OF PARENTING TEENAGE BRAINS

Two different normative discourses surrounding the idea of parenting-as-external-frontal-lobe can be distinguished. I dub these discourses *moral repertoires*, based on the notion of interpretative repertoire (Wetherell, 1998; Wetherell & Potter, 1988), as they consist of a range of symbolic codes (in this case parenting norms, values, and ideas about adolescents) that are often performed or put forward without being questioned or made explicit. Each of these moral repertoires consists of

[a]The teenage brain has also found a willing audience in educators and educational policymakers. In popular media, the "un-finished" cognitive development of the teenage brain is often used to explain problems such as low test scores, a lack of motivation for reading books or registering for Greek and Latin at high schools, or to argue for or against specific educational policies, ways of teaching, and curricula change.

two parts. The first part contains normative ideas about parenting—action programs based on knowledge of the teenage brain and ideals of good parenting. The second part of a moral repertoire consists of a developmental perspective about the natural development of the teenage brain and what it needs, including an evaluation of character traits and behavior, which are deemed good, bad, beneficial, or harmful for this development. Through these repertoires, different epistemic and normative assumptions regarding parents, teenagers, and their plastic brains are mobilized. Both repertoires propose that parents should act as an external frontal lobe for their adolescents, but *how* to be a good frontal lobe and for what goal or value differs. The teenage brain gains traction precisely because it can be appropriated for these different ideals of good parenting, and combine them, while reconfiguring them in terms of the brain at the same time. In the first repertoire, parents are cast as a kind of guardians of external stimuli, whereas in the second repertoire, parents are seen as stimulating coaches. It appears that the teenage brain in both repertoires is made valuable as a biological justification for more parental control and setting rules, albeit discussion remains about acceptable degrees of parental control, how parents should exert it, and what parents can and should expect from their teenagers.

Moral repertoire 1: Parents as protective guardians of external stimuli
To the path of a matured brain and a responsible citizen, many risks loom around the corner. In this first repertoire, the teenage brain is often used as an *explanans* for indicating why certain prevalent—often seen as increasing—types of adolescent behavior are harmful for teenagers and their brains. The heightened emotionality and impulsiveness of the teenage brain are used in this repertoire to account and warn for irresponsible risk-taking behavior, as we read in this parenting magazine (Eerkens, 2007, p. 22):

> A teenager is inclined to take decisions that result in a short-term reward. Because of this, he quickly decides to do irresponsible things: eating too much, too fat or too sweet, practicing dangerous sports, like the newest fad to jump off of buildings or from roof to roof. As long as it gives an immediate kick. Teenagers do not see risks at that moment.

Parenting as external frontal lobe is concerned here with the protection from risks. The disrupting of the "natural" development of the growth of the developing brain of teenagers is seen as the greatest danger, and is used to urge parents to be more directive and protective. This repertoire is especially present in popular mass media, when publicly reported topical excesses such as street violence or binge drinking result in cries by commentators and experts for parents to take more responsibility and "finally start parenting again." Dangerous external stimuli for a developing teenage brain are those that physically hamper its (natural) development.

The most discussed danger for adolescent brains is addictive behavior: alcohol usage, smoking, and other drugs (cannabis, speed, the party drug gamma-hydroxybutyric acid), but also gambling, excessive eating, and "new" addictive risks such as gaming and social media use. Alcohol and drugs remain the most dangerous risks for a developing teenage brain, and as such are strongly condemned. Another perceived and topical danger for the teenage brain are digital technologies. Parents seem to struggle with smartphones and Internet use in their households. Problems such as cyberbullying, extensive gaming, Internet porn addiction, and heavy smartphone use are frequently featured topics in parental magazines and parenting books. Because of the PFC thesis, teenagers are said to be prone to obsessive and "monomaniac" behavior (impulsivity combined with focus on short-term rewards), therefore parents are urged in this repertoire to set clear rules for computer, Internet, and phone use to prevent addictive and obsessive behavior. Just as in the case of alcohol, knowledge about the teenage brain is used here as an unquestionable biological justification and necessity for parents to protect its development, by setting and upholding clear limits.

To protect the natural development of their adolescents' brains, parents should set and uphold rules by using strategies of disciplining. Two main strategies are proposed in this repertoire: prohibition and external motivation through giving rewards or punishment. In the case of alcohol, for example, postponing its use should be achieved by strictly prohibiting it without discussion, as alcohol and drugs are seen as too dangerous for the teenage brain to leave it as a matter of negotiation. Less dangerous risks such as Internet and smartphone use should be limited by setting rules through agreements with teenagers, consistently upholding these, and rewarding for good behavior. Stressing the emotional sensibility of teenagers, giving direct rewards is preferable over punishment as "young brains learn the most from rewards" (Korteweg, 2010). When upholding certain rules or limits, parents should "dare to seek confrontations, not avoid difficult conversations, and demand obedience" (Jansen Schoonhoven, 2013, p. 9). Or, as these self-declared teenage-experts ironically phrase it: "Treat them like dogs: reward good behavior. [...] Just like you'll catch flies with syrup, you'll catch teenage attention and good will with compliments" (Van der Wal & Dijkgraaf, 2013, p. 55).

Parents in this repertoire should lower their expectations as adolescents are not able to fully use their frontal cortex, as this pedagogue argues (Horsthuis, 2008, p. 9):

> For a long time, our expectations of teenagers were too high. We wanted to take them seriously, because they look so mature and because they themselves want to be taken seriously. But the fact is that you cannot expect them to be all that reasonable.

As a consequence, parents should be more proactive and "help teenagers with everything their brains have difficulties with" (Pardoen, 2008, p. 8). The maturing PFC thesis is used here to frame teenagers as a kind of *defective* adults: their brains are a "work in progress," not there yet, and their brains are just *wrong* compared to the right brains of adults. Their plastic brains are interpreted as *vulnerable*, by which the focus shifts to risks that might harm or disrupt the natural development of the teenage brain. The developmental perspective of natural brain development of adolescents is interpreted here as needing a particular kind of guidance, correction, or protection. As a result, parents as substitute brains become a kind of *guardian* against harmful stimuli from the outside world and responsible for natural and optimal development of their adolescents' brains.

Moral repertoire 2: Parents as stimulating coaches
In the second moral repertoire, the central question is: How to get the best out of your teenager? Being an external frontal lobe in this repertoire means that parents should give their teenagers some freedom to experiment, as the teenager is "enterprising, is open for many stimuli, focused on challenges, and thus needs to gain experiences" (Jolles, 2011, p. 16). The teenage brain is used here to talk about the role of parents in the process of adolescent self-actualization. The underlying developmental perspective is that of a developing adolescent brain that seeks new experiences and new input to mature. To reach the goal of a mature brain in adulthood, a natural development of adolescent brains includes exploring, experimenting, and acquiring different experiences.

For example, regarding social exploration and experimenting with social identities, it is often suggested in parent magazines that parents should cut their adolescents some slack and offer them the opportunity to acquire new experiences for themselves (Pardoen, 2008, p. 10):

> Give a teenager the chance to do things which you as an adult don't like (anymore), such as watching horror movies, big physical challenges, or other exciting stuff. A teenage brain not only fancies excitement, but by doing so it has an opportunity to learn to deal with problems and feelings of fear.

Similarly, parents should not protect or prevent their adolescent from acquiring negative or disappointing experiences. Regarding cognitive tasks, such as homework or planning, parents should help and support where possible—but not take over. Teenagers themselves must learn to plan their homework; helping them doing their homework is out of the question.

The plasticity of teenage brains is interpreted here as an *opportunity*; a unique phase in life in which the brains is able to grow and develop the most. Consequently, the teenager with his or her plastic brain is framed as a unique individual: "Teenage brains are not like unfinished adult brains; they are creative innovation and learning machines, with their emotional circuits focused on new experiences and exploring social contacts" (Van Hintum, 2012, p. 6). Parents in this repertoire still need to set rules and limits, but not for reasons of protection. Rather, such rules and limits are seen as necessary for the process of self-actualization. The idea is that teenagers need to have some limits or rules they can rebel against. So when setting rules and upholding them, parents should refrain from punishments or rewards, but focus rather on intrinsic motivation, negotiation, mutual agreements, and making decisions together. Even regarding alcohol, which led experts in the first repertoire to stress the need for strict prohibition, experts in this repertoire cling on to the need to let adolescents themselves make the right choices, as we read in this workshop about Crone's teenage brain at a high school in Amsterdam (Dugomay & Mok, 2009, p. 12):

> Reward or punishment for good or bad behavior doesn't work anymore. Important is rather to think along with the teenager, discuss different options together, clarify consequences and discuss what the (logical) effects could be of their actions—for example regarding traffic, alcohol and drug use. Try to stimulate development by letting them make choices from different options. That's how a teenager learns to make a rational judgment, and it brings ratio and emotion faster in balance. They still have to learn the chain of cause and effect. Dedicated, non-patronizing support of parents can help them with this.

Similarly, in contrast with the first repertoire, parents should not draw the conclusion that teenagers are just not ready for adult behavior, because their brains aren't finished. Instead of low expectations, parents in this repertoire are encouraged to actually raise their expectations as this developmental psychologist argues (Pont, 2010, p. 3):

> The right conclusion should be: precisely because the brains of teenagers are not mature yet, we have to stimulate and challenge them. Adult requirements make you mature, even if you're not. Just like you're not waiting to teach your child how to talk, until the moment that he actually can already.

Instead of focusing on behaviors such as irrationality or riskiness, the focus in this repertoire shifts toward more positively evaluated behaviors as characteristic for the teenage brain: behaviors that stress opportunities to excel. The maturing PFC thesis is interpreted here with an emphasis on the increased sociality, emotionality, and creativity of the teenage brain. Teenage behaviors that are deemed as unequivocally beneficial for their developing brains are especially behaviors that stress creativity and out-of-the-box thinking, sociality, (political) idealism, and sincerity. These character traits, framed as typical of a teenage brain, are seen as opportunities for adolescents as unique individuals that

TABLE 13.1 Normative and Epistemic Assumptions of Parenting as "External Frontal Lobe"

Acting as external PFC for the teenage brain	Moral repertoire 1: Parents as guardians of external stimuli	Moral repertoire 2: Parents as stimulating coaches
Action program: Parental norms	Prohibition, punishment, praise (disciplining)	Stimulation, support, steering (motivating)
Parenting goal	Natural/normal/undisturbed brain development	Teenage (brain) self-actualization
Teenager as	Defective adult	Unique individual
Diagnosis (interpretation of PFC thesis)	Focus on riskiness, emotionality (negatively evaluated character traits)	Focus on creativity, sociality, idealism (positively evaluated character traits)
Plasticity as	Vulnerability	Opportunity

should be cherished and encouraged by parents. Parents in this repertoire thus become kind of stimulating coaches, only curbing the freedom of their adolescents in order for them to flourish and excel (Table 13.1).

The two moral repertoires make visible how the teenage brain can be mobilized by experts to propose different and sometimes conflicting parental norms. The teenage brain appears to be versatile enough to reconcile different epistemic and normative assumptions regarding teenagers and parenting. The distinction between the two moral repertoires is analytical; in practice, experts in parenting books or magazines easily switch between the two moral repertoires and suggest parents both be more strict and let go at the same time. The message is that acting as external frontal lobe entails balancing between protecting teenage brains from dangerous stimuli and stimulating teenage brains to flourish. How do we explain the emergence of these different repertoires of parenting norms? What can they tell us about contemporary parenting ideals? And does parenting as external frontal lobe actually change the ideals of good parenting?

PARENTING TURMOIL: THE TEENAGE BRAIN AND IDEALS OF GOOD PARENTING

Neuroscience research of adolescent brain development can be seen as the latest "evidence" in a tradition of science-based parenting advice. As the historians Depaepe, Simon, and Van Gorp (2005), Wubs (2004), and Wesseling (2002) show, science-based advice from parenting experts over the past century—consisting of advice regarding parenting goals/ideals and ways of parenting—reflects not only societal developments, such as industrialization, modernization, and individualization, but are also closely tied to changes of theories, methods, and practices of the scholarly discipline of (developmental) psychology.

The psychologist Stanley Hall is famous for his first explanation of adolescence as a biological phase in life. Less well known are his "pedagogical imperatives," which stressed disciplinary and instructional techniques, set against raising by tradition or neglect (Lesko, 2001, p. 88). After the period of so-called normative pedagogy, in which different religious and ideological views (the Dutch pillarization of society) defined parenting goals (Depaepe, et al., 2005; Wubs, 2004), the behaviorist approach toward parenting was succeeded by a psychoanalytic perspective, with the famous Dr. Spock as the most well-known proponent (cf. Spock, 1946). Prior to the 1970s, Dutch parenting advice mainly centered around character development and the societal responsibility of parents to make sure their children became responsible and virtuous citizens (Wubs, 2004, p. 210). From the 1970s onward, parenting goals became increasingly formulated in terms of developmental psychology. Self-development and (emotional) well-being of the child were no longer seen as a condition for healthy development toward a normative goal (such as a virtuous or disciplined citizen), but became parenting goals themselves (Wubs, 2004, p. 216). Parenting experts focused on practical methods and techniques of parenting instead of explicating parenting ideals. An example is the notion of parenting styles (cf. Baumrind, 1971), which became popular in the social sciences by the 1980s and 1990s. During the 1980s and 1990s, developmental psychology remained a strong source of authority for parenting experts, though a new form of evidence emerged as well: genetics. An example is Judith Harris' *The Nurture Assumption* (1998), in which she argued that genetic factors and peers are far more influential for the development of the child than parents and that the role of parents and of parenting should be seen as minimal (Harris, 1998). The teenage brain can be seen as the latest evidence in this "biological turn" of parenting advice, but with quite different normative inferences.

Experts in the two moral repertoires rarely explicate parenting ideals, but they do frequently use the teenage brain to differentiate between good and bad forms of parenting, which they describe in Baumrind's notion of parenting styles. Parents who adopt a more laissez-faire kind of attitude and embrace liberal and more lenient parenting styles are strongly rejected or condemned. In the first repertoire, "bad" parents are those who stress a mutual agreement between parent and adolescent (described as liberal and democratic parenting styles) instead of exerting a certain amount of control through disciplining or prohibition.

In the second repertoire, bad parents are those who have too low expectations of their adolescents and thus don't take the needs of the adolescent (brain) seriously enough (described as neglectful or permissive parenting styles). Contrary to the explicit rejection of "bad" ways of parenting, views on what actually is good parenting mostly remain implicit.

If we look at the emerging image of parents as external frontal lobe, we can see a striking resemblance to the so-called *authoritative parenting style* (Baumrind, 1971, 1991; Maccoby & Martin, 1983). The authoritative parenting style can be described as "a balancing act" or a golden mean between self-actualization (of the adolescent) and the protection of dangers. This balancing act is seen by many contemporary Dutch pedagogues and sociologists, as well as parents, as typical for the contemporary Dutch ideal of parenting (Brinkgreve, 2012; De Winter, 2005). It is a parenting style in which parents actively set boundaries and explain them with the goal of promoting adolescent's responsibility (De Winter, 2005, p. 11). Parents who are urged to act as external frontal lobes are expected to combine setting clear limits (strong control) while acknowledging and nurturing the needs of the adolescent (in this case: the needs of his or her brain). The teenage brain seems to function as credible support to this difficult balancing act. However, the emergence of two moral repertoires—with different action programs and ideas about how and why to set limits—shows that the teenage brain does not give an inconclusive answer, but rather is interpreted through this very ideal of authoritative parenting itself.

When we compare parenting as external frontal lobe with earlier debates on parenting, a continuity can be observed. Just as parenting goals in the 1970s became formulated in terms of developmental psychology, current parenting goals are formulated in terms of the developing teenage brain. The idea of parenting as external frontal lobe, including the very concept of parenting style itself, is a *procedural* notion: it doesn't include a substantial, normative ideal of good parenting. As the two moral repertoires show, the teenage brain is appropriated to tell parents *how* they should intervene, or *when*. The *why* is formulated in terms of protecting or stimulating teenage brains. *What* parents actually should teach their adolescents to become is not described in normative ideals (such as character building or responsible citizenship). Rather, the implicit end goal for many experts and parents using the teenage brain is "normal" or natural brain development as a prerequisite for adulthood. The teenage brain is heralded by parents as justification for the idea that teenagers "really" are different: their children are not "abnormal." The maturing PFC thesis is used in both moral repertoires to frame adolescents as a distinct category of people with shared biological characteristics, assuming that there is a single brain type that would be common for all members of this category (O'Connor, Rees, & Joffe, 2012). What is deemed normal or healthy brain development is not questioned. Teenagers in any case seem to be everything that adults don't want them to be: impulsive, irrational, irresponsible, and impulsive (implying that adults are the opposite: controlled, rational, responsible, and balanced). In effect, adolescents become symbolically distanced from adults, and the implication is that they also should not be treated as adults—thus increasing the need for parents to exert themselves more. By rooting teenage behavior in the brain, the differences between adults and adolescents become normalized, and as a consequence, the focus turns to their parents; it's they who have to learn how to take the brain development of their adolescents into account.

Next to this historical continuity, there is also a clear discontinuity to be observed between earlier parenting debates. Some skeptics infer from the teenage brain a diminishing role for parents—resonating the debate around *The Nurture Assumption* by Judith Harris (1998). These critics assume a deterministic role of the teenage brain as neurobiological explanation, leading parents to think they can't do anything about it (Van der Heijden, 2013). However, as we have seen, the teenage brain is foremost a *plastic* brain. The teenage brain as model of growth and development is often described in terms such as "work in progress," "immature," "unfinished," "undeveloped," "malleable," "in transition," "in motion," "flexible," "dynamic," "unbalanced," "imperfect," "sensitive," and "vulnerable." Though not all references to teenage brains explicitly use the term plasticity, these descriptions of teenage brains equal the developmental dimension of brain plasticity, the forming of the structure and functions of the brain in a particular sensitive period of growth. Where Hall (1904) spoke of adolescents as "becoming," the phrase now has its neuroscience equivalent in plasticity.

The notion of brain plasticity "opens" up the brain for interventions (Abi-Rached & Rose, 2013). As we have seen, plasticity of the teenage brain is interpreted differently in the two moral repertoires. In the first repertoire, the plastic teenage brain is seen as particular vulnerable, especially for behaviors that undermine the natural development toward a balanced rationality: addictions (alcohol) but also emotional behavior, or any other behavior that makes teenagers "out of control." In the second repertoire, the plastic teenage brain is presented as having "unique" flexibility or creativity, which allows them to learn and grow. The concept of the plastic, teenage brain enables and invites parents, experts, and others to do something with it: in this case, protect its vulnerability or stimulate its unique flexibility. In contrast to the fierce debates in the 1990s around Harris' *The Nurture Assumption*, the story of the teenage brain shows that it is now safe to acknowledge biological characteristics of behavior without leaving parents empty-handed. The biological imperative of the teenage brain is to view adolescent behavior as a result of brain development, a natural process that requires specific parental action programs and responsibilities.

CONCLUSION: POPULAR NEUROSCIENCE AS ETHICS THROUGH DIFFERENT MEANS

In this chapter, I have explored how the teenage brain is made valuable by parenting experts to address parenting issues and ideas concerning good parenting. I have shown how the teenage brain is used as answer to the timely question of how much control or freedom parents should exert or give their adolescents. Parents are urged to act as the external frontal lobe of their adolescents. Two distinct moral repertoires emerged around this idea of parenting as external prefrontal lobe, each with different normative suggestions regarding the expectations parents should have, the degree of control that should be exerted, and the way of exerting control. Different versions of the plastic teenage brain were mobilized in each moral repertoire to justify its action programs. Together, these two moral repertoires urge parents to both control and let go, in order to foster and nourish the development of their children's brains. As this "balancing act" between different parenting values can be seen as typical for contemporary ideals of good parenting, I argue that the teenage brain in Dutch parenting discourse does not so much change existing ideas of good parenting, but that it is rather appropriated through the lens of this ideal of authoritative parenting. Just as parenting goals in the 1970s were formulated in terms of developmental psychology, contemporary parenting goals are described in terms of the teenage brain. As the latest evidence in the history of science-based parenting advice shows, the teenage brain does differ from parenting evidence in the 1990s: instead of diminishing the role of parents in favor for determinist biological explanations of development, the plastic teenage brain comes with specific prescriptions and action programs.

The popularity of the teenage brain is a particularly successful example of the "valorization" or dissemination of neuroscience knowledge into society. The widespread appropriation of the brain in public discourse has become a topic of extended scrutiny in recent years. Studies on the public representation of neuroscience knowledge in the UK, the US, and Germany have shown that many neuroscientific ideas in public discourse perpetuate rather than challenge existing policy measures, the status quo or modes of understanding of self, others, and society (cf. Choudhury & Slaby, 2012; O'Connor et al., 2012; O'Connor & Joffe, 2013). Similarly, the case of the teenage brain indicates that we should not so easily assume a transformative potential of the neurosciences to change our conceptions of ourselves, our values, or our responsibilities toward others. At the same time, probing the actual use of neuroscience knowledge in society remains a fruitful venue for exploring ethical implications of the "neuro-turn": the increasing cultural interest in the neurosciences, both in the social sciences and the humanities, as in society at large. A neuroscience understanding of behavior doesn't speak for itself: technical brain facts have to be *made valuable*, and as such receive the potential to influence or destabilize our norms, values, and ideas of the good. As this chapter has shown, the teenage brain is not only valued for its promise to finally solve "the mystery of adolescence," it also gets imbued with the normative power to address what it means to be a good parent. Simply discarding such popular understandings of the neurosciences as neuromyths, "neuro-hype," "neurotrash" (Tallis, 2009), or just "bad" or overinflated interpretations of science, ignores that the popularization of neuroscientific knowledge in society is an important part of the flourishing of neuroscience as knowledge system itself, playing a constitutive role in mobilizing support, funding, and engagement with the wider audience (Choudhury & Slaby, 2012; Heinemann, 2012). Moreover, it obscures proper understanding of the different ways neuroscience knowledge is made valuable in different societal and scholarly contexts.

References

Abi-Rached, J. M., & Rose, N. R. (2013). *Neuro: The new brain sciences and the management of the mind*. Princeton, NJ and Oxford: Princeton University Press.
Baumrind, D. (1971). Current patterns of parental authority. *Developmental Psychology Monographs, 4*, 1–103.
Baumrind, D. (1991). Effective parenting during the early adolescent transition. In P. A. Cowan, & E. M. Hetherington (Eds.), *Advances in family research* (Vol. 2). Hillsdale, NJ: Erlbaum.
Borup, M., Brown, N., Konrad, K., & Van Lente, H. (2006). The sociology of expectations in science and technology. *Technology Analysis & Strategic Management, 18*(3/4), 285–298.
Brinkgreve, C. (2012). *Het Verlangen naar Gezag: Over vrijheid, gelijkheid en verlies van houvast*. Amsterdam: Atlas Contact.
Choudhury, S. (2010). Culturing the adolescent brain: What can neuroscience learn from anthropology? *Social Cognitive Affective Neuroscience, 5*(2–3), 159–167.
Choudhury, S., & Slaby, J. (2012). Introduction: Critical neuroscience: Between lifeworld and laboratory. In S. Choudhury, & J. Slaby (Eds.), *Critical neuroscience: A handbook of the social and cultural contexts of neuroscience*. West-Sussex: Wiley-Blackwell.
Crone, E. A. (2008). *Het puberende brein*. Amsterdam: Bert Bakker.
Crone, E. A. (2012). *Het sociale brein van de puber*. Amsterdam: Bert Bakker.
Depaepe, M., Simon, F., & Van Gorp, A. (2005). *Paradoxen van pedagogisering: Handboek pedagogische historiografie*. Leuven and Voorburg: Acco.
De Winter, M. (2005). *Oratie: Democratieopvoeding versus de code van de straat*. Utrecht: Universiteit Utrecht.

Dugomay, P., & Mok, D. (2009). *Een kijkje jonder de zinderende hersenpan van adolescenten: Hand-out bij de workshop Het puberende brein.* Amstelveen: Herman Wesselink College Amsterdam.

Eerkens, M. (2007). Wat gebeurt er toch allemaal in dat koppie? *J/M Ouders, 10,* 22–25.

Epstein, R. (2007, April/May). The myth of the teen brain. *Scientific American Mind,* 57–63.

Giedd, J. N., Blumenthal, J., Jeffries, N. O., Castellanos, F. X., Liu, H., Zijdenbos, A., et al. (1999). Brain development during childhood and adolescence: A longitudinal MRI study. *Nature Neuroscience, 2*(10), 861–863.

Hall, G. S. (1904). *Adolescence: Its psychology and its relations to physiology, antropology, sociology, sex, crime, religion and education* (Vol. 2). New York: Appleton.

Harris, J. (1998). *The nurture assumption: Why children turn out the way they do.* New York: Simon & Schuster.

Heinemann, T. (2012). *Populäre Wissenschaft: Hirnforschung zwischen Labor und Talkshow (Popular science: Brain research between laboratory and pop culture).* Göttingen: Wallstein.

Horsthuis, A. (2008). Lach, luister, voel mee en wees helder. *J/M Puberspecial, 2,* 11–15.

Jansen Schoonhoven, M. (2013, January 31). Beste ouders: eis van uw kind gehoorzaamheid. *Trouw,* p. 21.

Jolles, J. (2011, December 10). Puberen? Verbreinen! *Het Parool,* p. 16.

Korteweg, N. (2010, April 3). Billenkoek voor het brein. *NRC Handelsblad.*

Lesko, N. (2001). *Act your age! A cultural construction of adolescence.* London and New York: Routledge.

Luciana, M. (2011). Development of the adolescent brain: Neuroethical implications for the assessment of executive functions. In B. J. Sahakian, & J. Illes (Eds.), *Oxford handbook of neuroethics.* Oxford: Oxford University Press.

Maccoby, E. E., & Martin, J. A. (1983). Socialization in the context of the family: Parent–child interaction. In P. H. Mussen, & E. M. Hetherington (Eds.), *Handbook of child psychology: Vol. 4. Socialization, personality, and social development.* New York: Wiley.

O'Connor, C., & Joffe, H. (2013). How has neuroscience affected lay understandings of personhood? A review of the evidence. *Public Understanding of Science, 22,* 254–268.

O'Connor, C., Rees, G., & Joffe, H. (2012). Neuroscience in the public sphere. *Neuron, 74* (2), 220–226.

Pardoen, J. (2008, October/November). Pubers kunnen niet anders dan puberen. *Leef! Magazine,* pp. 8–10.

Pickersgill, M. (2013). The social life of the brain: Neuroscience in society. *Current Sociology, 61*(3), 322–340.

Pont, S. (2010, February 4). Untitled. *Het Parool,* p. 3.

Ryberg, J. (2014). Punishing adolescents: On immaturity and diminished responsibility. *Neuroethics, 7*(3), 327–336. <http://dx.doi.org.ezproxy.ukzn.ac.za:2048/10.1007/s12152-014-9203-6>.

Spock, B. (1946). *The common sense book of baby and child care.* New York: Duell, Sloan, and Pearce.

Swierstra, T., & Rip, A. (2007). Nano-ethics as NEST-ethics: Patterns of moral argumentation about new and emerging science and technology. *Nanoethics, 1*(1), 3–20.

Swierstra, T., Van Est, R., & Boenink, M. (2009). Taking care of the symbolic order: How converging technologies challenge our concepts. *Nanoethics, 3*(3), 269–280.

Tallis, R. (2009, November 10). Neurotrash. *New Humanist.* Retrieved from <https://newhumanist.org.uk/2172/neurotrash>.

Van der Heijden, M. (2013, April 13). Ruzie over het Puberbrein. *NRC Handelsblad.*

Van der Wal, M., & Dijkgraaf, J. (2013). *Het enige echte eerlijke puberopvoedboek.* Amersfoort: BBNC Uitgevers.

Van Hintum, M. (2012, September 15). Het idee van de "rijpende" hersenen is toch wat simplistisch. *De Volkskrant,* p. 6.

III. THE NEUROSCIENCES IN SOCIETY. SOCIAL, CULTURAL, AND ETHICAL
IMPLICATIONS OF THE NEURO-TURN

Wesseling, L. (2002). Deskundige waarschuwingen tegen deskundigen: een gemeenplaats uit de psychologiserende opvoedingsvoorlichting. In F. Van Lunteren, B. Theunissen, & R. Vermij (Eds.), *De opmars van deskundigen: Souffleurs van de samenleving* (pp. 147–160). Amsterdam: Amsterdam University Press.

Wetherell, M. (1998). Positioning and interpretative repertoires: Conversation analysis and poststructuralism in dialogue. *Discourse and Society, 9*(3), 387–412.

Wetherell, M., & Potter, J. (1988). Discourse analysis and the identification of interpretive repertoires. In C. Antaki (Ed.), *Analysing everyday explanation: A casebook of methods* (pp. 168–183). Newbury Park, CA: Sage.

Wubs, J. (2004). *Luisteren naar deskundigen: Opvoedingsadvies aan Nederlandse Oouders 1945–1999*. Assen: Koninklijke van Gorcum.

Toward Neuroscience Literacy?—Theoretical and Practical Considerations

A. Bergmann, A. Biehl and J. Zabel

University of Leipzig, Leipzig, Germany

INTRODUCTION

Neuroscience is the leading discipline of the 21st century. At least this is what you can conclude if you give credence to the reports of acknowledged German neuroscientists in a memorandum on neuroscience called "Das Manifest" (Elger et al., 2004). Only time will tell whether they are right. What seems to be undisputable is the fact that "neuro" is ubiquitous. Bear, Connors, and Paradiso (2012) comment on this in a more or less jocular fashion, crediting neuroscientists and their branch with fame and public awe that some decades ago used to belong to rocket scientists.

Within the last 20 years, neuroscience has developed into an interdisciplinary field of research in which physicians, biologists, psychologists, computer specialists, and others jointly attempt to explain the structural and functional relations in the nervous system. There has been rapid development in the technological conditions, and several new branches have emerged including neuroethics, neuroanthropology, neuroeconomics, and neuropedagogy. This rise of neurosciences has led to important new but also controversial perspectives on the human brain, e.g., the debate on free will caused by contradictory research results on unconscious cerebral initiative (Haggard & Eimer, 1999; Libet, 1985; Soon, Brass, Heinze, & Haynes, 2008). There is a broad discussion on

© 2017 Elsevier Inc. All rights reserved.

recent research findings and their socioscientific impact in the neuroscientific community as well as in the general public. Questions have been raised about emerging technological opportunities, philosophical challenges, ethical challenges, and demand for political action.

In order to participate responsibly in the public discourse, it is becoming more and more important to understand the basic neuroscientific assumptions, explanations, and methods used, including their limitations. Neuroscience literacy (Howard-Jones, 2010; Zardetto-Smith, Mu, Phelps, Houtz, & Royeen, 2002) is the term used to describe this skill. Unfortunately, one rarely finds an example of a reflective and neuroscientifically educated citizen. The public's understanding of neuroscientific ideas and results is basic at best.

For example, a large-scale questionnaire ($n = 2158$) on brain-related issues showed that the public recognizes the benefits of brain research for insights into human nature and for the improvement of quality of life. However, the questionnaire showed there is a lack of knowledge of basic neuroscientific research results. Common misconceptions included that we use only 10% of our brain capacity and that our brain works like a computer (Herculano-Houzel, 2002). One reason for the misconceptions may be the complexity of neuroscientific research. Not everybody can and is able to extensively concern themselves with molecular, cellular, systemic, or cognitive aspects of neuroscientific research (Bear et al., 2012). Moreover, the public's understanding of neuroscientific research results is further complicated by media distribution. Newspaper articles, podcasts, and web pages reporting about so-called brain gyms and cyberattacks on the brain raise the public's awareness; on the other hand, they create false expectations, misunderstandings, and fear.

Promoting neuroscience in a proper manner will solve this problem. Organizations such as the *DANA Foundation* or the *Society for Neuroscience* are eager to promote public understanding of neuroscience with the help of educational campaigns, forums, and publications of well-researched and comprehensible papers.

The science classroom, the most important area to promote neuroscience literacy, is not prepared to face this challenge. In the following we investigate what it means to be neuroscientifically educated and how understanding—popular, academic, and classrooms activities—can come together. It is our aim to conceptualize neuroscience literacy based on scientific literacy and to derive actions and measures to improve teaching neuroscience. Therefore we examine current science classroom conditions and determine the conditions that need to prevail in order to enable neuroscientific learning. We also examine the extent current science teaching meets these requirements. Special focus is placed on science teaching in Germany.

CONSIDERATIONS ON THE CONCEPTUALIZATION OF NEUROSCIENCE LITERACY

What does it mean to be neuroscientifically literate? To answer this question, we broadened the perspective and looked into the field of science education research to find aspects to create a definition. We consider the concept of scientific literacy suitable; firstly, it encompasses a broader understanding of science as a discipline and thus offers the possibility to specify neuroscience as a subbranch. Secondly, in almost 60 years the term scientific literacy has evolved to become a key concept of science education. So what does it mean to be scientifically educated?

According to Derek Hodson (2008) scientific literacy as a keyword for the scientific part of general education was coined during the science education debate of the 1950s and dates back to publications from Paul Hurd (1958) and Richard McCurdy (1958). Triggered by the Sputnik crisis, where Americans feared the development of an educational gap in the sciences, they radically reformed their educational system. The focus shifted to the aims of science education in order to educate scientifically proficient students as well as skilled workers in the branches of science and technology. Scientific literacy then comprised "an understanding of the basic concepts of science, the nature of science, the ethics that control scientists in their work, the interrelationships of science and society, the interrelationships of science and the humanities, and the differences between science and technology [...]" (Hodson, 2008, p. 4).

Nevertheless, according to Hodson, it took another quarter of a century until those ideas were taken up on a larger scale by the *American Association for the Advancement of Science* with the project "Science for All Americans." For them a scientific literate person was "one who is aware that science, mathematics, and technology are interdependent human enterprises with strengths and limitations; understands key concepts and principles of science; is familiar with the natural world and recognizes both its diversity and unity; and uses scientific knowledge and scientific ways of thinking for individual and social purposes" (Project 2061 AAAS, 1989, p. 4).

Almost contemporaneously to this theoretical development of didactic aims in science teaching, there had been attempts to create and implement practical concepts in science classrooms. Where the study of core concepts of the individual sciences such as chemistry and physics had been at the center of attention, the focus was shifted onto the connection of science, technology, and society. Wolfgang Gräber (2002) refers to the significant role of the Science-Technology-Society Movement (STS Movement), which was mostly based on the ideas of Norris Harms (1977) and John Ziman (1980, 1994). Robert E. Yager and

Rustum Roy (1993) summarized the goals of STS-oriented science teaching and claimed that it has to:

> Prepare students to use science for improving their own lives and for coping in an increasingly technological world; teach students to deal responsibly with technology/society issues; identify a body of fundamental knowledge that students should master to deal intelligently with STS issues, give students an accurate picture of the requirements of and opportunities in the many careers available in the STS field. (Gräber, 2002, p. 5)

The growing connection between the theoretical and practical educational debates within the international science education research community in recent years has led to diverse perspectives on the nature of science and its objectives on the one hand and its implementation and methodology on the other.

In their article "The Meaning of Scientific Literacy" Jack Holbrook and Miia Rannikmae (2009) succeeded in finding a modern interpretation of science education's aims while systemizing the area of research. Holbrook and Rannikmae differentiate between two major camps of science educators and their understanding of scientific literacy. These two opposing camps form the definitional continuum.

One camp advocates a central role for the knowledge of scientific content and sees the aim of science education still in the preparation of students for scientific careers. The second camp contextualizes scientific literacy with strong emphasis on its social usefulness:

> The first camp seems to be very prevalent among science teachers today. It builds on the notion that there are "fundamental ideas" in science that are essential and that there is content of science which is a crucial component of scientific literacy. [...] The second camp encompasses the longer term view and sees scientific literacy as a requirement to be able to adapt to the challenges of a rapidly changing world. [...] It recognises the need for reasoning skills in a social context, and above all, this view recognises that scientific literacy is for all, having little to do with science teaching solely focusing on a career in science, or providing only an academic science background for specialisation in science. (Holbrook & Rannikmae, 2009, p. 278)

According to Holbrook and Rannikmae, the definition of scientific literacy is strongly linked to the idea one has about the nature of science. They agree with the second camp, and in their definition of the nature of science they accordingly underscore the connection between science and economic, environmental, social, and political areas in certain moral and ethical aspects. When they discuss the decision-making process, they state that the nature of science is "one of interacting with all these areas leading to a decision in which the reasoning can be related to arguments on the importance of the science and the other aspects at the time the decision is being made" (Holbrook & Rannikmae, 2009, p. 282). Their overarching idea of science education is best described with the words "education through science," which is the exact opposite of what had been popular shortly after the Sputnik crisis when scientific literacy followed the slogan "science through education." Today, according to Holbrook and Rannikmae, a person who wants to be educated in the modern conception of science needs to manage the following tasks:

- Learn the science knowledge and concepts important for understanding and handling socioscientific issues within society.
- Undertake investigatory scientific problem solving to better understand the science background related to socioscientific issues within society.
- Gain an appreciation of the nature of science from a societal point of view.
- Develop personal skills related to creativity, initiative, safe working, etc.
- Develop positive attitudes toward science as a major factor in the development of society and scientific endeavors.
- Acquire communicative skills related to oral, written, and symbolic/tabular/graphical formats to better express scientific ideas in a social context.
- Undertake socioscientific decision-making related to issues arising from society.
- Develop social values related to becoming a responsible citizen and undertaking science-related careers (Holbrook & Rannikmae, 2009, p. 283).

Following this train of thought one needs to ask which concepts and aspects of neuroscience are of significance for social discourse. Besides aspects of molecular and cellular neuroscience (e.g., structure, types, and development of cells), the systemic and behavioral neuroscience (e.g., neural circuits, neuroregulation, and neuropathology) as well as cognitive neuroscience (e.g., consciousness, language, and imagination), which topics matter?

The World Health Organization (2001) and the National Research Council (1996) suggested that a special focus should be placed on the pathology of the human nervous system. They describe neuroscience literacy as "the knowledge and understanding of concepts and processes in neuroscience required for understanding issues related to diseases and disorders of the brain, as well as how humans interact with their environment and each other because of their unique nervous system characteristics" (Zardetto-Smith et al., 2002, p. 397). They also define a small set of outcomes of neuroscience literacy, which are for the most part strongly connected to health issues. For example, neuroscientific literate persons should be able to "make informed decisions about personal or a family member's health and care to support optimal nervous system function and participation across the life spectrum (infant to aged adult); [...]; participate in a knowledgeable fashion in government decisions regarding research initiatives and new treatments for various neurological diseases; and understand and critically judge neuroscience-related material in lay media, [...]" (Zardetto-Smith et al., 2002, p. 397).

This definition has to be seen as a valuable first approach to the question: Which topics matter? But it also hides a large part of the ethical considerations connected with neuroscience that are important for science education. Elisabeth Hildt defines neuroethics as based on the detailed description of the scientific-medical state of knowledge in terms of the brain and its functions. Neuroethics concerns itself with philosophical aspects and ethical implications as well as societal impacts of current developments of the neurosciences (Hildt, 2012, p. 10). Neuroethicists agree that there are two dimensions of neuroethics, each of which have a distinct perspective on neuroscientific research (Roskies, 2002).

"Ethics of neuroscience" is concerned with the ethical implications of brain research and clinical practice. According to Elisabeth Hildt (2012), there are possible issues that are relevant for the public, they include: What is a "normal" brain? Should people with neurodegenerative diseases be allowed to make decisions on their own? To what extent should neuroenhancers be allowed to optimize the functions of the brain?

The second dimension is referred to as "neuroscience of ethics." It discusses neuroscientific insights in connection to questions about human nature with regards to free will, self-determination, and autonomy.

Taking the definition for scientific literacy and the aspects of neuroethics into consideration we define neuroscience literacy as the ability:

- to understand neuroscience as a socially embedded science and to be aware of the consequences of neuroscientific research for personal life and society;
- to understand basic concepts of neuroscience and its research methods;
- to use this knowledge to separate science from science fiction in media;
- to explore the philosophical and ethical dimensions of neuroscience; and
- to be able to participate as a responsible citizen in an informed discourse about current socially relevant philosophical and neuroethical questions that influence public dialog and those that may arise from neuroscientific research in the near future.

This definition for neuroscience literacy imposes several requirements for science teaching on all levels, especially in secondary education. How can we design science lessons that integrate neuroscientific content, the general educational aims of science teaching, and opportunities for ethical reflection?

In accordance with Holbrook's and Rannikmae's idea of "education through science" and our definition of neuroscience literacy, we advocate well-balanced science lessons that combine basic biological facts, key concepts of neurobiology, and debates on philosophical and ethical issues concerning neuroscientific research. Therefore biological content should always be tied to one or more societally, personally, philosophically, or ethically relevant questions and case-based discussions. There are already some teaching strategies that recognize this consideration. Among international research, the works of Dana L. Zeidler (Zeidler & Nichols, 2009; Zeidler, Sadler, Simmons, & Hows, 2005) have to be mentioned. They are of particular educational relevance because of the integration of socioscientific issues in the science classroom. In the German science educational community comprehensive drafts for chemistry and physics can be found. They are subsumed under the keywords sociocritical and problem-oriented science teaching (Marks & Eilks, 2009; Menthe, 2006), but biology education currently lacks these teaching strategies. We espouse the idea that lessons should place special emphasis on central philosophical questions because biology has the human being and the conception of man as subjects of discussion. We call this educational concept critical-reflective biology education. Needless to say this is a considerable challenge, not least since it asks for qualified teachers who manage the complexity of neuroscience as a subject, who themselves are neuroscientific literate, and who are endowed with a positive attitude toward the ambivalent ethical dimension of neuroscience.

Having laid out the theoretical basis for an education that meets the demands of lessons promoting neuroscientific literacy, we need to find the status quo in German science classrooms to derive practical consequences for teaching strategies.

AN EDUCATIONAL GAP

The PISA study, a large-scale, in-depth assessment of secondary pupils' competencies in 2000, ranked Germany well below the OECD average (Organisation for Economic Co-operation and Development) in reading, math, and science. The German educational system saw the need for radical reform—comparable to the American reform after the Sputnik crisis. Therefore the curricula in the 16 German federal states were revised under the administration of the Kultusministerkonferenz (KMK), the *Standing Conference of the Ministers of Education and Cultural*

Affairs of the Länder [Federal States] in the Federal Republic of Germany. The focus shifted from input-oriented to output-oriented curricula that demanded learning situations aiming at the acquisition of competencies rather than knowledge. In addition, the 16 federal states were to be made comparable by a common framework for minimum requirements for the outcome of education, including science (KMK, 2005).

The shift to outcome-oriented curricula brought with it the concept of competence. In order to provide for a common understanding of the concept, the curricula were adapted in accordance with a competence model for science that was based on Franz E. Weinert's definition of competence and the definition for scientific literacy according to the OECD. Weinert's definition for competence refers to combinations of those "cognitive, motivational, moral, and social skills available to (or potentially learnable by) a person ... that underlie the successful mastery through appropriate understanding and actions of a range of demands, tasks, problems, and goals" (Weinert, 2001, p. 2433). The OECD described scientific literacy as "the ability to engage with science-related issues, and with the ideas of science, as a reflective citizen. A scientifically literate person, therefore, is willing to engage in reasoned discourse about science and technology" (OECD, 2015, p. 7).

Working with scientific matters, the most significant competencies are:

- Explain phenomena scientifically: Recognize, offer, and evaluate explanations for a range of natural and technological phenomena.
- Evaluate and design scientific inquiry: Describe and appraise scientific investigations and propose ways of addressing questions scientifically.
- Interpret data and evidence scientifically: Analyze and evaluate data, claims, and arguments in a variety of representations and draw appropriate scientific conclusions (OECD, 2015, p. 7).

In order to use these competencies the OECD stresses the need for content-related, process-related, and epistemic knowledge, in addition to an open-minded, interested, and acknowledging attitude toward science and its methods. Whereas content-related knowledge refers to scientific facts, terms, ideas, and theories about nature, procedural knowledge focuses on the scientific method and inquiry. This involves repeated measurements, error reduction, and control of variables, as well as the academic communication of scientific results. The third aspect of scientific literacy refers to epistemic knowledge, which deals with the nature of science itself. Epistemic knowledge according OECD is the "understanding of the rationale for the common practices of scientific enquiry, the status of the knowledge claims that are generated, and the meaning of foundational terms such as theory, hypothesis and data" (OECD, 2015, p. 6).

Unfortunately, the concept of the close connection between science and society from Holbrook and Rannikmae cannot be found here. Currently, the German educational ideal for science still pursues the strategy "science through education," which was established in the 1950s. It follows a very one-dimensional science education program, aiming at the accumulation of specific knowledge rather than at true-to-life discussions and situations from everyday life. This one-dimensional program promotes the few who may eventually work in the scientific sector.

Despite the introduction of a common framework for minimum requirements for the outcome of science education through the governmental KMK reform from 2005, the subcompetence "evaluation of ethical issues" has still not found its way into the educational system. This subcompetence takes into account that biological topics such as genetics, genetic engineering, environmental education, and neurobiology are very often linked to an ethical dimension that exceeds the subject knowledge significantly. Students should be able to recognize these ethical implications in the subject's contexts and to participate in the often highly controversial public discourse. In addition, students should be able to change their point of view and to evaluate ethically problematic situations (KMK, 2005).

From the perspective of modern science educators this change in the basic administrative documents was welcomed. However, the shift of focus to socially relevant ethical dilemmas in connection to biological issues has not found its way into the curricula of all federal states. This means that most federal states' lesson plans that have a practical impact on science teaching in German schools have not been updated yet. From the federal states that updated, the most ignored the recommendations of the KMK concerning the subcompetence "evaluation of ethical Issues." For example, this discrepancy is plain to see regarding neuroscience in biology.

What all curricula in Germany have in common is the instruction of basic neuroscientific contents that mainly belong to the cellular and molecular branches of neuroscience, e.g., senses and perception, structure and function of the central nervous system, structure and function of nerve cells and synapses, and nerve conduction and electric potentials.

About half of the curricula today include elements of the systemic and behavioral branches of neuroscience such as memory and learning, reflexes, stimulus response, influence of drugs and neurotoxins, and connections between the nervous and the endocrine system.

Only 3 of the 16 curricula include the initial steps for critical-reflective biology teaching, as mentioned above. The federal state of Berlin introduced the conception of man; in Brandenburg they introduced recent results of neuroscientific research; and in Hamburg it was the neurobiology and self-conception, consciousness, and free will. This clearly shows that the German curricula mainly concentrate on basic physiological aspects of neuroscience without any reflection or opportunity for reflection on neuroethical issues.

In addition to this, teachers and teacher trainees are often not able to discuss ethically relevant issues that exceed the scientific realm of their subject (Dittmer, 2006). Moreover, they rarely consider bioethical issues to be a part of their field of profession (Hartmann-Mrochen, 2011), and they themselves lack the competence to discuss bioethical and especially neuroethical dilemmas. Therefore it is questionable whether science teachers themselves are neuroscientifically literate.

CONSIDERING TEACHERS' LITERACY

Some research has already been carried out on the level of neuroscience literacy within the teaching community (Bartoszeck & Bartoszeck, 2012; Dekker, Lee, Howard-Jones, & Jolles, 2012; Howard-Jones, 2014). Paul Howard-Jones describes teachers gaining their knowledge of neuroscience from the mass media as well as science magazines and folk psychology as a major problem. Media contribute to the creation of misunderstandings and so-called neuromyths. Howard-Jones is following the OECD and defines the term neuromyth as a "misconception generated by a misunderstanding, a misreading or a misquoting of facts scientifically established" (OECD, 2002, p. 111). In his study titled "Neuroscience Literacy of Trainee Teachers" (Howard-Jones, 2014), 158 trainee teachers were required to fill out a questionnaire with 38 statements concerning the brain (16 general claims, 15 neuromyths, and 7 statements on the mind-brain-relation). Sanne Dekker et al. (2012) made a spot check of 242 primary and secondary school teachers that answered a 32 item online survey about the brain and its influence on learning that included 15 neuromyths. Amauri Betini Bartoszeck and Flavio Kulevics Bartoszeck (2012) acquired data by means of two questionnaires with 10 items each, the first focusing on the brain's influence on learning (83 teachers surveyed) and the second focusing on the molecular mechanisms of learning (42 teachers surveyed).

The results are unsettling. In Howard-Jones' study the teachers answered 9 out of 16 general questions correctly on average. They failed to recognize two-thirds of the neuromyths, which are relevant for science classrooms. The study of Dekker et al. (2012) had similar results: the teachers thought 49% of the number of neuromyths to be true and they answered incorrectly every third general question about the brain. The same holds true for the study of Bartoszeck & Bartoszeck (2012).

In an ongoing study at the University of Leipzig, 107 biology students in the teacher's program evaluated a number of neuromyths. According to the results, 69% agreed that we only use 10% of our brain, 57% believed that listening to classical music during pregnancy promotes intelligence of the unborn child, and 91% thought smartphone applications such as brain gym rejuvenates the brain. The most prevalent neuromyths among teachers are:

- We only use 10% of our brains.
- Memory is stored in the brain like in a computer. Each memory is stored in a tiny part of the brain, like a hard drive.
- Individuals learn better when they receive information in their preferred learning style.
- Differences in hemispheric dominance can help explain individual differences among learners.
- Exercises that improve coordination of motor-perception skills also improve literacy skills.
- Environments that are rich in stimulus improve the brains of preschool children.
- Children are less attentive after sugary drinks and snacks.

In conclusion there is a strong need to improve the neuroscience literacy of teacher trainees and in-service teachers.

To improve neuroscience education we could wait for a top-down revision of the curricula coming from the federal states or we could take a bottom-up approach, where teacher trainees and in-service teachers learn a critical-reflective teaching strategy to integrate neuro-philosophical and neuroethical debates into science classrooms and to deal with neuromyths in a proper manner.

SCIENCE EDUCATION IN BIOLOGY CLASSROOMS THROUGH "EVERYDAY MYTHS"

Pupils cannot be regarded as *tabulae rasae* entering our science classrooms. By now, educators agree that the idea of a "giving and taking of knowledge" as a simple transfer is outdated (Marsch, 2009). Instead, learning is regarded as an active and self-determined constructional process in which pupils draw on their wealth of experience, which is derived from daily routines, everyday language, the mass media, peers, and family. This experience generates ideas and notions about reality, which have to be considered by teachers of all subjects. These notions consist of an individual argumentative structure and are relevant for an individual's judgment and behavior (Combe & Gebhard, 2007). Therefore they have to be considered as alternative notions, rather than false beliefs. They present chances for learning, but also obstacles that have to be overcome.

In order to work with pupils' notions about science, two methodologies have evolved. Harald Gropengießer's theory of experientialism (2007) attempts to explicate and reflect metaphors pupils, educators, and scientists use to make sense of biological phenomena, e.g., the cell as a factory (Gropengießer, 2007) or the brain as a computer or society (Goschler, 2008). This approach refers to the "embodiment theory" and the "conceptual metaphor theory" of John Lakoff and Mark Johnson (1980, 1999), who realized that metaphors play a central role in comprehension. With regards to the understanding of neuroscience, the authors concentrated on topics such as memory, the visual process, and nerve conduction (Gropengießer, 2001; Sundermeier, 2009).

For our purpose, the second methodological approach "everyday myths" is more useful due to its connections to the bioethical discourse and to narrative approaches to science teaching (Zabel & Gropengießer, 2015). The "everyday myths" approach is based on the cultural philosophic and cultural psychologic works of Ernst Eduard Boesch (1980) and Ernst Cassirer (1972), as well as theoretic interconnections between different conceptions of education, transformation (Koller, 2012; Marotzki, 1990), and experience (Husserl, 1963; Waldenfels, 1997).

Arno Combe and Ulrich Gebhard suggest interactions between individuals and their environment as the process of parallel subjectification and objectification:

> Objectification is the "objective", systematic perception, description and explanation of reality. Subjectification, on the other hand, is concerned with the symbolic meaning of things which is expressed in subjective notions, phantasies and connotations. These symbolisations have to be considered to be an integral and mostly indispensable part of everyday language. They are used in a systematic and mostly unconscious way. (Combe & Gebhard, 2007, p. 21)

Boesch illustrates the relation between subjectification and objectification with an example of an architect designing and planning a house. The house does its normal task of offering shelter and protecting against threats. The architect needs certain skills, as well as theoretical knowledge about statics and mathematics. The thought process concerned with designing the house and the concept of the house (shelter and protection) are seen as the objective relation. As soon as the architect moves into the house, it becomes his home and is filled with subjective meaning, expectations, and emotions. The individual's need for meaning can only be satisfied by combining the objective (the house) and subjective (the home) relations to the individual's world.

Ulrich Gebhard (2007) states that these kinds of subjectifications also become relevant in the science classroom, e.g., when pupils speak of "monster tomatoes" in the discourse on genetic engineering. "Everyday myths" can be found in symbolically charged, biographical notions, and stories in which wishes, personal values, fears, and meaning are concentrated (Gebhard, 2007, p. 120). They exceed the scientific dimension and mainly occur as implicit and intuitive rather than explicit knowledge. Due to their extent of meaning, as they comprise aspects of the self-image, conception of man, and a world view, they influence the values, interests, and behaviors of a person (Combe & Gebhard, 2007, p. 63).

The combination of an individual's "everyday myths" and the scientific objectifications provides the transformation of scientific insights into everyday thinking (Gebhard, 2007), especially in science classrooms that are enriched with ethical and philosophical questions. In the context of bioethical issues the intuitive part of moral reasoning is emphasized. Intuitions do not give better judgments, but they influence thought and action and therefore need to be considered in the process of reflection (Gebhard, 2007, p. 122; see also Dittmer, 2005). This also links to Jonathan Haidt (2001), who ascribes a significant influence to the intuitive system in moral reasoning.

Following the ideas in Blumenberg's book *Paradigms for a Metaphorology* (1986), we, the authors, demand that teachers and pupils speak both the objective and the subjective language. Both languages are important for the successful and biographically valuable construction of meaning. Gebhard concludes that a reflective attitude (argumentation, communication, change of perspective, and evaluation) and the discussion of individual associations, intuitive judgments, and emotional reactions are necessary to improve the pupils' competence of ethical reflection (Gebhard, 2007).

Several studies confirm the benefits of considering "everyday myths" in biology classrooms. It was shown that the explication of everyday myths is a suitable method to reflect on the conception of man. Moreover, the explicit reflection of everyday myths strengthens the self-concept, motivation, and interest development of pupils, e.g., in science lessons on genetics (Born, 2007; Gebhard & Mielke, 2002; Monetha, 2009). This method is built on the method of philosophizing with primary school pupils. It requires teachers to open up a space within discourse for spontaneous ideas, fantasies, and associations. Respect and empathy are crucial for positive results. An open atmosphere for the free exchange of opinions and ideas is the starting point to explore the scientific facts, as well as their philosophical and ethical dimensions. This will meet the criteria for the critical-reflective teaching in science classrooms.

In an ongoing study focusing on the role of moral intuitions in pupils' decision-making we are investigating their "everyday myths" about neuroscience. We designed a thought experiment where year 9 and 10 pupils discuss the moral dimension of three imaginary neuroscientific research proposals as members of an ethical commission. The research proposals contain elements of current existing neuroscientific research, e.g., the development of brain-computer interfaces, and fictional elements taken from the mass media such as mind-reading through techniques of neuroimaging. We analyze, interpret, and systematize the pupils' statements on the basis of the grounded theory methodology (Breuer, 2009). So far we have been able to derive four essential "everyday myths" that structure the pupils' discourse:

Mechanized Man

According to the pupils, neuroscientific research will soon be able to transform man via the use of brain-based technologies. The notion of transhuman lifeforms is objected by pupils; for example, one of our participants said:

> Pupil 1: If you have such a thing (BCI), I don't want to call you a cyborg as in Star Trek, but in some way you are ..., you are not 100% human anymore. Okay, you could also say so if you have an artificial joint or something like that.
> Pupil 2: Somehow you become a terminator.

Demystified Man

The debate on the binding problem between brain and mind also influences pupils' perspectives on neuroscience. The possibility of solving the binding problem through neuroscientific research is regarded as problematic as people should have a right "not know," i.e., there should be areas that are not included in the scientific inquiry:

> Pupil 3: Since you are playing around with the mind and soul in this project, I don't like it. In my opinion the mind or soul of a person shouldn't be reduced to a scientific concept, because then a large part of what makes a human is lost.

Manipulated Man

Pupils derived possibilities to manipulate the nervous system and therefore humans themselves through the explanation of structural and functional conditions of our nervous system. Neuroimaging techniques are considered particularly "powerful." As one pupil commented:

> Pupil 4: I gave Project 2 the least funding. The main reason for me was ... well, I believe that the risk to manipulate man is at its highest in this project, because thoughts and stuff can be traced back to biological and chemical reactions.

Manipulated Society

The pupils expressed that the next logical step from the manipulation of a single person is the abuse of neuroscientific research to control and manipulate whole societies:

> Pupil 5: If it is possible to read everyone's mind, you are frightened by the possibility that your thoughts can be revealed. Any opposition will be suppressed automatically. You will have a totalitarian regime without any opposition.

On the whole, neuroscientific research is perceived as something dangerous and is clearly negative. In the group discussions the pupils rarely mentioned positive aspects of neuroscientific research, like medical treatment for people with neurodegenerative diseases. The corpus of "everyday myths" concerning neuroscience needs to be gradually increased and differentiated through further research. The aim is to integrate the "everyday myths" we found into our didactic concept of critical reflection in order to support teachers in teaching neuroscience.

CONCLUSION

Since neuroscience has become more important in societal discourse, there is a necessity to deal with neuroscientific information as a reflective citizen. Therefore it must find its way into science education. The science education community needs to promote responsible treatment of neuroscientific insight as part of general education. We have shown that the current administrative and curricular situation is not suitable to meet the requirements to promote neuroscience literacy. In conclusion the gap between neuroscience and science education, the curricular and administrative documents as well as the overarching concept of education needs to be closed.

In order to develop teaching strategies suitable for sociocritical reflection the concept of neuroscience literacy has served as a point of reference. We have shown that science teachers lack the reflective attitude to deal with neuroscientific and neuroethical issues because they are in need of profound knowledge about neuroscientific and neuroethical developments. Furthermore, teacher trainees and in-service teachers need to deal with bioethical issues; this means a reflective attitude on controversial matters. This creates the demand for improvements in teacher education, a stronger networking between neuroscientists, neuroethicists, science educators, science education researchers as well as educational administratives.

While we acknowledge that administrative changes can only come about on a long-term basis, we suggest the usage of "everyday myths" as a practical and more short-term strategy to promote neuroscience literacy for critical-reflective science teaching. Thereby we offer a valuable strategy to support pupils in the construction of meaning and to educate them to become responsible, reflective citizens who are able to participate successfully in our society.

References

Bartoszeck, A. B., & Bartoszeck, F. K. (2012). How in-service teachers perceive neuroscience as connected to education: An exploratory study. *European Journal of Educational Research, 1*(4), 301–319.

Bear, M., Connors, B. W., & Paradiso, M. A. (2012). *Neurowissenschaften: Ein Grundlegendes Lehrbuch für Biologie, Medizin und Psychologie*. Berlin and Heidelberg: Springer Spektrum.

Blumenberg, H. (1986). *Die Lesbarkeit der Welt*. Frankfurt am Main: Suhrkamp.

Boesch, E. E. (1980). *Kultur und Handlung: Einführung in die Kulturpsychologie*. Bergen, Stuttgart, and Wien: H. Huber.

Born, B. (2007). *Lernen mit Alltagsphantasien: Zur expliziten Reflexion impliziter Vorstellungen im Biologieunterricht*. Wiesbaden: VS Verlag für Sozialwissenschaften.

Breuer, F. (2009). *Reflexive Grounded Theory: Eine Einführung für die Forschungspraxis*. Wiesbaden: VS Verlag für Sozialwissenschaften.

Cassirer, E. (1972). *An essay on man: An introduction to a philosophy of human culture*. New Haven, CT: Yale University Press.

Combe, A., & Gebhard, U. (2007). *Sinn und Erfahrung: zum Verständnis fachlicher Lernprozesse in der Schule*. Opladen and Farmington Hills, MI: Verlag Barbara Budrich.

Dekker, S., Lee, N. C., Howard-Jones, P., & Jolles, J. (2012). Neuromyths in education: Prevalence and predictors of misconceptions among teachers. *Frontiers in Psychology, 3*(429), 1–8.

Dittmer, A. (2005). Vom Schattenboxen und dem Verteidigen intuitiver Urteile. *Ethik und Unterricht, 2*(05), 34–40.

Dittmer, A. (2006). Zur Diskussion gestellt—Wissenschaftsphilosophie am Rande des Faches? *Mathematischer und Naturwissenschaftlicher Unterricht, 59*(7), 432–438.

Elger, C. E., Friederici, A. D., Koch, C., Luhmann, H., Von der Malsburg, C., Menzel, R., et al. (2004). Das Manifest. Elf führende Neurowissenschaftler über Gegenwart und Zukunft der Hirnforschung. *Gehirn & Geist. Das Magazin für Psychologie und Hirnforschung, 6*, 30–37.

Gebhard, U. (2007). Intuitive Vorstellungen bei Denk-und Lernprozessen: Der Ansatz "Alltagsphantasien." In D. Krüger, & H. Vogt (Eds.), *Theorien in der biologiedidaktischen Forschung* (pp. 117–128). Berlin and Heidelberg: Springer.

Gebhard, U., & Mielke, R. (2002). Alltagsmythen und Selbstkonzepte zur Gentechnik. In R. Klee, & H. Bayrhuber (Eds.), *Lehr-und Lernforschung in der Biologiedidaktik Band 1* (pp. 75–88). Innsbruck: Studienverlag.

Goschler, J. (2008). *Metaphern für das Gehirn: Eine kognitiv-linguistische Untersuchung*. Stuttgart: Frank & Timme GmbH.

Gräber, W. (2002). "Scientific Literacy"—Naturwissenschaftliche Bildung in der Diskussion. In P. Döbrich (Ed.), *Qualitätsentwicklung im naturwissenschaftlichen Unterricht. Fachtagung am 15. Dezember 1999* (pp. 1–18). Frankfurt am Main: DIPF.

Gropengießer, H. (2001). *Didaktische Rekonstruktion des Sehens*. Oldenburg: Didaktisches Zentrum der Carl von Ossietzky Universität.

Gropengießer, H. (2007). Theorie des erfahrungsbasierten Verstehens. In D. Krüger, & H. Vogt (Eds.), *Theorien in der biologiedidaktischen Forschung* (pp. 105–116). Berlin and Heidelberg: Springer.

Haggard, P., & Eimer, M. (1999). On the relation between brain potentials and the awareness of voluntary movements. *Experimental Brain Research, 126*, 128–133.

Haidt, J. (2001). The emotional dog and its rational tail: A social intuitionist approach to moral judgment. *Psychological Review, 108*(4), 814–834.

Harms, N. C. (1977). *Project synthesis: An interpretative consolidation of research identifying needs in natural science education.* Boulder, CO: University of Colorado.

Hartmann-Mrochen, M. (2011). *Zwischen Notengebung und Urteilsfähigkeit: Einstellung und Vorstellungen von Lehrkräften verschiedener Fachkulturen zum Kompetenzbereich Bewertung der Nationalen Bildungsstandards* (Dissertation, Universität Hamburg).

Herculano-Houzel, S. (2002). Do you know your brain? A survey on public neuroscience literacy at the closing of the decade of the brain. *Neuroscientist, 8*(2), 98–110.

Hildt, E. (2012). *Neuroethik.* München and Basel: Ernst Reinhardt Verlag.

Hodson, D. (2008). *Toward scientific literacy: A teacher's guide to the history, philosophy and sociology of science.* Rotterdam and Taipei: Sense Publishers.

Holbrook, J., & Rannikmae, M. (2009). The meaning of scientific literacy. *International Journal of Environmental and Science Education, 4*(3), 275–288.

Howard-Jones, P. A. (2010). *Introducing neuroeducational research: Neuroscience, education and the brain from contexts to practice.* London: Routledge.

Howard-Jones, P. A. (2014). Neuroscience and education: Myths and messages. *Nature Reviews Neuroscience, 15*, 817–824.

Hurd, P. D. (1958). Science literacy: Its meaning for American schools. *Educational Leadership, 16*(1), 13–16.

Husserl, E. (1963). *Cartesianische Meditationen und Pariser Vorträge.* Hua I, La Hague: Martinus Nijhoff.

Koller, H. C. (2012). *Bildung anders denken: Einführung in die Theorie transformatorischer Bildungsprozesse.* Stuttgart: Kohlhammer.

Kultusministerkonferenz, KMK (2005). *Bildungsstandards im Fach Biologie für den Mittleren Schulabschluss.* München and Neuwied: Luchterhand.

Lakoff, G., & Johnson, M. (1980). *Metaphors we live by.* Chicago, IL: University of Chicago Press.

Lakoff, G., & Johnson, M. (1999). *Philosophy in the flesh: The embodied mind and its challenge to western thought.* New York: Basic books.

Libet, B. (1985). Unconscious cerebral initiative and the role of conscious will in voluntary action. *The Behavioral and Brain Sciences, 8*, 529–566.

Marks, R., & Eilks, I. (2009). Promoting scientific literacy using a sociocritical and problem-oriented approach to chemistry teaching: Concept, examples, experiences. *International Journal of Environmental and Science Education, 4*(3), 231–245.

Marotzki, W. (1990). *Entwurf Einer Strukturalen Bildungstheorie Biographietheoretische Auslegung von Bildungsprozessen in Hochkomplexen Gesellschaften.* Weinheim: Deutscher Studienverlag.

Marsch, S. (2009). *Metaphern des Lehrens und Lernens* (Dissertation, Freie Universität Berlin).

McCurdy, R. C. (1958). Toward a population literate in science. *The Science Teacher, 25*, 366–408.

Menthe, J. (2006). *Urteilen im Chemieunterricht—eine empirische Untersuchung zum Einfluss des Chemieunterrichts auf das Urteilen von Lernenden in Alltagsfragen* (Dissertation, Christian-Albrechts-Universität zu Kiel). Osnabrück: Der Andere Verlag.

Monetha, S. (2009). *Alltagsphantasien, Motivation und Lernleistung: Zum Einfluss der expliziten berücksichtigung von Alltagsphantasien im Biologieunterricht auf motivationale Faktoren und Lernleistung. Studien zur Bildungsgangforschung 26.* Opladen and Farmington Hills, MI: Verlag Barbara Budrich.

National Research Council (1996). *National science education standards.* Washington, DC: National Academy Press.

OECD (2002). *Understanding the brain: Towards a new learning science*. Paris: OECD.

OECD (2015). *Pisa 2015 draft science framework*. Paris: OECD.

Project 2061 American Association for the Advancement of Science (1989). *Science for all Americans: A Project 2061 report on literacy goals in science, mathematics, and technology*. Washington, DC: American Association for the Advancement of Science.

Roskies, A. (2002). Neuroethics for the new millenium. *Neuron, 35*(1), 21−23.

Soon, C. S., Brass, M., Heinze, H. J., & Haynes, J. D. (2008). Unconscious determinants of free decisions in the human brain. *Nature Neuroscience, 11*, 543−545.

Sundermeier, S. (2009). *Der Prozess der Sinneswahrnehmung: Historisch didaktische Rekonstruktion und Entwicklung einer fächerübergreifenden Lernumgebung*. Oldenburg: Didaktisches Zentrum der Carl-von-Ossietzky-Universität.

Waldenfels, B. (1997). *Topographie des Fremden*. Frankfurt am Main: Suhrkamp.

Weinert, F. (2001). Competencies and key competencies: Educational perspective. In N. J. Smelser, & P. B. Baltes (Eds.), *International encyclopedia of the social and behavioral sciences 4* (pp. 2433−2436). Amsterdam: Elsevier.

World Health Organization (2001). *ICF: International classification of functioning, disability, and health*. Geneva: Author.

Yager, R. E., & Roy, R. (1993). STS: Most pervasive and most radical of reform approaches to "science" education. In R. E. Yager (Ed.), *The science, technology, society movement 7* (pp. 7−13). Washington, DC: National Science Teacher Association.

Zabel, J., & Gropengießer, H. (2015). What can narrative contribute to students' understanding of scientific concepts, e.g. evolution theory? *Journal of the European Teacher Education Network, 10*, 136−146.

Zardetto-Smith, A. M., Mu, K., Phelps, C. L., Houtz, L. E., & Royeen, C. B. (2002). Brains rule! Fun = learning = neuroscience literacy. *The Neuroscientist, 8*(5), 396−404.

Zeidler, D. L., & Nichols, B. H. (2009). Socioscientific issues: Theory and practice. *Journal of Elementary Science Education, 21*(2), 49−58 (Spring 2009).

Zeidler, D. L., Sadler, T. D., Simmons, M. L., & Hows, E. V. (2005). Beyond STS: A research-based framework for socioscientific issues education. *Science Education, 89* (3), 357−377.

Ziman, J. (1980). *Teaching and learning about science and society*. Cambridge: Cambridge University Press.

Ziman, J. (1994). The rationale of STS education is in the approach. In J. Solomon, & G. Aikenhead (Eds.), *STS education: International perspectives on reform* (pp. 21−31). New York and London: Teachers College Press.

15

"Strangers" in Neuroscientific Research[*]

B. Bringedal[1,5], M. Christen[2], N. Biller-Andorno[2], H. Matsuzaki[3] and A. Rábano[4]

[1]LEFO - Institute for Studies of the Medical Profession, Oslo, Norway
[2]University of Zürich, Zürich, Switzerland [3]University of Oldenburg, Oldenburg, Germany [4]Carlos III Health Institute, Madrid, Spain [5]Columbia University, New York, United States

INTRODUCTION

New technological opportunities drive new types of research. The developments in digital technology have made it possible to collect, store, and analyze huge amounts of data, with significant impacts on science and society. "Big science" is the common term for such large-scale research, and the associated digitalization of research has simplified scientists' collaboration in many ways. It allows the creation of new types of research infrastructures—durable institutions, technical tools and platforms, and/or services that are put in place for supporting and enhancing research. Such infrastructures are increasingly set up as virtual research environments (VREs): web portals that provide services to users that are connected to underlying databases and repositories of various kinds. The technological developments promote collaboration among

[*]The title is inspired by David Rothman's book *Strangers at the Bedside* (Rothman, 1991). It also implies advantages of being an "external insider" who can discover new perspectives that the internal insider hardly becomes aware of. It does not imply that the members of the former Ethics, Legal and Social Aspects Committee and the current Ethics Advisory Board of the Human Brain Project are nonexperts in neuroscience—actually 8 out of 11 members of the current EAB have a primary background (PhD) in neuroscience or technology.

The Human Sciences after the Decade of the Brain.
DOI: http://dx.doi.org/10.1016/B978-0-12-804205-2.00015-X

© 2017 Elsevier Inc. All rights reserved.

many research institutions, research groups, and individuals, which facilitates the sharing of knowledge and contributes to more effective knowledge accumulation. The technological opportunities also involve new practical and ethical challenges tied to the size of the project, e.g., related to the collection, storage, analysis, and application of large amounts of data (usually referred to as "big data").

The European Commission's Human Brain Project (HBP) is an example of research that aims to create various VREs involving big data. The HBP is one of the world's largest initiatives in brain research and neuroscience, comprising more than 110 partners and, with matched partner funding for research, a 10-year budget of €1.2 billion.

Early on, there was an awareness of the need to pay attention to the inherent ethical, social, legal, and philosophical implications of the research.[a] Two ethics committees were established, later merged to one, as well as The Ethics and Society subproject assigned to explore social and normative issues emanating from the HBP research and to contribute to fostering responsible research and innovation by raising ethical awareness among the project participants.

The former Ethics, Legal and Social Aspects Committee (ELSA) and the Ethics and Society subproject complement ethics and ethically justified research. While the ethics committee was intended to provide independent views on ethical issues raised during the progress of the HBP, the research subproject identifies ethical issues qua the actual research.[b] In September 2015, ELSA was merged with the second committee that advised the HBP executive management: the Research Ethics Committee (REC). The two bodies were merged during the course of various organizational changes in the ramp-up phase of the HBP (see additional explanations in the text). All authors of the current chapter were members of the previous ELSA, while Christen, Rábano, and Bringedal are also members of the current EAB. The views expressed in the chapter are the views of the five authors only, and are not intended to reflect either the view of the (former or current) committee as a whole or the views of the leaders of the HBP.

In this chapter, we report our experiences of being involved in the ethical oversight of such a large-scale project and discuss some of the challenges from the perspective of experts that currently are mainly

[a]In fact, the importance of paying proper attention to ethics in this research was underscored as early as in 2010 by Dudai and Changeux (personal communication) in a meeting with the EC.

[b]An example is the "foresight lab," which "will be responsible for monitoring HBP research and exploring its social and ethical implications for European citizens, European industry, the European economy and European society." (See https://www.humanbrainproject.eu/ethics-and-society.)

active in ethics/sociological research.[c] We intend to highlight a few selected issues to promote an open discussion of how crucial ethical, social, and legal questions can be addressed and dealt with in such a large-scale project. Although many of the issues pertain to research promoting VREs and big data in general, our discussion is limited to the HBP.

We begin with a brief presentation of the HBP and the newly formed EAB, followed by a description of features (of the organization) of the research that pose particular ethical challenges. (For readability, we use "ethical aspects" as short for "ethical, legal, social, and philosophical" aspects.) Based on this description, we discuss how the issues could be dealt with. We suggest three overarching normative principles as guidance for all activity in the HBP in general and the work of the EAB in particular, and conclude the chapter by suggesting recommendations.

THE HUMAN BRAIN PROJECT

General Description

The HBP is a large-scale, long-term research project that includes 112 partners in 24 countries with a budget of €1.2 billion over 10 years. The project's central aim is "to build a world-class experimental facility to study the structure and functions of the human brain. This new information and communications technology (ICT) infrastructure will integrate neuroscience data and will be used to design brain-computer models to understand and simulate the human brain" (European Commission, 2014). This will "accelerate our understanding of the human brain, make advances in defining and diagnosing brain disorders, and develop new brain-like technologies" (the official HBP website: https://www.humanbrainproject.eu).

The HBP is one of the two winning projects of the European Commission's flagship initiative.[d] This initiative was formulated under the 7th Research Framework Program of the European Union in the field of Future and Emerging Technologies (FET). The intent was for FET flagship projects to be large-scale, ambitious research projects with a visionary goal in ICT. A total of 26 consortiums submitted projects. In 2011, six candidates were nominated to prepare a detailed

---

[c] The disciplinary background of the authors is as follows: BB: sociology; MC: empirical ethics/neuroinformatics; NBA: biomedical ethics/medicine; HM: sociology/science and technology studies; AR: neuroscience/neuropathology.

[d] The other is the Graphene flagship project; see: http://graphene-flagship.eu/.

proposal. In January 2013, the European Commission announced the two selected projects.

These FET flagship projects are big science projects with a strong focus on ICT, but the projects' activities are expected to extend beyond research, addressing aspects such as coordination, strategy development, mobility programs, international cooperation, road-mapping activity, training and education, outreach, communication, and PR activities (http://cordis.europa.eu/fp7/ict/programme/fet/flagship/doc/flagshipflyer-july2013_en.pdf).

The HBP structures its research along three major topics: First, *Future Neuroscience*, i.e., the aim to achieve a unified, multilevel understanding of the human brain that integrates data and knowledge about the healthy and diseased brain across all levels of biological organization, from genes to behavior. This also includes establishing in silico experimentation as a foundational methodology for understanding the brain. Second, *Future Computing*, i.e., the aim to develop novel neuromorphic and neuro-robotic technologies based on the brain's circuitry and computing principles. This includes the build-up of supercomputing technologies for brain simulation, robots and autonomous systems control, and other data-intensive applications. Third, *Future Medicine*, i.e., the aim to create an objective, biologically grounded, map of neurological and psychiatric diseases based on multilevel clinical data. This map should then be used to classify and diagnose brain diseases and to configure models of these diseases. It finally should lead to personalized medicine for neurology and psychiatry.

The HBP flagship project is coordinated by the Swiss Ecole Polytechnique Federale de Lausanne. Other important members include the German Ruprecht-Karls-Universität Heidelberg and the Forschungszentrum Juelich, the French Centre National de la Recerche Scientifique, the Swedish Karolinska Institutet, and the Spanish Universidad Politecnicade Madrid.

Controversy in the Neuroscientific Community—The Open Letter

In 2014, a debate emerged in reaction to developments within the HBP. The debate was initiated by a change of plans in the project, in particular when cognitive and systems neuroscience (cognitive architecture) allegedly was given diminished significance in the project. Within a few months, an open letter to the European Commission was signed by more than 800 scientists, who were not part of the HBP (see http://www.neurofuture.eu). The letter included a critique regarding the scientific approach ("overly narrow approach") as well as the governance structure ("lack of flexibility and openness of the consortium"). For a similar critique, see Frégnac and Laurent (2014).

The critique led to a mediation process, which focused on conflicts in the management.[e] "The open letter influenced the first review process that began with a screening but led to a full review due to the complexity and size of the project. It was no surprise to anyone that a full hearing would take place," says Kevin Grimes, research coordinator for the Human Brain Project Ethics Governance and Regulation (personal communication). The European Commission's response to the open letter included an independent review and a mediation process.

As a result of the external review and the mediation, cognitive architecture became part of the project again (including a new call for proposals), and significant changes were made in the governance structure. We return to the question of organization and governance later. One of the changes, however, concerned the role and structure of the ethics committees directly.

The Ethics Advisory Board

Ethical, social, legal, and philosophical aspects of the research were part of the project from the beginning. One of the subprojects, Ethics and Society, focused on research regarding these aspects. In addition, two external advisory committees were established; one was dedicated to research ethics (REC) while the second was intended to take a broader long-term perspective on ethical, legal, and social aspects (ELSA). One and a half years after the external ethical review of the HBP and the mediation report initiated by the open letter, the Board of Directors decided that the two committees should be merged into one Ethics Advisory Board (EAB).[f] The rationale for the decision was the recognition of overlap between the responsibilities of the two committees, that the REC could not and should not provide a formal evaluation of the projects, and the need to adapt ethical advice to a new governing structure. The members of the two committees considered the change an improvement and suggested they merge as soon as possible. The leadership of the ELSA and the REC, as well as the coordinators of the two committees, led the progress toward the merger.

[e]See the official HBP response to the open letter (https://www.humanbrainproject.eu/documents/10180/17646/HBP-Statement.090614.pdf).

[f]The idea of merging the two committees had come up earlier, independent of the open letter. One of the reasons was that the sharing of responsibilities between the Ethics and Society research project and the ELSA was unclear.

The current board was established in September 2015.[g] The majority of the members of the two previous committees were reappointed. The selection of members was primarily based on the assessment of core competencies needed in the group, while geographic distribution, balanced gender representation, and availability to attend meetings were also considered. The required expertise was assessed by the four chairs and cochairs of the ELSA and the REC in collaboration with administrative support in the HBP and based on the contents of the 12 HBP subprojects. Eight out of eleven members of the current EAB have a primary background (PhD) in neuroscience or technology.

The responsibilities of the EAB are similar to those of the former ELSA and REC. In both cases, the description of responsibility is described in general terms. The EAB, to a large extent, decides its specific tasks by itself—thus far based on deliberation in the committee of how the general mandate should be interpreted and operationalized. A standard operating procedure was recently agreed on.[h]

The EAB currently consists of 11 members with diverse disciplinary and professional backgrounds. The members are appointed for 3 years and may be reappointed for a second period. The general mandate of the EAB is to advise the direction of the HBP on specific ethical, regulatory, social, and philosophical issues raised by the HBP research. In this respect, the principle of subsidiarity will be upheld; i.e., the responsibility for ensuring compliance with ethical and legal principles and regulations (local, national, and European Union (EU) level) will lie with the research organizations and research groups who are actually undertaking the research. Thus the EAB will not duplicate the work of those organizations and procedures for vetting and approving research activities. However, the EAB is expected to advise on matters regarding the ethical review of research where conformity with relevant legislation and Horizon 2020[i] rules is not guaranteed by existing bodies and procedures.

This includes in particular issues related to data-sharing and research procedures, e.g., research involving the use of data, samples, or resources generated outside the HBP or carried out in non-EU countries (e.g., China and the United States).

[g]See https://www.humanbrainproject.eu/ethics-advisory-board?inheritRedirect = true.

[h]A standard operating procedure can be found at https://www.humanbrainproject.eu/documents/10180/1139903/EABSOP_2015-10-06-2.pdf/0132960f-77cb-4782-ba5d-946eca9c0e25.

[i]Horizon 2020 is the EU's new program for research and innovation that runs from 2014 to 2020 with a ~€70 billion budget. The research program is managed by the European Commission.

CHALLENGES

Before we identify and discuss some of the characteristics of the HBP that pose potential ethical challenges, the general tendency toward optimism bias (Kahneman, 2011) in many technological and scientific undertakings should be examined. Researchers, policymakers, and funding bodies frequently are disproportionally optimistic about what the research can accomplish. "Innovation in basic science is often a cause for wonder and excitement," says Jonathan Wolff (2014, p. 27), and "(t)hose associated with a new development are quick to point out the anticipated benefit: a cure for cancer or dementia, an end to unsafe water and hunger."

The optimism bias phenomenon is more likely to be pronounced in a big science project, since so much is at stake, in terms of resources, political expectations, more or less explicit promises, and academic expectations. This phenomenon is obviously not restricted to innovation and research; public policy frequently shares the same bias (see, e.g., Irwin & Wynne, 1996).

Disproportional or exaggerated focus on the positive achievements expected from research involves the risk of downplaying the potential negative effects (which, of course, may include the lack of effects). The selection process of the flagship project itself could also be seen to promote this tendency, as an attempt and/or prerequisite to win the competition. Several of the finalists had similar broad expectations. "Overselling" of, or overoptimistic expectations for, the research can have significant ethical implications, since adverse effects almost always are inherent and unavoidable, as are errors and negative results. To reduce the problem of optimism bias, explicit attention to "what can go wrong?" is required. We will elaborate on this in subsequent parts of the chapter.

Specific Challenges in the HBP[j]

The following description of features of the HBP research that may pose particular ethical challenges is not intended to be exhaustive; the description is limited and based on the experiences we as authors have encountered thus far as members of the advisory committees. Further, the categories are not mutually exclusive; characteristics can obviously belong to more than one category.

---

[j]The description builds on Christen, Biller-Andorno, Bringedal, Grimes, and Savulescu (2016); Christen, Domingo-Ferrer, Draganski, Spranger, and Walter (2016).

Big Data—Big Neuroscience

A salient feature of the HBP is its size. Several hundred researchers are involved, as well as huge amounts of data. This poses particular challenges for communication, coordination, and accountability. The HBP involves substantial geographic (research institutions in different countries and cities), economic (half of the funds for research are expected to come from local sponsors, while the other half is provided by the EU[k]), multidisciplinary and multicultural (including different cultures in different disciplines), and multinational collaboration. The last involves not only different cultures and values but also different legislation.

Some of the specific issues raised by the project's size are as follows: (1) "too big to fail," (2) unclear responsibilities, (3) important issues are lost in the structural complexity, (4) the relationship between massive public investments and the potential for private gain, and (5) implications of the public/private partnerships. When so much money has been invested in a project, the expectation of getting a substantial return on investment may bias progress evaluations. Any decision to terminate a flagship project early might also embarrass those who had given a greenlight for the funding so it can be assumed efforts will be made to stabilize the project in a way that smaller and less visible projects could not expect.

The larger a project, the more complex its management will become. As scientists are not usually trained as administrators and communicators, this task may be underestimated. However, good governance is paramount for the success of a project.

Another issue regards the interface of public and private interests. Projects such as the HBP receive significant public funding. Although public/private partnerships are quite fashionable today, a number of issues remain unresolved regarding fair distribution of investment and gain.

Further, big data projects require advanced ICT. The technology itself contributes to structuring the scientific research, as well as the communication between different researchers, subprojects, and stakeholders.

Big data also means big money. In this case, the European Commission's financial support is particularly generous since the project is one of the two FET flagship projects. In addition to the funding from the EU, there is also a substantial amount of local cofunding. The concentration of funds was one of the primary concerns of the critics (cf. the open letter). Clearly, the HBP takes up a substantial

---

[k]For research funding, 50% comes from EU FET and 50% from partners; for management activities, 100% comes from the EU.

amount of resources at the expense of other projects. Seen from the critics' perspective, the concentration of funding poses a threat to securing diverse approaches; however, the concentration of resources can be essential in order to realize the ambitious goals that are set for the research—not least, the goal of developing models and simulation tools to reach a unified understanding of the human brain.

Big money on the funding side is just one part of the ethical challenge. Big money in terms of the potential commercial use of research findings also involves ethical challenges. Where prospects of big profit are involved, there is always the risk of ignoring, or, at least, a lack of attention to, the potential unfavorable effects of the enterprise—be it in pharmaceutical research, technological innovation, or neuroscience. Recent years have seen a growing awareness of this issue in pharmaceutical research (Goldacre, 2012; Healy, 2012). There is, however, no reason to believe that this phenomenon is limited to pharmaceutical research only.

Not only profit motives can lead to (or, diminish proper attention to) misuse of data or findings. Social acknowledgment, scientific standing, and power—there are a number of motives why some individuals are willing to use data or scientific knowledge for unethical or illegal purposes. Pioneering research in an area, which characterizes the HBP, may in particular involve such risks, since it involves the prospects of gaining high academic and social status as well as significant commercial gains.

Organization and Information Flow

The project's size represents particular organizational and governance challenges. Essential information can be "lost in complexity" or responsibilities unclear. Good coordination and communication require unambiguous and well-known responsibility, decision, and information lines. Clear systems are essential—in each subproject as well as in the HBP as a whole.

Further, many scientists are involved in formal modes of collaboration that create a tradeoff between organizational coordination and individual freedom. In particular, the interests of individual members can be overridden in the pursuit of the collective goal (Shrum, Genuth, & Chompalov, 2007), which requires governing structures to manage such conflicts. In contrast to other types of formal collaboration in science (e.g., universities, faculties, institutes), big science projects usually lack a long "collaboration history" among members and are confronted with a fixed termination date, which generates management challenges, in addition to a requirement to communicate regularly with the funding organizations and the public. This means that much "nonscientific" expertise is needed, which was not present during the generation of the proposal, when the scientific goals dominate. Integrating management and communication structures among consortium members thus is a process largely performed after the project starts—and has to be done in parallel to the scientific work. This may lead to conflict regarding resources for securing such "nonscientific" activities.

Expectations Regarding the Aims of the Projects

The ambitions of the project are high. It aims at collecting a huge amount of data, both research published in scientific journals as well as clinical data from hospital records. The intention is to bring all these data together into several ICT platforms to make all the data available to the researchers and to simulate the human brain (brain-related processes). If this goal is reached, the next step is to use the model to study brain diseases and analyze how different therapeutic interventions affect the brain.

There is a risk that some researchers and other stakeholders will be tempted to exaggerate the potential achievements of the research. This may be due to the competition for funding, but there is also a driving force created by communication with the public and other stakeholders. It is much easier to communicate, and gain enthusiasm, for research that promises to solve some of the big challenges of our time, than to promote a new technology that—in principle perhaps—can provide the possibility of simulating a human brain on selected dimensions. The specific contents of the research are currently vague. At some point in the evolution of the project, more specific objectives regarding what can really be modeled and simulated will have to be established.

To what extent some of these goals are achieved is also an ethical question. It contains at least three aspects. First, economically: Since the project takes such a large part of the research budget,[1] it supersedes other projects. In principle, alternative uses of the research funds could provide more valuable knowledge. In economic terms, the question is whether the opportunity costs of this allocation of research resources are lower compared to alternative uses of the resources.

The second ethical element is epistemological: Is it feasible to integrate knowledge from such diverse disciplines and scientific studies into one and the same model? (Rose, 2014). In both cases, economically and epistemologically, it can be hard to assess the probability of success ex ante (which, of course, is an inherent problem in many scientific enterprises). There are, however, strategies, insights, and

[1]The HBP takes up a substantial part of not only EU research funds but also from other sources since half of the money should be provided by the partners. Thus HBP projects must compete with other neuroscience research groups for available European public and private research funds.

perspectives that might help to identify flaws and weaknesses in the scientific approaches. We will discuss these strategies in the following sections.

The third ethical element concerns data origin, quality, and storage (Christen, Domingo-Ferrer, et al., 2016). One of the main objectives of the HBP is to develop models that can identify clusters of data that serve as specific signatures of human neurological and psychiatric diseases. Such models require the use of huge numbers of multilevel data from patients originally obtained for clinical care. Accordingly, the Medical Informatics Platform of the HBP has to deal with the substantial challenge of optimizing the scientific quality and use of the patients' sociodemographic and clinical data while respecting the ethical-legal context in which these data were obtained (original informed consent) and/or anonymizing personal data at the highest standards recommended by the EU (Opinion 05/2014 on Anonymization Techniques). These issues are particularly relevant now, since within the next 2 years the EU is expected to adopt a new General Data Protection Regulation that poses explicit limitations to the use of personal data for biomedical research. Protection of personal data and responsibility, transparency, and provenance in the use of data related to human beings will undoubtedly be a key factor in the perception of the HBP by patients, relatives, and charities, and for their active and enthusiastic involvement in the project.

Finally, the expectations for the project may involve an element of competition between the three largest economies of the world, Europe, the United States, and China. Currently, similar neuroscience research initiatives being conducted in all three economies[m] and it would be naïve to ignore that government funding on this scale most likely involves an element of competition for an economic head start.

Specific Challenges for the EAB

Determining the Expectations for the EAB

The expectations for the previous REC and ELSA were formulated in the work package description: "The ELSA committee will support HBP management on issues of policy and strategy. The REC will support local research sites on regulatory issues and compliance, maintain an ethics data registry, and the responsibility for communicating the

---

[m]In the United States, the Brain Research Through Advancing Innovative Neurotechnologies (BRAIN) initiative (see http://braininitiative.nih.gov/) is comparable. In China, the Chinese Brain Project was recently announced to be funded from 2016 to 2030 (http://en.people.cn/n/2015/0630/c98649-8913112.html).

official project position on specific issues in research ethics" (Work package description December 12, 2013). The subsequent description of the responsibilities of the ELSA gave a detailed description of its composition, number of meetings, and other administrative requirements but very little about its mandate. In addition to supporting the management on policy and strategy, "strategic oversight of ethical, legal and social issues" is stated. The open mandate led to a period of deliberation in the committee in order to determine specific tasks, which were continued in the new EAB.

The expectations for the REC were more specific and broader: "A separate Research Ethics Committee (REC), independent of the ELSA, will help the partners ensure that HBP research meets the highest possible ethical standards and that it complies with all relevant European, national and regional, law, as well as with the deontological standards imposed by relevant professional bodies (...) The work (...) will include preparing and revising guidelines, responding to researcher queries, and mandatory reviewing HBP local research ethics applications prior to their submission to local Independent Review Boards" (Work package description December 12, 2013).

This statement is not clear on the distinction between an IRB function of research proposals and a monitoring, higher-level role. The members of REC agreed that the research institutions themselves should be responsible for the ethical reviews, while the REC's role should be to advise, monitor, and promote research ethics, especially for data collection and protection. The REC made the following clarification: "The HBP REC is an advisory committee which will endeavor to assist with inquiries in regard to ethics, as is our remit. ... The REC does not receive or proactively collect all the ethics material for all studies associated with the HBP. We see our role as giving advice and helping resolve any queries in regard to ethics which are brought to us by ELSA or Sub-Projects of the HBP or in issues identified by REC members" (Office of Management of Work Package 12.5).

The task description of the EAB is as general as that for the ELSA and the REC. The EAB is expected to advise the Board of Directors of the HBP on "specific ethical, regulatory, social and philosophical issues raised by research that is being undertaken or planned under the auspices of the Human Brain Project" (SOP, EAB, 2015). A vague broad mandate clearly involves the possibility of defining the EAB's specific responsibilities independently. A vaguely stated mandate also involves the possibility of developing and changing the tasks during the project. The strength is flexibility and involvement by the members, but the weakness is that too much time is spent on deliberation and clarification, in contrast to implementation.

Independence and the Composition of Members

The independence of the previous ELSA and REC and the current EAB has been a manifest value since its start-up. Financial independence, meaning that no member receives an honorarium for his or her work, promotes such independence. The voluntary commitment implies, of course, that the members of the committee work under a significant time restriction. This involves a tension between expectations and what the EAB can accomplish in practice.

In principle, financial dependence can be a hindrance for an objective and/or sufficiently distant perspective. However, it seems unlikely that an honorarium for ethics consultants should compromise independence, unless the honorarium is unrealistically high. From a more cynical perspective, the volunteer contributions of ethicists signal how ethics is valued in the project or, more pragmatically, as a way to save money.

More important, in order to promote independence, EAB members must not hold a financial interest in the research itself, either as researchers or as funders. A slightly more subtle version of interest is the following: Since most members are in the same research fields, in broad terms, there can be a tradeoff between the need for well-informed experts as members on the one hand, and influence from individual academic interests, or intellectual bias toward certain scientific perspectives, on the other. Such potential influences on the EAB's work can, and should, be reduced through a diverse representation of members.

Generally, ethics committee should be aware of their members' potential conflicts of interest. The members, and/or their employers, may have academic or financial interests in certain parts of the research. Although the members of ELSA, REC, and EAB were selected on the basis of their personal capacities, there is reason to pay attention to their positions outside the ethics committee as well—since such roles implicitly or explicitly influence their judgments.

Clearly, influences from the different roles that an individual holds are unavoidable. In order to be aware of any unjustified influence, however, explicit attention to this fact is crucial. It should be noted that professional positions and experiences not only pose potential conflicts of interest; they can also be beneficial for the work of and ethics committee, since systems and tools from other contexts can improve the work of the committee.

In conclusion: the diversity of professional backgrounds represents a challenge with respect to identifying potential conflicts of interest and unjustified influence on judgments, while, at the same time, representing a strength due to the transferability of systems to improve research ethics.

In addition to the members of the ethics committee, REC, ELSA, and the current EAB are supported by people who are employed in the project. The ethics manager, who is part of the Ethics and Society subproject, plays an important role in terms of contributing to the preparation of cases, the agenda, and the minutes, as well as serving as the connection between the EAB and the Board of Directors (BoD) as a nonvoting member of the BoD. The experiences thus far are that the ethics manager has significantly promoted information flow and communication compared to the first year when the committees lacked this function. However, there is the potential for unjustified or too much influence on the work of the EAB by the ethics manager, due to the power generated by the information privileges as well as the double, if not triple, roles he or she holds in the EAB, in the BoD, and as part of a research subproject. The perspectives of the BoD and the EAB in many cases differ and could turn out to be antagonistic. There is a need to be aware of the potential conflict of interest if such a situation arises (Box 15.1).

Further, there are four "ex officio members" of the EAB, three of whom are on the HBP payroll. Similar to the ethics manager, these

---

BOX 15.1

FEATURES THAT POSE POTENTIAL ETHICAL CHALLENGES IN THE HBP

- Size
- Organization
- Distribution of responsibility
- Optimism bias
- Exaggerated expectations
- Concentration of funding
- Too big to fail
- Data origin
- Data storage
- Informed consent procedures
- Information flow
- Role of EAB
- Role of ethics management
- Communication between EAB and the scientists
- Communication between EAB and the management

---

members are nonvoting, but they contribute to the discussions that in principle and in practice influence the work in the EAB.

ADDRESSING THE CHALLENGES

Ethics work can be characterized as an oscillation between practical, empirical challenges and principled thinking. In the following, we suggest three basic principles on which the strategies to meet the specific challenges should or may build on. Those principles reflect the opinion of the authors; they do not represent an ethical framework to which all EAB members have formally agreed on. Our intention is to propose a general account, in order to promote a principled approach to the challenges in this kind of research. Clearly, the proposal is open to suggestions and amendments. As most of the work in ethics (committees), the perspectives and strategies must continuously develop in close connection to the challenges as such become evident. At the same time, general principles can prevent unjustified ad hoc solutions, as well as contribute to stronger awareness of inherent challenges (Box 15.2).

Principles

Primum Non Nocere—*First Do No Harm*

The golden rule of medicine can guide more than medical treatment. The positive intentions of research—producing new knowledge for the benefit of humans—cannot come without the downside of potential negative or harmful effects. The assessment of the balance between positive and negative implications tends to be in favor of the positive, due to optimism bias. To counteract this tendency, explicit attention to

---

BOX 15.2

GUIDING PRINCIPLES

- *Primum non nocere*—a precautionary attitude
- Weighing benefit and harm ("first do no *net* harm")
- Transparency

"what can go wrong?" is helpful. Such a precautionary attitude[n] is important in order to pursue a careful and cautious approach and to promote guards against hubris caused by optimism bias.

A precautionary attitude involves a necessary epistemological correction, since it challenges scientists—and others involved—to be concerned with observations that count against what one is eager to prove. Karl Popper's principle of falsification goes well with a precautionary attitude. Originally, Popper distinguished between a scientific and nonscientific statement according to whether the statement is, in principle, possible to falsify through empirical testing (Popper, 1959/1999). It is not the task of scientific inquiry to prove that a particular empirical statement is correct, but rather to search for evidence to its rejection. We are not concerned with demarcation between science and nonscience in this context, but the epistemological attitude it expresses. The difference between searching for verification versus falsification reflects a fundamental difference in attitudes to knowledge. Further, the falsification attitude can be a safeguard against optimism bias.

It is almost impossible to predict the total effects of research and innovations, especially the longer-term effects. For this reason, the precautionary principle is advocated. On a global level, the principle is included in virtually every policy document on environmental protection, sustainable development, and public health (Andorno & Biller-Andorno, 2015).

In European law, the principle is operationalized as follows: "The precautionary principle in public decision making concerns situations where following an assessment of the available scientific information, there are reasonable grounds for concern for the possibility of adverse effects on the environment or human health, but scientific uncertainty persists. In such cases provisional risk management measures may be adopted, without having to wait until the reality and seriousness of those adverse effects become fully apparent" (Von Schomberg, 2012, p. 147).

In public health, the principle is formulated as follows: "(W)here there are significant risks of damage to public health, we should be prepared to take action to diminish those risks, even when the scientific knowledge is not conclusive, if the balance of likely costs and benefits justifies it!" (Horton, 1998, p. 252).

To be willing to take action despite insufficient evidence is less straightforward than it might seem at first sight, as the debate on

---

[n]Although we discuss *primum non nocere* as a variant of the precautionary principle, they are not equivalents. One important difference is that the precautionary principle generally concerns groups of people, while *primum non nocere* regards the individual patient.

precautionary principles readily demonstrates. Precautionary principles are accused of being antiscientific, conservative, and outright irrational—as potential benefits of scientific and technological advances are sacrificed on its altar (Harris & Holm, 2002).

Since the potential negative implications of research and innovation can be seen as arguments against innovation altogether, a principle of precautions seems too absolute. A precautionary attitude, however, is in place. The challenge is to strike a balance between benefit and harm; the duty to avoid harm is not the same as the duty to abstain from carrying out research (or any action) altogether—clearly also because no action can involve more harm than the action itself. This introduces a second general principle we build on, namely, the need to weigh the benefits and the drawbacks.

Weighing Benefits and Harm

Jonathan Wolff argues in favor of the precautionary attitude not least because it involves a more pragmatic attitude to risk, compared to the precautionary principle (Wolff, 2006). Any action involves risk; thus, the task is not to avoid risk altogether but to weigh the potential positive implications against the negative ones. This requires an explicit assessment of the potential beneficial as well as potential harmful implications.

Risk is a product of hazard and probability. There is a substantial difference between severe, perhaps fatal, outcomes of low probability and minor problems of high probability. As a guiding principle, identifying high-risk areas—those that should be marked with red lights—is useful.

When weighing benefits against harm, it is important to address the question, "Whose benefit, whose harm?" Whether the decision maker and those affected by the decision are one and the same or not is important since one person's benefit can be another person's harm (Luhmann, 2005). The same holds true for harm and risk. In a risk situation, the one who runs a risk, for example, in order to obtain a preferred state in the future, need not be the same one who bears the costs. One HBP-related example would be the differences regarding risk-taking in controversial areas such as neuroenhancement to improve individual performances of healthy subjects, or neuroeconomics research that could inform neuromarketing (Voarino, 2014).° Those who

---

°Neuro-marketing is a new discipline that uses expertise in medical neuro-technology to study brain responses to marketing stimuli and to attain specific business goals. Although its applications are still preliminary, it seems likely that research in this area will provide subtle means of brain manipulation to provoke desired behaviors without consumers being aware that they are being manipulated (Breiter et al., 2015).

exploit commercial opportunities and laypeople tend to have different views, especially when it comes to potential long-term threats emerging from the uncontrolled application of neuroscientific findings. This makes it difficult to reach a consensus because someone's risk is often another's danger, and perspectives differ according to the respective position. Thus the evaluation of risk associated with the outcomes of scientific research in the HBP, and the willingness to accept risk decisions, are most of all social problems.

As a starting point, scientific knowledge never comes free from social interests or implications. Disregard of public concern and laypeople's perspective in the scientific enterprise entails a normative problem; as such disregard undermines the possibility of establishing a democratic knowledge society (European Commission, 2007). It is also important to acknowledge laypeople's specific knowledge, based on their everyday lives, and needed for assessing long-term ethical/social implications (Myskja, 2007) in terms of benefits and harms. The inclusion of lay expertise is thus crucial to establish empirically well-informed ethical governance of science and technology.

The combination of the "first do no harm" principle and the weighing of harms and benefits can be combined into a principle of "first do no *net* harm." Since all action potentially involves negative and positive outcomes, the task is to choose the ones where the harm is as little as possible, while the benefit is maximized.

Transparency

Transparency not only has the advantage of enhancing quality of research but also increases our attention to what can go wrong. When the research content is shared within a wider community, epistemological and ethical implications are laid open to broader scrutiny. Transparency has the potential benefit of involving diverse individuals and approaches, which can be particularly helpful in large collaborations such as HBP, since it is impossible for the few to maintain oversight over, or possess insight into, all dimensions of all subprojects.

Clearly, transparency is not always justified; in some situations, it can involve a breach in fundamental ethical principles, such as when information includes confidential medical records. The principle of transparency means that nondisclosure is explicitly qualified (e.g., by reference to the rules of confidentiality) in contrast to a requirement to qualify openness explicitly.

Why Principles—And Why These Three?[P]

Advice to researchers and management in the HBP concerns a wide range of issues. They include short- and long-term consequences; implications for individuals, smaller groups, or for societies; and consequences in terms of ethical, legal, or social aspects (or a combination of the three). To prevent inconsistent choices of the issues that are selected for discussion or inconsistent advise to the researchers and other stakeholders, the determination of fundamental principles is crucial. This way, the EAB itself, as well as the researchers, the managers, and the public, will be aware of what basic principles we build on when actions are taken, or advice is given.

The three principles we suggest in this chapter are chosen for three reasons. First, it is necessary that the principles are on a level of generality that make them relevant for a wide number of issues that can come up. This way, only a small number of principles are required.

Second, it is necessary that the principles at least are internally consistent. Ideally they should support each other mutually, such as when the principle of transparency can serve as a hinder for unwarranted exaggeration of potential gain from the research. In this case, the principle of transparency enforces the principle of balancing harm and gain.

The third reason for the choice of the particular principles is that they should reflect common values in the European research community as well as in the general public. First, do no harm is the essential ethical principle in medicine, hence clearly also a core value in this context. Since the extent of unforeseen consequences is very high in the HBP, a guiding principle should concern how to relate to uncertainty. Combining the recognition of great uncertainty with the human tendency to optimism bias, we should be concerned with precaution as well the need to weigh negative and positive effects (harms and benefits). Finally, the choice of the principle of transparency is supported by its being a widely shared value in our culture. Transparency is also an essential principle because of its strength as one of the most effective safeguards against misconduct or other unjustified or illegitimate actions.

Finally, the determination of basic principles also helps to identify specific measures required in order to do our work. If, for example, transparency is a principle, measures to secure openness are required. In the next section we describe the measures we have developed so far.

[P]Clearly, the choice of principles is not cut in stone. They are suggestions, and as such considered to be open for substitution and/or completion.

Measures

The choice of measures is based on two elements. First, it is crucial that the guiding, normative principles are clear (they were stated in the previous section). Second, the measures should be based on a description of the challenges identified in the research (provided in the first part of the chapter). Let us underscore that this description is preliminary; thus it is important that new measures are included in accordance with how the research develops. It is also crucial that measures are determined as a result of the experiences along the way. Thus what we propose here is what we think is needed and/or useful at this stage.

To avoid harm, risk assessment systems are necessary—in all research projects and for general governance. Risk assessments are already part of the governance activity; however, the explicit weighing of benefit and harm is a necessary prerequisite to avoid an unjustified over- or underestimation of benefits and risks. Further, the explicit attention to optimism bias is required.

The work of the EAB includes concrete systems to register and follow up on concerns among the scientists or other stakeholders. An online submission application, called Point Of Registry (PORE), can be used to contact the ethics committee, while a system with "ethical rapporteurs" is established to promote contact between the committee and the scientists. The appointment of an ombudsperson is also discussed. All three systems are described in detail in a forthcoming paper.

To promote collaboration, information sharing, and ethical values in general, transparency and explicit and public reason-giving are necessary. The members of the ethics committee publish scientific papers and discussions on different aspects of their work, of which the current chapter is an example.[q] These publications are a way to facilitate public participation and to open up a wider discussion of values and measures, while it also serves as reason-giving for, or qualification of, the EAB's decisions. Encouragement of transparency and public debate should be part of the HBP generally, and systems to support this should be developed.

External ethical reviews are also a way of promoting awareness, transparency, and information sharing. As part of the review carried out after the first year, the external committee met with all principal investigators to discuss specific ethical challenges in all subprojects. Representatives of the ELSA and REC committees were invited to these

---

[q]Such publication needs to follow a certain etiquette. In addition to the Vancouver system for scientific publications, specific attention to the relation between the contents of the chapter and the views of the committee as a whole is required. A description of this etiquette is currently in progress.

BOX 15.3

MEASURES FOR REDUCING INHERENT CHALLENGES

- Identify high-risk areas (where are the red lights?)
- Risk assessments that include weighing of benefit and harm
- Streamlined information and decision routes
- Clear responsibility
- Must be well-known to all stakeholders (Internet sites)
- Regular external ethical reviews (EAB members present in the discussions with PIs)
- Data origin always declared
- Declaration of conflict of interest as routine for all scientists and decision-makers and other stakeholders (including EAB members)
- Data protection systems
- Ethical rapporteurs
- Other systems for communication between the EAB and scientists (PORE)
- Ombudsperson
- Transparency measures: All is public unless there are important reasons; public and explicit reason-giving

discussions, which was a valuable information source for the committees. If not exactly the same model, a similar way of identifying ethical challenges in each subproject should be carried out yearly (Box 15.3).

Finally, attention to the social implications of HBP research requires specific measures. The tradition in research ethics is to focus on data protection and informed consent (data origin, storage, and protection). While recognizing the importance of these issues, there is also a need to pay specific attention to social implications. The issue should be part of the risk assessments (who bears the risk, who gains?), and the EAB should have a specific responsibility in identifying social aspects of the research.

CONCLUSION

So, how can an ethics committee contribute to the goal of promoting strong research that will ultimately have some public benefit? The first message is that there is a need to translate general mandates into concrete measures, such as the ethical rapporteur system that promotes information flow and collaboration between the committee and the researchers.

The vague goals of the ELSA/REC and the subsequent EAB involved a slow start-up; a clearer goal could have sped up the work. However, the opportunity to define the specific goals and measures by the committees themselves promoted a wider discussion of their role and tasks. This opened up the discussion of the fundamental question of how ethics can contribute. The answer to this question needs to evolve from practical activities of the ethical committee and the sharing of these experiences. This requires vivid exchange and collaboration between the EAB and the researchers, as well as an open discussion and publication of experiences.

Further, "strangers" in a research community can provide fresh perspectives on the research, precisely because of their outsider position. However, there is a fine line between being ignorant and providing a fresh outsider approach. Striking a balance requires continuous collaboration and candid information-sharing between the EAB and the researchers.

Another issue to negotiate is the balance of a close and trusting relationship with researchers and critical distance to see potential flaws. If researchers cannot trust their ethical advisors, important information might be withheld. However, if the ethicists feel obliged to be loyal to "their researchers," the ethicists might misinterpret this as trying not to stand in the way and being disruptive. It is important to reach a common understanding that the role of the members of the EAB is to look for potential problems and bring up criticism, albeit in a constructive and respectful way. This can be an important contribution to avoiding pitfalls and, at worst, scandals that can jeopardize the success of a project.

Finally, if the EAB is to have any real impact, structural systems must be established to support the board's role. Without such systems, the ethics discussions and recommendations can easily serve as the frosting on the cake—as an ethical alibi without any practical significance. An interesting thought experiment might be the following: How could we tell in a few years if the EAB has worked well or has had "real impact"?

In this chapter, we identified the EAB's role as consisting of three parts: (1) to identify the inherent ethical challenges, (2) to provide a normative framework for EAB contribution, and (3) to suggest measures including structural systems that support the goals. Examples of supporting structures include the presence of EABs at external ethical reviews, regular meetings with the BoD, a system with ethical rapporteurs and their contact persons in the EAB, and the promotion of transparency through publications.

Acknowledgments

Thanks to Karin Blumer, Jean-Pierre Changeux, Kathinka Evers, Kevin Grimes, Abdul Mohammed, and Henrik Walter for valuable comments.

References

Andorno, R., & Biller-Andorno, N. (2015). The risks of nanomedicine and the precautionary principle. In B. Gordijn, & A. M. Cutter (Eds.), *In pursuit of nanoethics*. New York: Springer.

Breiter, H. C., Block, M., Blood, A. J., Calder, B., Chamberlain, L., Lee, N., et al. (2015). Redefining neuromarketing as an integrated science of influence. *Frontiers in Human Neuroscience, 8*, 1073. Available from http://dx.doi.org/10.3389/fnhum.2014.01073.

Christen, M., Biller-Andorno, N., Bringedal, B., Grimes, K., & Savulescu, J. (2016). The ethical challenges in simulation-driven "big neuroscience." *American Journal of Bioethics Neuroscience, 7*(1), 1–13.

Christen, M., Domingo-Ferrer, J., Draganski, B., Spranger, T., & Walter, H. (2016). On the compatibility of big data driven research and informed consent based on traditional disease categories—the example of the Human Brain Project. In L. Floridi & B. Mittelstadt (Eds.), *Ethics of biomedical big data*. Law, Governance and Technology Series, vol 29, pp. 199-218. Springer International; Switzerland.

European Commission (2007). *Taking European knowledge society seriously: Report of the Expert Group on Science and Governance to the Science, Economy and Society Directorate, Directorate-General for Research*. Luxembourg: European Commission.

European Commission. (2014, July 18). *No single roadmap for understanding the human brain*. Retrieved from <https://ec.europa.eu/digital-agenda/en/blog/no-single-roadmap-understanding-human-brain>.

Frégnac, Y., & Laurent, G. (2014). Where is the brain in the Human Brain Project? *Nature, 513*, 27–29.

Goldacre, B. (2012). *Bad pharma: How drug companies mislead doctors and harm patients*. London: Fourth Estate.

Harris, J., & Holm, S. (2002). Extending human lifespan and the precautionary paradox. *Journal of Medicine and Philosophy, 27*(3), 355–368.

Healy, D. (2012). *Pharmageddon*. Berkeley, CA: University of California Press.

Horton, R. (1998). The *new* new public health of risk and radical engagement. *The Lancet, 352*, 251–252.

Irwin, A., & Wynne, B. (Eds.), (1996). *Misunderstanding science?* Cambridge: Cambridge University Press.

Kahneman, D. (2011). *Thinking, fast and slow*. New York: Macmillan.

Luhmann, N. (2005). *Risk: A sociological theory*. New Brunswick, NJ: Aldine.

Myskja, B. K. (2007). Lay expertise: Why involve the public in biobank governance? *Genomics, Society and Policy, 3*, 1–16.

Popper, K. (1999). The logic of scientific discovery. London: Routledge (Original work published 1959).

Rose, N. (2014). The Human Brain Project: Social and ethical challenges. *Neuron, 82*, 1212–1215.

Rothman, D. (1991). *Strangers at the bedside: A history of how law and bioethics transformed medicine*. New York: Basic Books.

Shrum, W., Genuth, J., & Chompalov, I. (2007). *Structures of scientific collaboration*. Cambridge, MA: MIT Press.

SOP, EAB. (2015). *Human Brain Project*. Retrieved from <https://www.humanbrain-project.eu/documents/10180/1139903/EABSOP_2015-10-06-2.pdf/0132960f-77cb-4782-ba5d-946eca9c0e25>.

Voarino, N. (2014). Reconsidering the concept of "dual-use" in the context of neuroscience research. *Bioéthique Online, 3/16*, 409–427.

Von Schomberg, R. (2012). The precautionary principle: Its use within hard and soft law. *Symposium on the European Parliament's Role in Risk Governance. European Journal of Risk Regulation, 2*, 147–156.

Wolff, J. (2006). Risk, fear, blame, shame and the regulation of public safety. *Economics and Philosophy, 22*, 409–427.

Wolff, J. (2014). Precautionary attitude. Special report: Synthetic future: Can we create what we want out of synthetic biology? *Hastings Center Report, 44*, S27–S28.

16

At the Push of a Button, Narrative Strategies and the Image of Deep Brain Stimulation

O. Hayes

University of Tübingen, Tübingen, Germany

INTRODUCTION

New technologies certainly have the potential to polarize the broad public. Adding to that intrinsic property, they also offer news reporters the opportunity to coin extravagant headlines. It's no wonder then that science communication is a hot topic as well in practice as on the meta-level that helps to analyze and support the practice. When disseminating knowledge to the broad public, the educational mission appears obvious. But when the information transmitted is about medical research, science communication takes on a whole other relevance. Even without being personally affected by a given topic, there is a sense of urgency about what is being done in research and what is eventually possible in terms of therapy.

This is particularly true of print-media reports about deep brain stimulation (DBS), commonly dubbed "brain pacemaker." The analogy to the artificial cardiac pacemaker gives an idea of DBS's regulating function. This therapy, used among others to alleviate symptoms of Parkinson's disease, calls for attention by uniting both a high-profile condition and a high-tech neurosurgical procedure. This emerging therapy faces the challenges these traits bring with them: as much a

The Human Sciences after the Decade of the Brain.
DOI: http://dx.doi.org/10.1016/B978-0-12-804205-2.00016-1

© 2017 Elsevier Inc. All rights reserved.

chance as a pitfall, the hype around neurosciences seems to promise a brave new world. Based on examples from German-speaking media, we will see which strategies are applied in the communication about DBS. We will observe how cultural imagery plays a role in the representation of this therapy and to what extent they uphold and/or fail in their function. To do so, we will first investigate the representation of time, so as to be able to understand the idea of immediacy relayed by a mechanistic view of the therapy, body, and disease. To frame these mechanistic notions, we will define the situation of observation that is broadcasted in the media. Thus we will underline the metaphorical spectrum open to the representations of DBS and examine the narrative shortcuts this generates, culminating in the vision of an instantaneous miracle cure.

PUTTING DBS IN PLACE

Some news reports focus on spectacular moments that seem to summarize a topic in a nutshell. But do they really? When dealing with neurosciences, a certain fascination easily channels the attention of a lay public on key moments without necessarily giving away other sides of the story. In the following we will look at the time component of DBS and DBS representations.

People suffering from Parkinson's disease may resort to DBS as a therapy since its approval for that indication in 1998 in the European Union (Coenen, 2014) and 2002 by the FDA (2002) for the United States. Patients eligible for DBS need to correspond to a profile susceptible of responding well to the therapy (Clausen, 2011). Other conditions for which patients may resort to DBS are essential tremor, dystonia, obsessive-compulsive disorder (OCD), and epilepsy. Further applications are currently the object of research, such as chronic pain, treatment-resistant depression, and Tourette's syndrome, among others.

For the individual patient, DBS needs time. Before even given access to this therapy, a patient has to be selected. Not all people suffering from Parkinson's are eligible for DBS. Inclusion and exclusion criteria are a major concern in medical research in general, as the selection of patients impacts on the success statistics of any new therapy, thus decisively influencing its official approval. Additionally, the selection also benefits the patients as those selected are more susceptible to responding to the therapy; for DBS and Parkinson's, this will depend on the characteristics of the motor symptoms, on the duration of the disease, on a good response to L-Dopa treatment, on the side effects of the drug treatment, among others (Clausen, 2011). The selection also helps to avoid major negative events, which in the case of DBS can go as far as suicide if some psychiatric factors have been ignored (Lieb & Schläpfer, 2006). The screening of patients is only one of the steps before actual surgery. Parkinson's disease is known to be incurable and degenerative. Medication aims at improving quality of life and is adjusted to the needs of the patients according to their individual symptoms, and must be adapted as the disease evolves. Therapy for Parkinson's disease can be considered "palliative" (Schneider, Novak, & Jech, 2015). Likewise, DBS relieves Parkinson's symptoms but doesn't act on the disease itself.

DBS implies brain surgery. Before surgery the patients' usual medication must be interrupted. The main operation is constituted by the implantation of the electrodes in the basal ganglia, during which the patient has to be conscious so as to give feedback about the effects, thus permitting to verify the accuracy of the placement in the brain target. Another element of the operation consists of placing the pulse generator, to which the electrodes are connected, under the skin, usu-ally below the clavicle. This is done either subsequently to the main operation or in the following days. The surgical operation has been improved and its duration shortened: Dubiel (2006/2009) mentions ten hours for his operation in 2005, whereas 10 years later one can expect about half that time. After the patient recovers from surgery, the stimulation parameters need to be tried out and set. DBS also implies an individual tuning of stimulation parameters. Only then can the medication be adjusted, as a complement. One of the consequences of successful stimulation is a reduction of the medication. It is interest-ing to note that DBS is used in combination with medication. Reducing medication thanks to DBS remediates to many problems due to growing side effects of the standard drugs used for Parkinson's disease. All in all, from the surgery to the completion of the settings and adjustments, and not taking, e.g., physical therapy into account, we are looking at a process that lasts several months from surgery to first stable settings. For example, Schmalfeldt (2010) defines the end of the period for major programming to be about 5 months after the surgery.

Testifying to the length of the process, he also writes: "Programming of the neurotransmitters is a process. It can't be done in one visit, it takes several. As a result, I took several day trips [. . .]. Wearying!" His sense of time transpires from highlighted comments of his dairy, as here a month and a half later: "And yet another trip to Nashville for programming." And again a month later: "And yet, another trip!" (Schmalfeldt, 2010, p. 81, p. 97, p. 102). However, if the process is a big part of the patient's experience, the sense of time is not the focus of media reports.

"AT THE PUSH OF A BUTTON": SWITCHING DBS ON AND OFF FOR OBSERVATION

News reports point out the spectacular results of DBS-based therapy by comparing the on and off state of patients. This is particularly interesting in two respects. First, this cannot be done instantly with a drug-based therapy, as drug effects wear off over a period of time of several hours to several days. Second, the comparison would hardly be made without an observer. This can be the patient either him- or herself, for the sake of experiencing the impact of the stimulation, e.g., a patient staging himself, as in the video "DBS—3 years after surgery."[a] The observer can also be a physician, for the purpose of assessing the therapy or to document the patient's state for the medical record. In the course of the diagnostic examination, videos are routinely made of Parkinson's patients to chart their motor symptoms. These videos of patients before and after, or in an on and off state, are used for presentations and shown at conferences. They might occasionally appear in online information material about DBS, on the websites of institutions that implement DBS for instance. A patient's story on the website of the University of Pittsburgh Medical Center gives an example, with video sequences of the evaluation of a candidate for DBS, in which the tremor is particularly visible, and on which the physician additionally comments; a text screen announces that "[a]fter the procedure, his neurological team activates the stimulator," followed by a short sequence with tremor, after which another text screen explains "David's symptoms immediately improve," followed by several sequences of the patient without tremor, smiling, and succeeding in doing the diagnostic exercises and gestures.[b] Using precisely that imagery, the GEO reporter Jürgen Broschart uses two pictures of himself, before and after surgery (sic!), as well in his article (Broschart, 2015) and in the online video.[c] In these photos, he demonstrates his handwriting with a flashlight and a long shutter opening with the words "vorher" and "nachher" (before and after). The aforementioned is hardly recognizable due to the trembling. This kind of exhibition of Parkinson's symptoms and effects of DBS, with a comparison of "on" and "off" stimulation, are in many cases an obvious base for written media reports.

[a]Cf. video "DBS—3 years after surgery" at https://www.youtube.com/watch?v = 17ch1guvoLA.

[b]Cf. video "David's Story" at http://www.upmc.com/services/neurosurgery/brain/treatments/movement-disorders-and-epilepsy-treatments/pages/deep-brain-stimulation.aspx.

[c]Cf. video "Das geplante Wunder" at http://www.geo.de/GEO/info/newsletter/abo/video-das-geplante-wunder-80673.html.

For instance, a quite typical portrayal, taken from a major German-language newspaper, the *Frankfurter Allgemeine Sonntagszeitung*, states: "For that matter, the effect of the brain stimulation is more than impressive: at the push of a button, even a severe tremor can be simply turned off. If the pacemaker is switched off again, the trembling comes back"[d] (Heier, 2008, p. 57).

But in fact, a Parkinson's patient has no interest in turning off the stimulation without good reason, as this leaves the patient with but the little medication that completes the electrical stimulation, i.e., mostly without therapy altogether. Such an action artificially puts the person in an extremely vulnerable physical and mental state, exposed to the full blast of the disease's symptoms. So although they have have the possibility to turn off the device, most patients never make use of it.

One exception is presented by Dubiel (2006/2009) who wrote about his experience as a DBS patient and gives such a reason: The stimulation, while helpful for some symptoms, impairs other faculties, as he explains very clearly. Dubiel represents an exception not only as an articulate elaboration on the patient's perspective, but also inasmuch as he used DBS as a functional switch between the two states—on and off stimulation—with respective bodily capacities. The result sounds like an empowering compromise, even if the stimulation's side effects are debilitating in everyday life: "In time, Dubiel learns to deal with the unwanted side-effects by adjusting the pacemaker amplitudes. If he wants to enunciate clearly, he sets the amplitude very low knowing this will lead 'to relative immobility and depression' [...]. In the alternative, if he wants to walk any distance, he sets the amplitude much higher knowing that his speech will become inaudible. [...] having to choose between talking and walking is less than ideal" (Baylis, 2013, p. 521). However, Dubiel tells vividly about the drastic situation in which he was brought to turn off the device for the first time, and how his speech impairment was temporarily done away with, enabling him thereafter to resort to switching DBS off if needed, especially in professional circumstances. Turning DBS off the first time was done under constraint, which shows that it hardly comes as a voluntary act for patients.

Dubiel also uses the expression "at the push of a button" himself (Knopfdruck), thereby reflecting on the fascination due to the easiness of the procedure, and on how trivial and somewhat frivolous it seems to be able to switch between states in such a manner, especially considering the depressive state common in Parkinson's disease.

So if turning DBS off doesn't come easily, if switching it off means overcoming the apprehension of being fully vulnerable to the disease to

---

[d]In the following, translations from German are by the author of this contribution if not indicated otherwise.

no particular end except observation, how come that very experience is the one depicted in news reports?

Whether the agent is one's self or somebody else, the observation, i.e., the act of turning the DBS-device off, originates in the observation itself. Solely the will to observe causes the act of switching the DBS device off. This is obviously necessary in a medical context when assessing and adapting the stimulation parameters. But newspaper stories don't necessarily pick that setting, or any setting at all: they focus on the moment of the switch (in either directions, depending on the story), and that moment is characterized by the state of the patient before and after. In the hypothesis that a reporter meets the patient (s)he is telling about—and not base they article on secondary material, like videos—the journalist's observation coconstitute the very act of switching the device off and on again. The observation causes and shapes its own object.

This fundamental insight has been pointed out in various scientific disciplines: In ethnology, the presence of the observer within the studied environment and its impact on the observations is a basic principle in research methods (Lévi-Strauss, 1955). In philosophy, Barad (2003, p. 815) concurs: "The notion of agential separability is of fundamental importance, for in the absence of a classical ontological condition of exteriority between observer and observed it provides the condition for the possibility of objectivity. Moreover, the agential cut enacts a local causal structure among 'components' of a phenomenon in the marking of the 'measuring agencies' ('effect') by the 'measured object' ('cause'). [...] On an agential realist account of technoscientific practices, the 'knower' does not stand in a relation of absolute externality to the natural world being investigated—there is no such exterior observational point."

In quantum physics, Niels Bohr expands on the role of the observer in his agency regarding the observed. Thus "not only concepts but also boundaries and properties of objects become determinate, not forevermore, but rather, as an inseparable part of, what Bohr calls a *phenomenon—the inseparability (differentiated indivisibility) of 'object' and 'agencies of observation'*. Concepts do not refer to the object of investigation. Rather, concepts in their material intra-activity enact the differentiated inseparability that *is* a phenomenon. [...] Being is not simply present, there to be found, already given" (Barad, 2010, p. 253).

It is therefore not overly surprising to remark that, when reporting about DBS, the news reporter/observer creates a situation that wouldn't come into existence without its agency: Except for changing the stimulation parameters, there is no reason for "pushing the button" but mere observation.

THE METAPHORICAL MEANINGS OF ON AND OFF

"On" and "off" states are medical terms that seem to be directly accessible to lay people: To be on or off medication, meaning being in a therapy-induced state (or not), i.e., differentiating a state in which drugs are effective from one in which they have—for example—worn off. The latter is medically called an "off condition." But with DBS, the medical terms "on" and "off" coincide with turning an electrical device on or off. Thus "switching off" the DBS device can simply be understood in terms of everyday life. Beyond this casual level of language that merges medical terms and lay people's understanding, on and off also offer a fundamental metaphor—in accordance with the concepts of Lakoff and Johnson (1980)—of activity and passivity, consciousness and unconsciousness. These metaphors are so rooted in the language that they channel our understanding in a positive or negative way even without any kind of awareness.

In an analysis of over a hundred German-language newspaper articles from 1973 to 2014, from the dailies *Frankfurter Allgemeine Zeitung* and *Neue Zürcher Zeitung*, to the weeklies *Spiegel* and *Stern*, the terms "at the push of a button" or "switch" and "switching on" or "off," are used incidentally, with no reference to medical terms nor context explaining the use of a metaphor. This implicitly points to the fact that understanding the metaphor is taken for granted, such as in the title of an article "Peace at the Push of a Button" (Bidder, 2008), which goes on to describe: "For 15 years he was plagued by the typical trembling of Parkinson's disease. Now he can turn it off at the push of a button." Without an extensive common cultural background, referring to a mechanical understanding of the body, the readers couldn't understand nor indeed accept such a succinct "explanation" of a therapeutical technique. Another example even passes from a literal meaning to a metaphorical use and back, showing no explicit difference in the meaning: "A patient who was completely passive before the treatment, suddenly wanted to visit the cathedral of Cologne after the pacemaker was switched on, Schläpfer described of a change. [...] Through switching off the 'interference,' the blocked brain areas are nearly freed. [...] The effects of such a desynchronisation lasted for a few days for some patients, even after the pacemaker was switched off, said Peter Tass, scientist in Jülich" (Heier, 2008, p. 57).

Of course, a reflection about the mechanistic metaphor can hardly be chosen as an issue on its own in a simple news report. One article nevertheless ends on a warning about the conception and implicit verbal representation of DBS: "Could physicians one day relieve even more diseases that concern the control circuits of the brain with this method? Zumsteg cautions against exaggerated expectations. The correlations are generally much more complex, and the conception of a 'switch in the brain' turns out to be all too often much too naïve" (Klott, 2008, p. 8).

A GOOD STORY NEEDS A PLOT POINT

The "much too naïve" representation doesn't come out of the blue. News reports follow a classical narrative structure, especially if they are based on a patient's story. Like in a fictional narration, the main character has to be introduced, then the conflict is presented—here the disease and/or coping with the disease—and finally the story comes to a resolution. A standard narrative blueprint is to use a turning point to tell a story that concentrates the before and after in a pregnant moment. Choosing one particular moment or scene usually enables to render the whole story. Many medical narratives focus on the physician's intervention and not on the experience of the patients.[e] Relying on the agency of the doctor to intensify a story also goes along with a long biographical tradition in the history of medicine (Kühl, 2016), which has shaped both the public image of the physician and of his agency. Three examples can demonstrate this focus in media representations. For instance, organ transplantation acutely relies on donation. To increase organ donation, awareness campaigns prove to be influential tools (Felt, Fochler, & Winkler, 2010; Scheuher, 2016). The popular image of organ transplantation essentially revolves around the surgical operation and doesn't dwell much on the patient's case history, the time on the waiting list, pre- and postoperation treatments. Similarly, in vitro fertilization (IVF) appears mostly represented either through the test-tube fertilization in the laboratory or the moment of the implantation, without much thought about the couple's previous examinations and various medical supervisions in the process until then, nor the hormonal stimulation in the forerun to the implantation itself, nor the multiple attempts until a pregnancy is achieved—nor incidentally the rates of success (King et al., 2014; Tain & Robertson, 2002). A last example, that makes the focus on the physician's intervention obvious is the pop-cultural representation of birth. The medicalization of childbirth has a counterpart in the media (Luce et al., 2016). The crucial moment is the appearance of the baby—or of its head—and more often than not an episiotomy or the cutting the

[e]For the following examples and the recognizing of a pattern, I would like to thank Dr. Irene Poczka for our fruitful discussion.

umbilical cord plays a major part, whereas giving birth might be defined by a mother as beginning with the first contractions and ending with the placental expulsion, and will most certainly be experienced as a process more than a single moment.

By analogy, it thus seems plausible to assume that DBS stories will follow a similar narrative structure, focusing on the physician's intervention, i.e., implantation of the electrodes and possibly setting of stimulation parameters. What may come as surprising is that DBS narratives are altogether reduced to the climax, whereas better known medical procedures,[f] such as the abovementioned, concentrate a story that can be completed by the public's general knowledge or own experience. DBS generally is a new concept to the broad public, and most people are also unfamiliar with specifics of Parkinson's disease and neurological treatments. Most people who are not affected, either personally or by someone close, have no particular notion or knowledge of chronic diseases nor of the daily experience and management it involves, such as movement training, medication, adaptation of activities, and tricks for handling activities of daily living in and outside the home. One can therefore assume that recipients generally have a fairly blank point of view from which to understand a news report not only about an awe-inspiring disease—"Parkinson's" is feared, but knowledge about the disease is often limited—but also about a new therapy, both of which may appear quite abstract in comparison with a patient's concrete experience in dealing with his or her symptoms on a day-to-day basis. It is only one step from an abstract impression to the aspiration for a radical solution, an all around, complete, perfect cure, one that will make the disease simply disappear (turn it off at the push of a button).

Patients may develop a personal view according to their concrete experience: inescapability from the disease, dealing with certain symptoms at certain times, handling the medication, dealing with one's surroundings, adapting one's activities, receiving help from one's environment, and organizing help from professionals (e.g., physical or occupational therapy). But the general public (readers, viewers) can resort to a sensational coverage of the subject with a fascinating twist. It goes without saying: such an exclusively mechanistic conception of therapy doesn't cover the whole story, even in neurology.

The DBS stories don't speak for themselves as the technique is uncommon and neurology is surrounded by a certain fascination.

[f]A search in the archives of the Frankfurter Allgemeine Zeitung gives 179 hits for IVF and 18 for DBS; similarly, the databank LexisNexis gives 1718 hits for "In Vitro Fertilisation" and 414 hits for "tiefe Hirnstimulation" in German-speaking publication (as of July 12, 2016). This illustrates the respective mediatization of these medical subjects in mainstream publications.

III. THE NEUROSCIENCES IN SOCIETY. SOCIAL, CULTURAL, AND ETHICAL
IMPLICATIONS OF THE NEURO-TURN

They are staged so as to appear self-sufficient ("at the push of a button"), but they don't actually reveal the intermediary steps that therefore become utterly invisible—to the observer, incidentally, not to the protagonist.

Of course, reduction is a necessity when communicating the possibilities and constraints of medical interventions to a usually unaffected lay readership of the news report. However, the problem is not reduction as such, but the way complexities are reduced and stories are told. An alternative, or rather an opening, of these narrative structures would ensure a corrective. Instead, on the whole the same images are used repeatedly. A diversity is nevertheless possible, as an example shows: the Jülich research center has used the image of "desynchronizing" the neural activity when communicating to the press and the public.[g] "But there can be other ways. We know today that diseased parts of the brain—as with Parkinson's—are groups of cells which are pathologically synchronized. The brain cells fire together as one, very much like an audience clapping rythmically at the end of a concert. The one provokes an encore, the other has disastrous consequences: excessively synchronized brain cells cause chaos. [...] We are trying to give back to the brain its natural, unsynchronized Rythme" (Heier, 2008, p. 57). Here, synchronized activity causes chaos and desynchronized brain activity is represented as natural. Although this representation doesn't fit the usual cultural imagery of nature, control, and order, it still uses these strong topics to make an unusual point. Through diversity of images, a complex technique can be communicated in different ways and thereby ensure a better general understanding on the whole. Because "on" and "off" correspond to several cultural patterns (mechanistic, medical, metaphorical), the overlapping strengthens the one familiar and unchallenging image.

"AND THIS WORKS, LIKE THAT, INSTANTLY"

The culmination of the "push of a button" representation of DBS effects can be observed in a video example entitled "Andres Lozano: Parkinson's, depression, and the switch that might turn them off."[h] Aside from literary narrative structures, Lozano, as a neurosurgeon, is logically particularly aware of the role he himself plays in the patient's treatment. Consequently, his narrative will naturally correspond to a classical structure revolving around a turning point, that of the "push

[g]Cf. http://www.deutscher-zukunftspreis.de/de/nominierte/2006/team-4.

[h]Cf. Video at http://www.ted.com/talks/andres_lozano_parkinson_s_depression_and_the_switch_that_might_turn_them_off.

of a button." The video is that of a Tedtalk (TEDxCaltech), which presents format constraints such as the conciseness of the talk, obliging the speaker to get to the point and put the story in a nutshell.

In his talk Lozano announces he will show an example of a Parkinson's patient. He mentions that she has implanted electrodes in her brain (story before the turning point), that the video will show what the symptoms are like when the stimulation is turned off, and then what happens when the stimulation is turned on. As such, he simply demonstrates the effect of DBS. But the narrative structure and the reception of the example go far beyond the mere illustration, as we will see further. The patient presented in "off" state has very impressive Parkinson's symptoms. Her shaking (tremor) looks extremely impairing. She is barely able to achieve basic coordination exercises such as alternately touching the physician's finger and then her nose, giving up altogether with her right hand and succeeding with the left while exhibiting severe tremor. She is then made to hold a position in which the tremor is particularly remarkable. As soon as the stimulation is turned on, the tremor stops immediately. Her relief is audible and visible as she sighs and opens her eyes. Her features become clearer while the camera zooms in on her. Assuming empathy from the viewer, the scene is incredibly touching and at the very least impressive. So much so that the audience of the talk takes it in, and after a few seconds, breaks into applause.

Several elements are not made visible in that staging of DBS. For one, the patient is made to hold a strenuous position as a resting position, which will doubly accentuate the tremor: holding out both arms horizontally might overtax her tonus capacities and a voluntary movement will tend to minimize tremor. She could have her arms hanging on the sides of her body, but it wouldn't be as spectacular. Furthermore, she is obviously an excellent respondent to DBS as no flagrant symptoms are visible after turning the stimulation on. This not only implies that she has surgery and parameter settings over and done with, but also that her complementary medication most probably could be reduced. This means minimal medication and maximal therapeutical effect due to DBS that can actually be shown "at the push of a button," which is the purpose of the video. Last of all, of course, the DBS was turned off for observation. The "off" state shows how far the untreated disease has come, but usually, the patient is constantly under treatment. The "off" state is an absolute exception that almost never occurs, if not for observation.

Keeping the lay recipient in mind, watching the video, one discovers a patient with full-blown Parkinson's symptoms, where DBS is turned on and the symptoms disappear. In a nutshell. The focus on that moment—artificially created for the observation—is underlined by

Lozano's staging of the turning point, where he announces the action ("we're now going to turn it on ... it's on, just turned it on") and then goes on to comment the precise moment the stimulation is turned on with "and this works, like that, instantly," accompanying his words with a snap of the fingers. In doing so, he condenses the DBS narrative into the very moment when stimulation is turned on, concurring with the narrative that suggests that DBS works at the push of a button.

CONCLUSION

With an exclusive focus on the turning point, the story becomes: DBS therapy kicks in instantly. The climax takes on a life of its own and becomes the whole story. The invisible parts are blanked out, giving an image of "on and off" that blends with notions of "before and after," of an instantaneous change from unrestrained disease to complete cure. As a narrative strategy, the turning point structures a story, but in this case, it obstructs the visibility of other moments. Instead of constituting the quintessence or a summary of the whole story, the "push of a button" becomes a reduction that has to be put straight when potential patients inquire about DBS, not to mention the broad public's fascinated image of DBS in particular and neurosciences in general. Other associations with DBS also imply fears, such as that of alienation, changes of personality, or manipulation (Hayes, 2016). Either way, a simplistic and overly optimistic representation of DBS as well as all-embarassing fears might well have prejudicial consequences for the image of the therapy and of the whole field in the long run.

Repetition of narrational motives consolidates familiar patterns from which it is increasingly difficult to depart. Therefore it is urgent to reflect on the narrative traditions, as this may enable to contemplate alternative motives of narration and of thought. Metaphors and symbols stand for broad concepts, but the risk is to isolate them from their full meaning, as lay people cannot be expected to fill in the blanks. Trying out other ways of telling the story would mean bringing a broader understanding of DBS and neurosciences to the general public without having to operate within standard—but limiting—patterns.

Both hype and invisibility blur the representations of DBS. Hidden narrative frameworks channel and compromise the diversity of thought about DBS, about the image of disease and cure, especially within neurology and neurosurgery. As a result, notions about DBS in particular, but maybe also about the brain, brain functions, and neurosciences in general, become limited to a given set of representations. Seeing the framework means being able to question familiar representations that we take for granted. Paying attention to the—usually unreflected—frameworks and taking a critical stand on predetermined notions (images, pictures, etc.) might well be the only way for us to reflect on representations of scientific findings and medical practice after the neuro-turn. Ultimately, not all is black or white, on or off.

Acknowledgments

This work was drawn from research funded by the German Federal Ministry of Education and Research (BMBF) (promotional reference 01 GP 1306A).

References

Barad, K. (2003). Posthumanist performativity: Toward an understanding of how matter comes to matter. *Signs: Journal of Women in Culture and Society, 28*(3), 801–831.

Barad, K. (2010). Quantum entanglements and hauntological relations of inheritance: Dis/continuities, spacetime enfoldings, and justice-to-come. *Derrida Today, 3*(2), 240–268, Edinburgh University Press.

Baylis, F. (2013). "I am who I am": On the perceived threats to personal identity from deep brain stimulation. *Neuroethics, 6*(3), 513–526. <http://www.ncbi.nlm.nih.gov/pmc/articles/PMC3825414/>.

Bidder, J. (2008, February 19). Ruhe auf Knopfdruck. *Focus-Online.* Retrieved from <http://www.focus.de/gesundheit/ratgeber/gehirn/news/tiefe-hirnstimulation_aid_235872.html> Accessed 10.01.16.

Broschart, J. (2015 June). Das geplante Wunder. *GEO Germany,* pp. 76–86.

Clausen, J. (2011). *Technik im Gehirn. Ethische, theoretische und historische Aspekte moderner Neurotechnologie.* Köln: Deutscher Ärzte-Verlag.

Coenen, V. A. (2014). *Die tiefe Hirnstimulation—eine Erfolgsgeschichte: Neue operative Behandlungsoptionen.* Retrieved from <https://www.uniklinik-freiburg.de/fileadmin/mediapool/07_kliniken/nch_stereotaxie/pdf/artikel-hirnstimulation-forumsanitas2014.pdf> Accessed 10.01.16.

Dubiel, H. (2009). *Deep in the brain: Living with Parkinson's disease* (P. Schmidtz, Trans.). New York: Europa Editions (Original work published 2006).

FDA. (2002). *Activa® Parkinson's Control System P960009/S7.* Retrieved from <http://www.fda.gov/MedicalDevices/ProductsandMedicalProcedures/DeviceApprovalsandClearances/Recently-ApprovedDevices/ucm083894.htm> Accessed 10.01.16.

Felt, U., Fochler, M., & Winkler, P. (2010). Coming to terms with biomedical technologies in different technopolitical cultures: A comparative analysis of focus groups on organ transplantation and genetic testing in Austria, France, and the Netherlands. *Science, Technology, & Human Values, 35*(4), 525–553.

Hayes, O. (2016). Mind control? Fear and media portrayal of "brain pacemakers." In I. Dixon, S. Doran, & B. Michael (Eds.), *There's more to fear than fear itself: Fears and anxieties in the 21st century.* Oxford: Inter-Disciplinary Press. <http://www.interdisciplinarypress.net/product/theres-more-to-fear-than-fear-itself-fears-and-anxieties-in-the-21st-century/>.

Heier, M. (2008, January 20). Gehirn unter Strom. *Frankfurter Allgemeine Sonntagszeitung* (No. 3), p. 57.

King, L., Tulandi, T., Whitley, R., Constantinescu, T., Ells, C., & Zelkowitz, P. (2014). What's the message? A content analysis of newspaper articles about assisted reproductive technology from 2005 to 2011. *Human Fertility (Cambridge), 17*(2), 124–132.

Klott, A. (2008, August 24). Stromstösse gegen das Zittern. *Neue Zürcher Zeitung.*

Kühl, R. (2016). Wie umgehen mit den "großen" Ärzten? Entwicklungslinien und Perspektiven der medizinhistorischen Biographik. In Paul Thomes (Ed.), *Zwischen Narration und Methode: Neue Impulse in der historischen Biographieforschung*. Aachen: Shaker.

Lakoff, G., & Johnson, M. (1980). *Metaphors we live by*. Chicago, IL: University of Chicago Press.

Lévi-Strauss, C. (1955). *Tristes tropiques*. Paris: Plon.

Lieb, K., & Schläpfer, T. E. (2006). Deep-brain stimulation for Parkinson's disease, correspondence. *The New England Journal of Medicine, 355*(21), 2256.

Luce, A., Cash, M., Hundley, V., Cheyne, H., van Teijingen, E., & Angell, C. (2016). *"Is it realistic?" The portrayal of pregnancy and childbirth in the media*. BMC Pregnancy and Childbirth (16, p. 40).

Scheuher, C. (2016). What is being done to increase organ donation? *Critical Care Nursing Quaterly, 39*(3), 304–307.

Schmalfeldt, B. (2010). *Deep brain diary: My life as a guy with Parkinson's disease and brain surgery volunteer*. USA: Self-Published, ISBN: 1452801444.

Schneider, J., Novak, D., & Jech, R. (2015). *Optimization of Parkinson's disease treatment combining anti-Parkinson's drugs and deep brain stimulation using patient diaries*. Conference Publication of Engineering in Medicine and Biology Society (EMBC), 2015 37th Annual International Conference of the IEEE. Retrieved from <http://ieeexplore.ieee.org/lpdocs/epic03/wrapper.htm?arnumber = 7319133> Accessed 02.02.16.

Tain, L., & Robertson, G. D. (2002). The hospital, the woman and the physician: The construction of in vitro fertilization trajectories. *Population, 57*(2), 373–404.

Author Index

Subject Index

Printed in the United States
By Bookmasters